IT ADMINISTRATOR'S
TOP 10 INTRODUCTORY SCRIPTS
FOR WINDOWS

IT ADMINISTRATOR'S
TOP 10 INTRODUCTORY SCRIPTS
FOR WINDOWS

JEFF FELLINGE

CHARLES RIVER MEDIA, INC.
Hingham, Massachusetts

Acquisitions Editor: James Walsh
Production: Datapage Technologies, Inc.
Cover Design: The Printed Image

CHARLES RIVER MEDIA, INC.
10 Downer Avenue
Hingham, Massachusetts 02043
781-740-0400
781-740-8816 (FAX)
info@charlesriver.com
www.charlesriver.com

This book is printed on acid-free paper.

Jeff Fellinge. *IT Administrator's Top 10 Introductory Scripts for Windows*.
ISBN: 1-58450-212-6

Library of Congress Cataloging-in-Publication Data
Fellinge, Jeff.
  IT Administrator's top 10 introductory scripts for Windows / Jeff Fellinge.
     p. cm.
  ISBN 1-58450-212-6 (acid-free paper)
  1. Microsoft Windows (Computer file) 2. Operating systems (Computers)
3. Programming languages (Electronic computers) I. Title: IT
administrator's top ten introductory scripts for Windows. II. Title.
QA76.76.O63F445 2004
005.13—dc22
                    2003022749

Printed in the United States of America
04 7 6 5 4 3 2 First Edition

CHARLES RIVER MEDIA titles are available for site license or bulk purchase by insti-
tutions, user groups, corporations, etc. For additional information, please contact the
Special Sales Department at 781-740-0400.

Requests for replacement of a defective CD-ROM must be accompanied by the original
disc, your mailing address, telephone number, date of purchase and purchase price.
Please state the nature of the problem, and send the information to CHARLES RIVER
MEDIA, INC., 10 Downer Avenue, Hingham, Massachusetts 02043. CRM's sole obliga-
tion to the purchaser is to replace the disc, based on defective materials or faulty work-
manship, but not on the operation or functionality of the product.

To my wife Kellie, with whom I share my dreams and adventures.

Contents

Acknowledgments

I have written technical articles for *Windows & .NET* magazine for several years because I enjoy not only learning (and playing!) with new and cool technology, but also sharing my experiences with others with similar interests. I have felt for a while that writing a book would be something that I really wanted to pursue. First, I needed to find that *right* topic—something that I was passionate about and could get very creative with. The notion of a top 10 scripting book hatched in a hallway conversation in late 2002. For me, this was a book *I* always wanted—a blow-by-blow walk-through of complete scripts doing real work. Plus, I dig the creativity; scripting definitely taps deep into your creative energy to pull out technical solutions to any particular problem. For me, this was an ideal topic for my first book. Marching this idea up the mountain from a concept to a paperback has been an intense journey for me and one that I could not have made without the support and encouragement of my colleagues and friends.

I would like to first thank Nathan McCoy—a colleague, friend, and programmer extraordinaire—for his generous time and effort in helping make this a much better book. Nathan challenged my logic, edited my syntax, and helped show an IT guy some of the finer aspects of programming. When working together at Amazon.com, Nathan and I talked of "care factor," that inner drive to make something come out right. Nathan has the highest care factor of anyone I know.

I would also like to thank Ethan Wilansky for his initial encouragement and recommendation that I enter the very creative and rewarding world of technical writing. From technical reviews to brainstorming sessions, Ethan has extended me a tremendous amount of support throughout my writing career.

Thank you also to the entire *Windows & .NET* magazine staff who helped, critiqued, pushed, and taught me to become a better writer. Thank you, Michele Crockett and Amy Eisenberg; by listening to my ideas and accepting my work you have helped grow my writing in so many ways.

An idea can only be transformed into a book with the faith and confidence of a publisher. I would like to thank the team at Charles River Media, and especially James Walsh, who helped successfully guide me through the process of book writing. Thank you also to Laura Lewin and the staff of Studio B, who provided me with answers to my many, many questions along the way.

I'd also like to thank my colleagues at aQuantive, Inc. who tolerated and listened to me raving on and on about some neat or cool new scripting gizmo or wizardry I had discovered the previous day. The aQuantive team is one of the most talented and passionate teams I have worked with, and I enjoy working with each of you every day. Thanks also to Greg Meyer whose encouragement helped me pursue this personal goal.

Most of all, I give thanks to my wife Kellie, who tolerated six months of "imbalance" as I worked these two jobs. Kellie supported and encouraged me beyond words, and I am grateful and fortunate to share life with such a wonderful and caring person.

Happy scripting!

Jeff Fellinge
Labor Day, 2003

Preface

What an exciting time to work with computers. Finally, the tools have matured to a point that has expanded—no, exploded—people's ability to really start making cool things happen.

Twenty years ago you had to be a computer science major or serious propeller head (hacker) to understand and program your Apple®, Commodore®, or IBM® computer for anything beyond the most rudimentary tasks. Mainframes and minis offered much more power, but availability and access to them was greatly restricted to large corporations and universities. Ten years ago the Internet was still slowly ramping up toward its widespread future. Unix® ruled the Internet, and academia and government waded into the new world of internetworked computing. Still, the tools for programming consisted mostly of C compilers, which required abstract and laborious methods to program even the simplest functionality, let alone a solution to the problem you were trying to solve.

Fast-forward over the last 10 years through the trials and tribulations of the dot-com world. HTML, CFML, ASP, Java®, JavaScript®, VBScript, PERL, and CGI, known to many, offer many solutions with which to tackle work problems, and each of these successful languages excels in a genre. But the emergence of these languages has exposed a new truth: *you don't need to be a software developer to write code and make computers work.*

Even while Microsoft® continued to dominate the desktop, a second revolution began; Linux emerged as a Unix variant that ran well on the Intel® platform. Alien to Microsoft Windows® users, Linux relied heavily on the command line. The Linux GUI, X Window, was better and attracted new users of this operating system, but it too lacked the polish and refinement of the Windows interface (not to mention the familiarity), and Windows users continued to point and click their way into corporations around the world. What Linux had, though, was the decades of experience among Unix users and the elegance of truly powerful and stable computing. It wasn't fancy, but the command line provided access to the entire operating system, and many administrators coming from a Unix background wished for a similar level of automation in their Windows platform.

Today, many Windows administrators are fortunate to work on the latest versions of Windows: Windows® XP and Windows Server™ 2003, the most advanced operating systems to come out of Microsoft. These new operating systems have highly functional GUIs, and Microsoft spent a lot of time improving the stability of the operating systems as well as their networking capabilities. Gone are the days of rebooting to simply change an IP address, and the infamous Blue Screen of Death (BSOD) rears its ugly head much less often. But even with these advances and other improvements, Windows system administrators have remained mostly in the GUI. Consider the following example. Companies connected to the Internet or internetworked with other offices employ systems administrators who are responsible for maintaining and supporting hundreds or thousands or tens of thousands of servers, users, accounts, mailboxes, and other network or computer objects and widgets. A new employee is starting tomorrow? Great—Start > Control Panel > Administrative Tools > Active Directory User and Computers > Add. Type the first name, tab. Type the last name, tab. Type the password, tab. Repeat, tab. Times 10. Or times 100. You get the idea.

Progressive companies may have solved some of these resource-intensive tasks through custom-developed software, such as a Web interface into which you type new employees' first and last names and the account is automatically created. But who actually developed this tool or others like it? Did the company outsource it? Does the company employ a dedicated software engineering team to develop internal applications such as this? Today, many systems administrators have the skills that, when coupled with currently available tools, allow them to write these scripts themselves.

These tools are becoming easier and easier to build, and development tasks are now accessible to the IT systems administrator. New standards and programming technologies have opened up even some of the more historically complex programming tasks to *nondevelopers*. Sure, coding solutions takes time, depending on the complexity of the problem, but the tools available today to build these applications are much more advanced than they were even three years ago.

IT'S ALL THROUGH SCRIPTING

Scripting *can* make your life easier. Scripting *will* make your life easier. It's not difficult to learn. You don't have to have a computer science degree or work as a Web developer for five years. New programming languages and toolkits are available and well documented to remove you—the systems administrator—from the programming housekeeping and help you focus on programming solutions to your specific tasks.

The goal of this text is to show how easy and beneficial scripting can be using practical, real-world scripts that a Windows systems or network administrator would find useful on the job. The scripts themselves, which are powerful yet not overly complex, are intended for the systems administrator who is looking for a toolbox from which to build powerful tools. These tools unlock some of the most fascinating treasure chests in Windows® 2000, XP, and Server 2003.

To keep the scripts relatively short and easy to understand, we have included in many examples only the code that makes the magic happen and not robust error handling or optimized code.

LEARNING BY DOING

The Internet is teeming with scripting information and knowledge—almost to the point of inundation. The examples in this book are designed to demonstrate the scripting of a particular task, not necessarily an academic tutelage of development. For example, a script example may contain 10 lines of code to extract *stuff* from Microsoft Active Directory®. This basic code gives you an idea of how and what can be done, and from that you can extend your knowledge in any direction using any of the many available reference materials or books.

The chapters in this book are laid out with this method in mind. Chapters 2 and 3 rocket through just a few of the basics to set the simplest foundation of VB-Script, JScript®, Windows Script Host®, Windows Management Instrumentation®, and Active Directory Services Interface (ASDI). The intent of these chapters is to introduce you to the style of this text as well as to get you thinking about the basic vernacular of Windows programming, because many of the references you will come to rely on speak in this tongue. Chapters 4 through 13 each present a different script designed to solve a particular technical or business problem. The chapters begin with a scenario for a task that you need to accomplish. The scenario explains why one particular scripting technology was chosen over another and shines light onto the thought process behind the script.

Scripting adds a degree of freedom and creativity to your job. When you have built your own script, you will find it satisfying to add a new feature or tweak it a bit. As you read the scenarios, think of how you would approach and solve the problem. When you read the script and walk through it, think about alternative methods for solving the problem. Once you begin to challenge the scripts, you have graduated and then are able to create some truly powerful tools.

1 How to Use This Book

```
list-process.vbs - Notepad
File  Edit  Format  View  Help
'Scripting helps with routine tasks like highlighting
'a running process on any remote computer

strComputerName=WScript.Arguments(0)
WScript.Echo "Listing Running Processes for Computer: " & strComputerName
WScript.Echo "Process Name : Memory (MB)"
WScript.Echo "-------------------------"
Set colProcess=GetObject("winMgmts://" & strComputerName).InstancesOf("Win32_Process")
For Each objProcess in colProcess
    If objProcess.Name = "cmd.exe" Then
        WScript.Echo objProcess.Name & " <---- Highlight this Process"
    Else
        WScript.Echo objProcess.Name & " : " &_
        FormatNumber(objProcess.WorkingSetSize /(1024 * 1024),1) & " MB"
    End If
Next
```

```
Console
C:\>list-process.vbs corsair
Listing Running Processes for Computer: corsair
Process Name : Memory (MB)
-------------------------
System Idle Process : 0.0 MB
System : 0.2 MB
smss.exe : 0.3 MB
csrss.exe : 3.1 MB
winlogon.exe : 10.9 MB
services.exe : 3.3 MB
lsass.exe : 2.3 MB
svchost.exe : 3.0 MB
svchost.exe : 20.8 MB
svchost.exe : 1.6 MB
svchost.exe : 3.4 MB
spoolsv.exe : 3.6 MB
DefWatch.exe : 1.0 MB
MDM.EXE : 2.2 MB
Rtvscan.exe : 10.4 MB
nvsvc32.exe : 1.6 MB
svchost.exe : 3.3 MB
explorer.exe : 12.1 MB
VPTray.exe : 2.9 MB
CTHELPER.EXE : 0.7 MB
msnmsgr.exe : 3.0 MB
ctfmon.exe : 1.5 MB
mmjb.exe : 34.3 MB
mmdiag.exe : 2.6 MB
mm_director.exe : 2.7 MB
mm_TDMEngine.exe : 4.7 MB
cmd.exe <---- Highlight this Process
logon.scr : 2.2 MB
wmiprvse.exe : 4.1 MB

C:\>
```

As previously mentioned, Chapters 4 through 13 each present a different problem that you, as a systems administrator or engineer, may encounter in your work environment. The problem is introduced in the beginning of each chapter, and any constraints and requirements are laid out. After a short discussion of the problem statement, requirements, and constraints, we look at possible tools or approaches to take to conquer the problem. Presenting a right answer is helpful, but discussing pros and cons of several alternative solutions is even more powerful. We describe the benefits of using various tools for our scripting engine as well as what language and methodology are appropriate. For example, the password change and archive tools demonstrate using Microsoft Excel® and its built-in Visual Basic® for Applications (VBA) to dump, collect, change, and archive member server local administrators and passwords. This task could have been performed in Microsoft Access® or perhaps a Web frontend with a Microsoft SQL Server® back-end. However, Excel was chosen because of its availability and familiarity for most IT staff.

After the analysis of the problem scenario, open the script from the CD-ROM in your favorite script editor and review it before walking through its explanation. This process is like that used in cooking school when the chef takes the cake out of the oven and shows it to everyone before actually making one.

ON THE CD

All of the scripts are presented on the accompanying CD-ROM. They are contained in folders by chapter number and called out by their listing (for example, Listing 3.1 would be found on the CD-ROM as \chapter 3\scripts\listing_3_1.vbs) or by name (for example, \chapter 9\scripts\ClientCheck.hta).

Reviewing the script on the CD-ROM in its entirety first gives you an overall sense of the scope and size of the script as well as a big picture of how all of the pieces fit into making it work. Is it a house or a stadium? The goal of reviewing the entire script first is not to overwhelm but to provoke questions and inspire you to think "Whoa, so this is how that is done." After mapping the course and selecting the tools, the chapter presents the actual script. The chapter decomposes the script into parts and explains the key areas that make the script tick.

These scripts are real-world scripts that perform actual, practical work. Look within each script for the functions and techniques to pull and insert into your own future scripts. These script components may even become the basis of your scripting library. Most of the technologies are cross-indexed in the Glossary for easy reference. This reference will prove handy months from now when you may think back to the JavaScript enumeration technique that walked through a machine's NetBIOS shares.

BOOK STRUCTURE

This text adopts many of the same conventions used in other popular books. For example, code is set in a monotype font. Notes, tips, and CD-ROM references are highlighted where useful and appropriate.

A note offers additional explanation of how something works or points out a particular caveat to an approach or technique.

Whereas a note may provide a caveat, a tip offers an optional or heads-up approach that may be a bit more crafty or eye opening to solving a particular problem.

The CD-ROM icon highlights references to code contained on the accompanying CD-ROM.

Sample code may look like the following:

```
'----------------------------
'somescript.vbs — WSH script
'----------------------------
WScript.Echo "Hello Scripters"
```

To execute the script, save it to the file `somescript.vbs`, and then from the command line type:

```
cscript somescript.vbs
```

The section *What You Need* introduces some of the tools that you can use to author and run the scripts in this text. You can create and save many of these scripts using a simple text editor like Microsoft Notepad. However, as your scripts become more complex, you will benefit from using a more sophisticated script editor.

AUDIENCE

Keep in mind that the intention of this book is to show you how to build script-based tools by using real-world problems. The concepts covered in this text assume an appropriate level of technical experience and knowledge. The scenarios have been written for beginner and intermediate systems administrators and engineers. For example, if you have built Windows NT®/XP/2000/Server 2003 systems from scratch or have installed and configured a Windows 2000 domain controller, then you will have little problem following the scenarios presented in this text.

These scripts use Windows 2000 Server and Professional, Windows XP, and Windows Server 2003 platform technologies and topologies. Therefore, to get the most out of the problem-solving aspect of the scripts, you should be familiar with these environments. Walking through a script about how to dump Active Directory information is not useful if you are unfamiliar with Active Directory.

You certainly do not need a background in development, although if you have dabbled previously in scripting or other languages, you will pick up more of the subtle nuances in the scripts. Notes and tips are scattered throughout the text pointing out development best practices or coding techniques that are standard, common, or just cool. When you have finished reading this book and the scripts within, you will have a good idea of what technologies are available to help solve your own problems within your own environments. From there, you may decide that JavaScript is for you, and you can invest in one of the many JavaScript-focused reference books. Similarly, you can purchase a specific book on any of the many technological topics covered in this text.

WHAT YOU NEED

The scripts in this book cover a wide variety of everyday (or otherwise reoccurring) tasks faced by the Windows systems administrator. For many of these scripts, you can get away with a test environment of two computers, as shown in Figure 1.1. When extracting account information, domain controllers behave differently from member servers. For this reason, set up a Windows 2000 or Windows 2003 Server domain controller and a second computer as a member server within that domain. Access to the Microsoft Developer Network® (MSDN) or Microsoft TechNET® Web resources is also important. Appendix A lists the Web addresses of many popular scripting resources and tools.

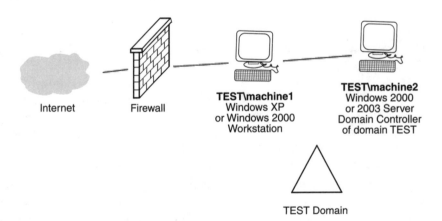

FIGURE 1.1   Sample test lab to practice your scripts. An Internet connection is useful for real-time research and reference during script writing.

ACCOUNTS AND PERMISSIONS

While you are learning scripting and *within your test environment*, test your scripts under an administrator or domain administrator account. Running your scripts under a privileged account removes security permissions from the list of reasons why your script isn't running. Certain tasks can be accomplished only by an administrator account (member of the local machine's administrators group). You may choose to run other scripts under a dedicated user account (for example, svc-SomeServiceAccount), especially if you decide to schedule the task to run repeatedly. (It may be more work to set up a dedicated service account to run the scheduled task, but if you run it under your account, you will likely be scratching your head three months later when you change your password and your account keeps getting locked out.) Script security is well documented in MSDN, and most objects have defined methods or properties that allow specification of the security context under which to run. After you have successfully mastered the hurdle of simply getting the script to do what you want it to, reexamine your script and remove any of the above constraints (such as the need to run as an administrator or domain administrator). As you get a sense of how the script technologies operate, you will soon be scripting security into your scripts straight from the start. When you use a privileged account when learning something new, remember to script in a test environment. You will be surprised how easy it is to make changes to directories, machines, the registry, and other critical infrastructure components, and you want to isolate your scripts to minimize any chance of harming a production or corporate environment.

TOOLS

Notepad is fine for creating or editing very short scripts or README files. For any larger amount of scripting, however, consider using an editor that offers at the very least syntax highlighting. Many of the more popular editors also offer tag or language completion, and some include integrated debuggers through which you can view variable values as you step through your script.

As with many of the approaches, you may have your own preference or fall in love with some other tool not mentioned here. Great! What you'll find is that you will become proficient in a particular tool, which offers a feature, however small, that you'll love, such as how it indents or what it can autocomplete for you. Available editors, in order of complexity from simple to state of the art, include Notepad, Microsoft Visual Basic Editor®, Macromedia® HomeSite™, and Microsoft Visual Studio .NET®.

Notepad

Notepad offers a rudimentary scripting editor. Most scriptwriters probably started out using this basic tool (or its predecessor, the MS-DOS® Editor). The feature list includes the ability to go to a specific line (by number) and insert the date and time. Beyond this, Notepad makes for a very basic text editor. Installed as the default on every Windows-based platform, its ubiquity makes it the dependable choice when scripting on an unknown machine.

Microsoft Office and the Microsoft Visual Basic Editor

Microsoft® Office core applications support Visual Basic for Applications (VBA) development and include the Microsoft Visual Basic Editor. When you record a macro, it is actually written into a module saved within the host Office file. Access the macro through the Visual Basic Editor by selecting Tools > Macro > Visual Basic Editor. (Alternatively, press Alt + F11). This full-featured editor supports VBA and includes syntax highlighting and inline debugging. The editor is fairly easy to get up to speed and works well for writing a quick script. Because VBA supports external scripting technologies such as Windows Script Host (WSH) or Windows Management Interface (WMI), you can use this editor and debugger to prototype or test your network and ADSI scripts and send the script output back to the easy-to-use and easy-to-save Office document. Figure 1.2 shows an example of the Microsoft Visual Basic Editor and its different components. A potential drawback to using the Microsoft Visual Basic editor is that it does not support JScript or VBScript—only VBA. This lack of choice does not present a challenge when you code a script from the start within an Office application, but it is something to remember if you want to port an existing script into an Office application.

The editor consists of various windows—the Project Explorer, the Code window, the Properties window, the Locals window, and three Watch windows.

Project Explorer: The Project Explorer lists the different objects of the application, such as the worksheets and any modules that you are using. If you are creating a new script, right-click Microsoft Excel Object's item and select Insert > Module to open a new module in the code window.

Code window: The Code window, the heart and soul of your VBA script, contains the actual development code. The Code window supports syntax highlighting, which automatically capitalizes and colorizes keywords in your code to make your code easier to read and debug. Also, if you erroneously enter a line, the editor highlights the mistake, allowing you to correct your mistake before even running your code for the first time.

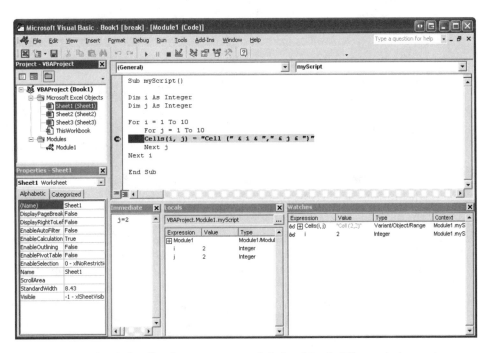

FIGURE 1.2 Microsoft Office features a powerful Visual Basic Editor to help create macros and other scripts used in Office applications.

Properties window: The Properties window displays the individual settings of the workbook or worksheet objects. For example, to change the worksheet name from the default *Sheet1* to *Inventory*, select the property and enter the new value.

Locals and Watch windows: The Locals and Watch windows are useful for debugging your code. An advantage of using the Microsoft Visual Basic Editor when coding is its ability to step through your code one line at a time (or to run to a specified line in code and then stop at a particular line, called a breakpoint). As you are stepping through your code (by consecutively selecting Debug > Step Into or by pressing F8), the Locals and Watch windows list your code's variables and their current values. As these values are updated by your code, they are shown immediately. This functionality is useful for debugging and determining why a program isn't running as it should.

The Locals window lists all variables identified in your code. The Watch window specifies any selected variable or expression in your code. Unlike with the Locals window, which shows all local variables, you must explicitly specify what object you wish to view in the Watch window. Additionally, you can choose to watch all sorts of expressions. For example, specifying `Cells (i, j)` displays the value of cells i and j as the variables i and j change.

Immediate window: You can set a variable to expression in real time in the Immediate window. This functionality is useful for tweaking a value or forcing a condition in the middle of a running script. For example, if you have a loop of a variable from one to 1,000 and you want to skip to the end, you can use the Immediate window to force the variable to a higher value, such as 995. Note that in this case, you will skip all intermediate processing and continue executing as if the variable was 995.

Even if you do not use an Office application as your scripting engine, you can still use the Microsoft Visual Basic Editor to create and debug your script. Once your script is working, copy the script into a text editor and run it from the console. Be aware that you must remove any Office-specific application calls that would not work outside the application. For example, you may need to substitute `Cells (x, y) = "Some Output"` to `WScript.Echo "Some Output"`.

Macromedia HomeSite

Macromedia HomeSite is an editor favored by many Web programmers. Its elegant yet powerful interface offers almost any feature the modern-day developer could desire. HomeSite provides syntax highlighting, a repository for building up your own script library, and tag completion. You'll find the tag completion functionality more useful for Web page development, but its syntax highlighting and orientation toward text- and script-based development is well done. Figure 1.3 shows the HomeSite interface. Drawbacks to HomeSite include its lack of debugging capabilities and its lack of support for VBA.

Microsoft Visual Studio .NET

Microsoft Visual Studio .NET is a powerful, integrated development environment (IDE). This complex, powerful tool can be used to create the most sophisticated applications. Although it is possibly overkill for more basic scripts, use Visual Studio's powerful features to assist with your script design.

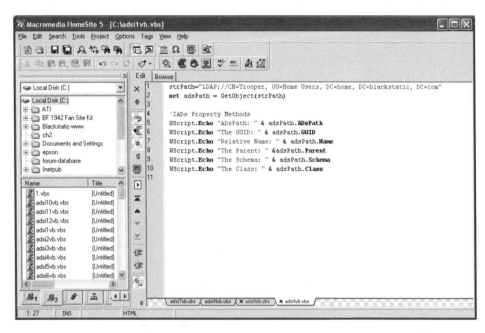

FIGURE 1.3   Macromedia HomeSite is a favored tool for scriptwriters and Web developers.

WHICH SCRIPTING TOOL IS RIGHT FOR ME?

Possibly one of most daunting challenges to starting anything new is simply figuring out how it all works—not necessarily the complexities of the problem itself, but understanding the toolset used to engage the problem. For that reason, get your feet wet and write (and run) some sample code using the previously mentioned editors. This practice is like gently shaking a ladder before ascending to ensure that it is stable and secure and that you can place your confidence in it. Once you have confidence that your scripts run, you can progress to more advanced tasks in no time.

SUMMARY

Chapter 1 introduced a few of the many tools that you can use to create and debug your scripts. Chapter 2 examines some of the different technologies used in creating scripts for the Windows platform, and we begin using these tools to demonstrate how these technologies work.

2 | Why Scripting?

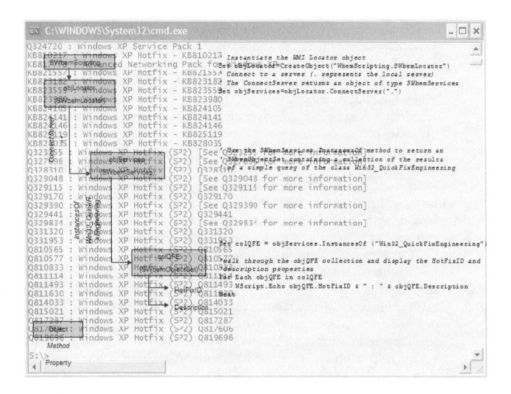

Quite simply, scripting saves time and reduces errors. Take a moment to think about your daily work tasks that you repeatedly perform, or project action items for which you must collect, review, or present system information. Take, for example, an engineer or administrator working in the information security field who is responsible for the ongoing monitoring and inspection of thousands of pieces of data. Suspicious data is often buried deep within huge volumes of legitimate information; the trick is to sort through all the good stuff to find the bad. Consider the following tasks, each of which involves managing large amounts of information at a detailed level.

- Monitoring what ports on what machines are exposed to the Internet.
- Auditing the user and group membership of sensitive servers.
- Establishing a reliable means of rotating the passwords on all servers when an IT employee who has access to the servers leaves the company.

So how do you go about finding what you know shouldn't be there? How do you apply the same action to a number of servers while minimizing mistakes? Manually scanning Internet traces, log files, and account memberships is tedious and mistake prone. Automation is the key, which leads to the next question of buying versus building a solution.

Commercial applications have been written to solve all of these problems; however, they cost money and may not perform exactly as you wish within your infrastructure. Also, the application may not output the report in the format you desire or include the data you require. Windows 2000 and Windows Server 2003 are the latest in a long line of Windows platforms, and over time, many small programs have been created that capably perform some of the previously listed tasks. Microsoft has bundled hundreds of these tools on the various product resource kits and support tools. Use these time-savers where possible, and remember that even these tools can benefit from custom scripting.

Scripting is not so much reinventing the wheel as harnessing what has already been created to work directly for you. Many times, you can script "wrappers" for existing binaries. These wrappers feed input and direct output to various text files or subsequent applications. Some scripts are a few lines long and others resemble small programs. Chapters 4 through 13 cover many direct examples of how scripting can help tacitly and tangibly make your life easier by automating tasks and ensuring that the tasks are carried out in a reliable and repeatable manner.

Let's return briefly to the three bulleted tasks mentioned previously and discuss some of the scripting options to assist with these processes.

- Monitoring what ports on what machines are exposed to the Internet.

To look for new machines or new services that may have opened up external ports to the Internet, consider using a command-line port scanner (such as NMAP). Port scanners can be configured to look for open ports on your entire external IP range assigned to you by your ISP. Run the scanner on a regular basis (such as biweekly), and compare the results to a previously generated baseline set of results. Scripts contain the subnets you wish to scan and start the port scanner. Scripts also parse the output of the reoccurring runs and look for new or changed ports from the known (and previously approved) baseline. All that you need to review is the report showing what has changed. The amount of actual information that you need to review is much less than when you review every machine and every port again and again. Use built-in filtering functionality included in *regular expressions* to reliably parse the output data of the port scanner.

- Auditing the user and group membership of sensitive servers.

You can audit user and group membership using ADSI by connecting to either a local machine or a domain controller. Consider writing a Microsoft Excel-based script that contains a list of all of the sensitive machines in your company. Walk through each of the machines, and output the groups and users with privileged access to that machine. Similar to the previously described network scanning example, you can repeatedly run this script to watch for changes in membership over time. Again, watching for deltas—or changes—in data is a much more reliable auditing practice than repeatedly inspecting the entire data set.

- Establishing a reliable means of rotating the passwords on all servers when an IT employee who has access to the servers leaves the company.

Password management is a subject approached by many commercial vendors; however, scripting can accomplish this task as well. Using Excel as the archiving tool, create a list of all servers in your network. Then, on a reoccurring basis (or when the need is triggered, such as with the exit of an employee with access to the passwords), run a script that iterates through each machine, sets the local administrator password, confirms that it is set, and then moves to the next server. Record the password in a single column next to the machine name. The next time you need to rotate passwords, insert a column and repeat the process. Skip unreachable machines, and you will build a password-archiving tool for which you will have access to all historical passwords.

These examples touch on some of the many possibilities that scripting affords. Chapter 1 introduced some of the tools that we use to demonstrate scripting in these examples. Now let's look at the languages we use.

LANGUAGES—WHAT DID YOU SAY?

The number of computer languages has grown tremendously over the last decade. In the not-so-distant past, you could count on a systems administrator or developer to be familiar with FORTRAN 77 or COBOL®. Now, however, development languages have splintered and fragmented into a whole slew of proprietary and application-centric dialects. Consider Web development alone—HTML, ASP, CFML, PHP, XML, VML, JavaScript, VBScript, ASPX, and others.

The Windows platform harbors its own language zealots who vocally endorse their favorite language as the most flexible, easiest, most powerful, or simply the best. In truth, the best language for you is the one that you can understand and most successfully leverage to meet your current, and hopefully future, business requirements. Not sure what to choose? Talk with your internal development staff, and ask about any standards that your company prefers to follow. A heavy Microsoft shop may prefer VBScript, but chances are if you talk with any past Web developers, they may have an affinity toward JavaScript (JScript in Microsoft parlance). Each of these languages offers benefits of which you will see a few in the pages to come. Consider, too, any local support you might be able to get to help you during your scripting, especially if you have development expertise inhouse—even in a different group. Many developers love the opportunity to expound the merits of their favorite language and usually will go the extra mile to help you solve a problem. All of the scripts in this book are written in three of the more popular Windows platform scripting technologies: VBScript, Visual Basic for Applications, and JScript.

But What about Batch and PERL?

This text does not cover batch files or programs written in PERL, nor does the text cover Unix and Linux shell scripting or tools (e.g., vi).

Batch Programming

Many of you may be using batch files (or have used them in the past) to accomplish some amazing things. However, this book focuses on Windows 2000, Windows Server 2003, and Windows XP scripting technologies. Some of you may be batch programming gurus and can replicate many of the scripts in this text using batch file programming. (By their very nature, as you will see, many of these scripts can be recreated in many different languages using a number of different tools.) So, if you are a batch file programmer extraordinaire, consider how any of these new technologies can complement and supplement your own toolbox—you'll be pleasantly surprised.

PERL

PERL is a popular language among Linux programmers, and it has made significant inroads into the Windows platform as well. In fact, Microsoft included a number of PERL scripts in the Windows 2000 Resource Kit. PERL is less distributable to Windows 2000 platforms because it requires a third-party PERL interpreter, which is not installed by default. However, for launching scripts from a single computer, PERL is an awesome language. Visit *www.activestate.com/Solutions/*

Programmer/Perl.plex for more information or to download ActivePerl®, the leading Windows PERL interpreter.

VBScript and Visual Basic for Applications (VBA)

VBScript and VBA are languages developed by Microsoft and are based on the popular and easy-to-learn BASIC language. VBScript is designed to be a lightweight and portable language, very suitable for independent scripting tasks and Web development. VBA offers more robust features and is integrated into the larger Microsoft applications, including Microsoft Office and Microsoft Visio®. In my experience, the tool you are using (such as Excel or the command line) usually drives your decision between VBScript or VBA. As you become more familiar with scripting in general, you can pursue the many texts or online documentation that describe in more detail the differences between VBScript and VBA.

VBScript is relatively easy to learn because of its friendly syntax. In addition, its support in all Microsoft scripting technologies and big brother support of VBA in Microsoft Office applications make it widely available to Windows users.

JavaScript and JScript

Ecma-262 is the official standard for the popular scripting language JavaScript, which was invented by Netscape and became a standard in 1997. The Microsoft full implementation of this standard is called JScript. JScript remains popular with Web developers and is supported by the most popular Microsoft scripting technologies, such as Windows Script Host.

JScript more closely follows a C programming style. (It requires the termination of code with a semicolon and shares a similar statement syntax.) Also, the much-hyped Microsoft C#® language used for Microsoft .NET® development borrows heavily from Java and JavaScript's background, which makes JScript a good stepping-stone toward C#.

It's funny when you talk with developers and ask them their preference—VBScript or JavaScript. Web developers tend to prefer JavaScript, and Microsoft-only programmers seem to favor VBScript. As you learn, you will come to find your own preferences.

PORTING

Choosing one language over another can be a difficult decision, but staying relatively flexible to using any language ultimately helps you be more receptive to the inevitable new technology. Worse-case scenario, you may be required to port your code from one language to another. Porting from VBScript to JScript (or vice versa) is a good exercise to learn the nuances of the language, although it is never really all that fun. You will likely end up coding helper functions or workarounds to limitations of the destination language.

Microsoft .NET has stirred the pot a bit in the release of Visual Basic .NET (VB.NET) and the depreciation of Visual Basic® 6.0 (VB6). This move has angered many Visual Basic programmers, because the conversion from VB6 to VB.NET is not trivial and requires quite a bit of redevelopment. Although scripting is not immune to these inevitable advances in technology, the scripts are usually smaller in size and generally simpler. However, keep in mind when creating a "killer script" that you may be required to port it someday to another language. Following a development structure (such as adhering to variable naming conventions) and liberally commenting your code dramatically aids future redevelopment or porting efforts.

PROGRAMMING FUNDAMENTALS

Regardless of the language in which you code, you must learn some basic software development principles. The rest of this chapter covers some of the programming fundamentals required to begin scripting effectively. If you already have some basic programming experience, you can skim this chapter or continue on to Chapter 3, which begins to introduce some of the more popular Microsoft technologies accessible through scripts. As mentioned in the Preface, this book is not an academic text on any one development language or process. Instead, it teaches you scripting through practical examples of some real-world scripts. However, that being said, some very basic building blocks are important to cover. Let's see these fundamentals exemplified through an actual script.

A script is a computer program created as a text file that, when run, is interpreted in real-time by a language interpreter engine. For command-line scripts, Microsoft offers Windows Script Host (WSH), which includes engines for both VBScript and JScript.

OUR FIRST SCRIPT

Open Notepad, and type the following line into a new text file:

```
WScript.Echo("Hello, World.");
```

Save this file as *HelloWorld.js*. (Make sure to override the extension so the file is not saved with a.txt extension.) Next, open a command window. (Click Start > Run, and type `cmd`.)

From a command line, execute the script by typing `CScript HelloWorld.js`. You should see the logo of the CScript engine followed by the words *Hello, World*.

Next, type `WScript HelloWorld.js`. Windows Script Host now processes this script through its GUI-based engine instead of the strictly command-line-based engine and shows the words *Hello, World* in a dialog box instead of in the console. Click OK to close the window.

WScript and CScript have similar functionality. However, you might find it beneficial to control which engine processes your scripts. For example, if you choose to dump a huge amount of data to the user, one line at a time, you may find the ability of CScript to write it to the open command window much more useful than the WScript dialog boxes. You can control WScript output for desired results, too, however. For example, to display a long summary of results, you can stuff a string with your message, and then display that string once in a dialog box. WScript and CScript are similar but do not offer the exact same functionality. (For example, WScript does not support the standard `in` and `out` functions, which we will cover a bit later.)

On your own machine, you can control the default behavior of Windows Script Host. From the command line, type `CScript /?` (or `WScript /?`) for a list of the WSH configuration parameters. To permanently turn off the WSH logo and make the console-based CScript the default script host, type:

```
CScript //H:CScript //NoLogo //S
```

Now try executing your new script by typing its name from a command line:

```
HelloWorld
```

You should see that the script is executed through the CScript script host engine without your specifying WScript or CScript. Of course, even with this standard set, you can always override the default by specifying a script host. Any user can set the default WSH engine to use (by default, Windows uses WScript unless overridden by the previous command), which is why you usually want to explicitly control how your scripts are called. For example, call your script with `CScript Hello-World.js` instead of simply typing `HelloWorld` at the console. Doing so prevents the accident of a user running your script under the wrong engine.

Porting Your Application to VBS

Copy the file *HelloWorld.js* to *HelloWorld.vbs*, and try to execute it by typing CScript HelloWorld.vbs. You will get a runtime error. Open the file, and remove the ending semicolon and parentheses so that the file contains the line:

```
WScript.Echo "Hello, World."
```

Save the file, and run it again. When you see the expected text, you have ported your first application from JScript to VBScript

The previous example really wasn't that exciting or difficult. However, the concepts and fundamentals used are the same as those used in any of the scripts in this text. As simple as the example is, it demonstrates the entire process of creating and running a script. Using a simple example lays a great foundation for building more complex scenarios.

ON THE CD

That being said, let's walk through the framework of a basic script, this first example written in both JScript and VBScript, as shown in Figure 2.1 (and contained on the CD-ROM as Listings 2.1 and 2.2 in the /chapter 2/scripts folder).

```
//********************************        '********************************
//* Script Framework                     ' Script Framework
//* framework.js                         ' framework.vbs
//* Created: Feb 2003, Jeff Fellinge     ' Created: Feb 2003, Jeff Fellinge
//********************************        '********************************
var i =22;                               Dim i
var j = '1';                             i=22
answer=sum(i,j);                         Dim j
WScript.Echo(answer);                    j="1"
j = 1;                                   answer=sum(i,j)
WScript.Echo(sum(i,j));                  WScript.Echo answer
                                         WScript.Echo concatenate(i,j)


//********************************        '********************************
//* sum/concatenate 2 variables          ' Calculate sum of 2 numbers
//********************************        '********************************
function sum(a,b)                        function sum(a,b)
{                                            sum = a + b
    result=a+b;                          end function
    return result;
}                                        '********************************
                                         ' Concatenate two strings
                                         '********************************
                                         function concatenate(a,b)
                                             concatenate= a & b
                                         end function
```

FIGURE 2.1 The framework of a simple script written in VBScript and JScript side by side.

Comments

This script begins with a comment section that introduces the script and defines any input parameters, functions used, created and modified dates, and authors. Be sure to comment your code and describe the functionality of key areas of your script. Liberal commenting of your code not only helps others understand your intentions, should they inherit your creation, but also helps you months after you have created the code when you inevitably need to revisit and update it. Comment syntax is shown side by side for both JScript and VBScript in Figure 2.2.

```
JScript                                    VBScript

//********************************         '********************************
//* Script Framework                       ' Script Framework
//* framework.js                           ' framework.vbs            '
//* Created: Feb 2003, Jeff Fellinge       ' Created: Feb 2003, Jeff Fellinge
//********************************         '********************************
```

FIGURE 2.2   Comment syntax for JScript and VBScript.

Comments are in-code explanations of what you are doing. You must differentiate your comments so that the interpreter does not try to execute them.

TIP

Refrain from using a "smart" word processor as your code editor. For example, Microsoft Office products automatically convert quotes and apostrophes to curly quotes and apostrophes, so a ' becomes a ' and " becomes a " or ". Although these characters look the same (even if copied to Notepad), they are different and may cause you scripting headaches.

JScript uses the C style comments of /* to begin and */ to end a comment, as follows:

```
/* This is a comment block. Every word between these
delineators will be interpreted as a comment, and this is
a useful way for specifying multiline comments. A problem
is that if they are very long, you may forget what is
a comment and what isn't. A code editor with syntax
highlighting can help prevent mistakes with long comments.*/
```

Alternatively, you can comment a single line with two forward slashes:

```
// This is a single-line comment in JScript.
// Multiple lines each need a separate delineator.
```

VBScript comments each line with an apostrophe:

```
' VBScript uses an apostrophe as its comment delineator.
' With this, too, you must comment each line individually.
```

Variables

Variables provide storage for the data that your program uses. In scripting, you declare your variable's name (and sometimes its type), initialize the value of the variable, and use the variable.

Variables represent many data types with which you are already familiar, such as integers (1,2,3) or strings ("Hello"). Other data types with which you may be less familiar are Boolean values (true and false) and objects. Later in this chapter, we discuss objects, but for now, understand that you can represent objects through variables, too.

Declare Your Variables

The next section of our example script declares the variables. JScript and VBScript do not require that you declare variables, but it is a good habit to do so. If you change to a strongly typed language, you are required to declare your variables and explicitly define their type. Figure 2.3 shows how to declare variables and set their values for both JScript and VBScript. Another advantage of declaring your variables is that declared variables appear in the Locals window when you debug and undeclared variables do not.

```
JScript                              VBScript

var i = 22;                          Dim i
var j = '1';                         i = 22
                                     Dim j
                                     j = "1"
```

FIGURE 2.3   Variable declaration differences between JScript and VBScript.

JScript and VBScript are weakly typed languages, which means that once a variable is declared (with the keyword var), it can hold many different literal (value) types. Notice, too, that with JScript, we can declare the variable and its literal on the same line. JScript and VBScript automatically determine the type of the variable (string, number, or Boolean). You want to be certain that it does so correctly. Look at the previous example in Figure 2.1. Notice that we have changed the type of the variable j from the string "1" to the number 1. Also, the function uses the operator

+, which not only sums two numbers but also concatenates a string and another string or number.

When we run this script, we get the following output:

```
C:\>framework.js
221
23
```

which may not be what we wanted, depending on the type of the variable. In this example, the value is easy to see, and control. If you are returning values from an object or working with otherwise unknown variables, you want to be sure you control their type. To aid in troubleshooting, use the JScript typeof operator to determine the type of a variable or expression.

VBScript uses the keyword Dim to declare variables. Also notice in the example in Figure 2.1 that VBScript differentiates between the sum operator (+) and the concatenate operator (&). Regardless of data type, VBScript tries to evaluate a variable as a number and sum two variables if + is used and always concatenates if & is used. In VBScript, we must create two separate functions to sum or concatenate the variables and achieve a result similar to the one earlier in the *framework.js* example. Table 2.1 shows the different language data types of VBScript and JScript.

TABLE 2.1   Language data types of VBScript and JScript

VBScript	*JScript*
empty	undefined
null	null
Boolean	Boolean
string	string
object	object
byte	number
integer	
currency	
long	
single	
double	
date (time)	
error	

Naming Convention

Following a naming convention for your variables dramatically increases the readability of your code and reduces programming mistakes. Apply your naming convention not only to your variables but also to your functions and other things that you come across as you delve further into programming. A naming convention increases readability and eases maintenance when you (or someone else) need to revisit the code in the future. For example, the variable named strName implies not only that the variable contains a name but also indicates that it is a string. Compare this example with iRow, which indicates that it is an integer representing the row number of something. You might be tempted to add the value 1 to iRow, but you would certainly think twice before you added the number 1 to strName. Although some commonly accepted naming conventions are available, many people choose their own, which they have found to work for them. In the end, choose and stick with a naming convention for your variables that works for you. Adopting some or all of the parts of a mainstream naming convention standard is advantageous, because it will increase the readability of your code. One of the more famous naming conventions is Hungarian Notation, established by Dr. Charles Simonyi who worked at Microsoft. Hungarian Notation introduced the notion of prefixing the descriptive name of the variable with its type (among other things). Internally, Microsoft widely adopted this naming convention. Although other naming conventions have emerged over the years, many programmers continue to use variants of Hungarian Notation.

The examples and scripts in this book follow a fairly generic (and simple) naming convention. Each variable is descriptively named and title capped, and the variable's data type is prefixed to this name (such as strFirstName, bExists, iRow, dtCreated). Procedures (subroutines and functions) are descriptively named and title capped; for example, ListAllDetail or Tabulate. Global and public variables are denoted with a g_, so, for example, if the variable objWsSetup was declared Public, it would be named (and referred to) as g_objWsSetup to represent its global nature. Table 2.2 shows most of the naming-convention prefixes and example variable names used throughout this text.

Arrays

Use an array to define a single object that contains multiple items. VBScript supports both single and multidimensional arrays. While JScript doesn't support multidimensional arrays, you can create arrays of arrays to accomplish similar functionality. Configuring and using an array involves declaring it, setting its value, and retrieving its value. JScript offers superior array manipulation features such as the push and pop methods, which allow you to add new elements to an array or remove elements using a single command.

TABLE 2.2   Variable naming conventions used in this text

Data type	Prefix	Example
integer	i	iRow, iCol, i
string	str	strName, strOutput, strServer
Boolean	b	bGoHelp, bIsRandom, bIsUser
object	obj	objState, objWsSetup, objShell
date	dt	dtToday, dtCreated, dtPwdExpires
array	a	aGroup, aDrives, aList
XML document	xml	xmlDoc, xmlRoot, xmlConfig
regular expression	re	reValidFile
collection of objects	col	colPrinters, colQFE
enumerator	e	eGroups, eLogicalDisk
	g_	global

Arrays in VBScript

To use an array in VBScript, you must first declare a variable and then assign it to an array. Create a static array of a specific size (for example, `Dim aList(10)`) if you know that you will not need to resize it in the future. Or define a dynamic array by omitting the size (for example, `Dim aList()`) or by defining the array elements explicitly using the Array function (such as `Dim aMonth` followed by `aMonth=Array("January", …, "March")`), which declares the array as dynamic and implicitly defines the size of the array. In VBScript, an array size must always be defined either at the declaration of a static array or later using the `ReDim` function for a dynamic array. For example, `Dim aList(10)` defines an array of 10 elements that can be used immediately, but `Dim aList()` defines a dynamic array, and before the dynamic array can be used, it must be sized using `ReDim`. A dynamic array can be resized by subsequent calls of the `ReDim` statement. Use `ReDim` to clear the array and set a new size, or use the `ReDim Preserve` parameter to retain previous array values. Set array element values explicitly by assigning the indexed element to a value (for example, `aMonth(3)="April"`). Listing 2.3 shows how to define and iterate—or cycle—through a VBScript array using a For…Next loop in conjunction with the `Ubound` function, which returns the upper bound of the array.

ON THE CD

LISTING 2.3   Working with single-dimension arrays in VBScript

```
Dim aList(3)
Dim aMonth
aMonth= Array("January","February","March")
Redim Preserve aMonth(6)
aMonth(3)="April"
aMonth(4)="May"
aMonth(5)="June"
For i = 0 to Ubound(aMonth)
    WScript.Echo(i & ") " & aMonth(i))
Next
```

VBScript supports multidimension arrays. Their construction is similar to that of single-dimension arrays. Simply define the array dimensions upon declaration (for example, Dim aMatrix(3,2)). Set the values by referencing the row and column index of the array. Like single-dimension arrays, use the Ubound function to return the upper bound of the array. However, Ubound also supports an optional parameter, dimension, which is used to specify the dimension upper bound to be returned. Listing 2.4 shows an example of looping through both dimensions of a 3x2 array.

ON THE CD

LISTING 2.4   Working with multidimension arrays in VBScript

```
Dim aMatrix(3,2)
aMatrix(0,0)="r0c0"
aMatrix(0,1)="r0c1"
aMatrix(1,0)="r1c0"
aMatrix(1,1)="r1c1"
aMatrix(2,0)="r2c0"
aMatrix(2,1)="r2c1"
for iRow = 0 to Ubound(aMatrix,1)
    for iCol = 0 to Ubound(aMatrix,2)
    WScript.StdOut.Write(aMatrix(iRow,iCol) & "  ")
    Next
    WScript.StdOut.Write(vbcrlf)
Next
```

Arrays in JScript

A JScript array must first be created by using the code new Array(x), where x is an optional definition of the array size. Unlike in VBScript, you do not have to specify the size of the array upon initialization, and the JScript array length property returns the actual number of elements defined in the array. Array elements in JScript can be defined as literals or by index value, as shown in Listing 2.5. Iterate through elements in a JScript array using a `for` loop from the beginning of the array to its `length`. Both VBScript and JScript arrays are zero-based, meaning that the first element has an index of 0. Array element values can also be explicitly set by their index number, also seen in Listing 2.5.

ON THE CD

LISTING 2.5 Working with single-dimension arrays in JScript

```
var i;
var aList1=new Array();
var aList2=new Array(10);
var aMonth=new Array("January","February","March");
aMonth[3]="April";
aMonth[4]="May";
aMonth[5]="June";
for(i=0;i < aMonth.length;i++) {
    WScript.Echo(i + ") " + aMonth[i]);
}
```

Unlike VBScript, JScript does not support multidimension arrays. Multidimension array functionality in JScript can be simulated by using an array of arrays. (Note that this technique can also be used in VBScript if its multidimension array functionality does not meet your particular needs.) Creating a three-row, two-column array (a 3x2 array) actually requires creating four arrays. The first array, `aMatrix`, consists of three arrays (one per row). The first element of the primary array (`aMatrix[0]`, where 0 equals the row number) is actually another array that represents the first row. The first element of this row array is the first column (`aMatrix[0][0]`), and the second element represents the second column. Using this technique requires additional boundary checking, because, unlike a matrix, in a JScript array of arrays, each row could have a different number of columns. Listing 2.6 demonstrates setting the values of row and column elements of a 3x2 matrix and then displaying the results. Actually, JScript arrays are full-fledged objects with myriad useful methods and properties. Although JScript arrays may seem more complicated at first than their VBScript counterparts, learning to use these objects effectively most definitely adds to your scripting abilities.

ON THE CD

LISTING 2.6 Working with multidimension arrays in JScript

```
var iRow,iCol;
var aMatrix=new Array();
aMatrix[0]=new Array();
aMatrix[1]=new Array();
```

```
aMatrix[2]=new Array();
aMatrix[0][0]="r0c0"
aMatrix[0][1]="r0c1"
aMatrix[1][0]="r1c0"
aMatrix[1][1]="r1c1"
aMatrix[2][0]="r2c0"
aMatrix[2][1]="r2c1"
for(iRow=0;iRow < aMatrix.length;iRow++) {
    for(iCol=0;iCol < aMatrix[iRow].length;iCol++) {
    WScript.StdOut.Write(aMatrix[iRow][iCol] + "  ");
    }
    WScript.StdOut.Write("\n");
}
```

Main Body of Code

Your script's main body of code is where the action is done. Ironically, depending on the modularity of your code, your main body could be (and will be as you become more proficient with scripting, code flow, and best practices) one of the shortest sections of your scripts as you begin to leverage and use functions for any repetitive work. In our example, the main body of code calls functions, sends in variables, and displays the result in the output. Using Windows Script Host, we are calling the echo *method* of the WScript object, as shown in Figure 2.4. Methods provide actions to use with *objects*. Both methods and objects are covered in the next sections.

JScript	VBScript

`answer=sum(i,j);`
`WScript.Echo(answer);`
`j = 1;`
`WScript.Echo(sum(i,j));`
	`answer=sum(i,j)`
	`WScript.Echo answer`
	`WScript.Echo concatenate(i,j)`

FIGURE 2.4 The main body of code calls functions to perform repetitive or specialized work.

Functions

Functions are isolated bits of code that are called to perform specific or repetitive operations. A program's main body of code or even other functions can call a function and pass it input parameters. The function then operates on these parameters before returning the result and code execution to the calling program. Common examples of functions include an output function that writes output to a file or to the console in a particular format, or a function that list all computers in an organizational unit or all users in a group. You will find that the more coding that you can offload or modularize into functions, the more portable and reusable your code will be. Ideally, once you create a function to, say, manage a log file, you can reuse that code in many other scripts. The syntax for setting up functions for both JScript and VBScript is shown in Figure 2.5.

```
JScript                                      VBScript


//********************************            '********************************
//* sum/concatenate 2 variables               ' Calculate sum of 2 numbers
//********************************            '********************************
function sum(a,b)                            function sum(a,b)
{                                                sum = a + b
    result=a+b;                              end function
    return result;
}                                            '********************************
                                             ' Concatenate two strings
                                             '********************************
                                             function concatenate(a,b)
                                                 concatenate= a & b
                                             end function
```

FIGURE 2.5 Functions perform specialized work and are commonly reused across different scripts.

Notice that because of the differences in operators, two functions are required for the VBScript code, one for summation and one for concatenation. Also notice that a JScript function can return any variable as the result of the function to the calling program, whereas the name of the function must be assigned the return value in VBScript.

Functions return a result of an operation and can be used in expressions. VBScript includes another statement, Sub, which operates similarly to a function, except that it does not return a value and cannot be used in an expression.

SCRIPT BUILDING BLOCKS

An object is a programming construct that groups methods and properties into a single unit. Methods can perform actions or do work, and properties are information about the object that you can read and sometimes write. You can work with external programming entities via objects and their associated methods and properties.

Expressions and *statements* make up the skeleton of the programming language, providing the syntax to build your program. Expressions include *initializers* and *operators*. Statements include *flow control* and *conditionals*. In the following sections, we see some of the major categories of these building blocks. Detailed examples of their use are illustrated throughout the scripting code in the remainder of this book. Many online and hardcopy references provide a detailed explanation of these programming concepts. Create a bookmark to the scripting content in the Microsoft Developer Network (MSDN) at *http://msdn.microsoft.com*. Search MSDN for JScript and VBScript, and expand the Reference section for detailed information on the syntax of these languages.

Links to these resources at the time of publication of this text follow:

MSDN JScript Reference

http://msdn.microsoft.com/library/en-us/script56/html/js56jsoriJScript.asp

MSDN VBScript Reference

http://msdn.microsoft.com/library/en-us/script56/html/vtoriVBScript.asp

Objects

Objects are elements that represent the physical or logical pieces of the environment. You will find yourself using objects throughout your scripts. Objects are the data types that allow you to interact with your operating system and other applications by using application programming interfaces (API). The API consists of the methods, properties, and documentation for using a particular object. For example, Windows Script Host includes 14 objects that allow your scripts to interface with your local or remote computers' network, environment, file system, run commands, and change administrative options. You can also create or get objects from a variety of other sources, which enables you to use WSH to instantiate objects from IIS, WMI, ADSI, or other supported interface that Microsoft exposes.

Figure 2.6 shows the relationship of objects, methods, and properties by comparing the architecture of a script side by side with the script syntax.

ON THE CD

The example shown in Figure 2.6 is contained on the CD-ROM as \chapter 2\ scripts\listing_2_7.vbs).

Throughout this book, we introduce and discuss new and different objects. However, JScript and VBScript have their own language-centric objects, listed in Table 2.3. Most of the time in your scripting, you will use COM to use objects exposed from other applications or libraries.

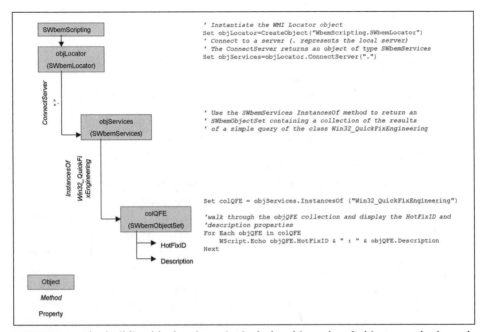

FIGURE 2.6   The building blocks of a script include a hierarchy of objects, methods, and properties.

TABLE 2.3   Objects built in to both VBScript and JScript

JScript	
ActiveX	function
array	global
arguments	math
Boolean	number
date	RegExp
debug	regular expression
enumerator	string
error	VBArray
VBScript	
class	match
debug	regular expression
err	

Object Models

An object model consists of a group of objects that work together to provide services. For example, the Windows Script Host object model defines interaction with the Windows operating system through a set of instructions to help you script Windows OS administration tasks, such as running programs, logging an event, or manipulating the system registry. Another example of a model is the *Common Information Model* (*CIM*) that is used by Windows Management Instrumentation (WMI) to manage Windows resources. CIM provides the framework to physically and logically describe networked objects. WMI is a Windows-based implementation of the CIM syntax that allows Windows computers to access information from any computer (Windows or other) that supports CIM.

The most common object model with which you work during your scripting is the Component Object Model (COM). For example, you can use COM automation to create a new ActiveX object representing an Excel worksheet from your console-based script without actually opening Excel. Through COM, your script creates a new Excel Worksheet object, which then exposes any of the Excel programming methods and properties of that particular object. Automation is the ability (provided by COM) for scripting languages to work with COM objects via methods and properties. Component developers create "automation compatible" interfaces—methods and properties—that are compatible with scripting languages so you can automate these components (objects).

Or vice versa, from within Excel you can use COM to create a new Windows Script Host FileSystemObject to read a text file.

Functions

Functions are independent pieces of code that perform a specific, often isolated action. Functions are usually built into the programming language or can be created by you, the programmer. Many functions do what they sound like (for example, the VBScript function IsNumeric returns a Boolean True or False, depending if the object that is passed to it is a number). Some of these built-in functions may be familiar to you if you have a coding (or math) background, or if you are an Excel power user.

Methods

Methods implement the actions that an object understands and can initiate. Methods accept parameters like a function, but they differ from functions in that methods are directly tied to their parent object, whereas functions exist independently. Like functions, you can create your own objects and define custom methods. However, for the majority of your scripting, you reference existing objects and their defined methods.

JScript relies on methods much more heavily than VBScript does. For example, most of the JScript math capabilities rely on the Math *object together with many* Math *methods. For example, to calculate the square root of a number in JScript, you use the code* Math.sqrt(number). *In this example,* Math *is the object, and* sqrt *is the method. The argument to the* sqrt *method is* number. *Contrast this to VBScript, which defines the function* Sqr *to calculate the square root. No object is needed; simply call* Sqr(number) *to return the square root of the argument,* number.

Objects support methods that enable them to perform actions. For example, to map a network drive to a computer, you can use objects and methods of the Windows Script Host object model, as the next JScript example shows.

First, create a reference to the WshNetwork object:

```
objWshNetwork=new ActiveXObject("WScript.WshNetwork");
```

The WScript object is the root object to the Windows Script Host object model. The WshNetwork represents the name of the object that we wish to manipulate, and a reference to it is assigned to the variable objWshNetwork. The WshNetwork object has eight methods—one of which is called MapNetworkDrive. You can access this method using the following code:

```
objWshNetwork.MapNetworkDrive("p:", "\\server\share")
```

Notice that we now use the variable name, objWshNetwork, to call the method. This practice is typical when working with objects and methods. The first step is to create a reference to the object, and the second step is to call the method using that reference.

Methods support both required and optional arguments. The method MapNetworkDrive supports two required arguments and three optional arguments. In our example, the only arguments we used were those specifying the local name of the mapped drive and the share's Universal Naming Convention (UNC) name. We could have specified to update the user's profile or to use a different username or password. A good reference guide or help text is essential for knowing all of the parameters supported by the different available objects and methods. Fortunately, Microsoft offers this assistance online at MSDN and through its many software development kits (SDK). The platform SDK is particularly useful; it covers many of the scripting objects discussed in this book. This SDK can be found at *www.microsoft.com/msdownload/platformsdk/sdkupdate/*. Although these SDKs may look daunting at first, you will soon be cruising through them, looking up methods and properties and using them in your own code.

Let's reexamine Figure 2.6 to walk through the objects and methods used. This VBScript example begins by creating a reference to the SWbemLocator object. Next, the ConnectServer method is called with the argument ".", which denotes to use the local server.

The second method used in our basic example is SWbemServices.InstancesOf. InstancesOf is one of 18 methods supported by the SWbemServices object, and it returns a collection of objects from a specified WMI class. Compare this method with the more flexible SWbemService.ExecQuery. ExecQuery also returns a collection of objects from a specified WMI class; however, it supports a SQL-like selection criteria using the *Windows Management Instrumentation Query Language (WQL)*. Let's say you want to query whether a specific Quick-Fix Engineering (QFE) update is installed on a machine. You could enumerate and search the entire collection returned by the InstancesOf method or, alternatively, by using the ExecQuery method you request to return only updates that satisfy your criteria.

To demonstrate this flexibility, in Figure 2.6, replace the following line:

```
Set colQFE = objServices.InstancesOf("Win32_QuickFixEngineering")
```

with the following line:

```
Set colQFE = objServices.ExecQuery("select * "_
& "from Win32_QuickFixEngineering where HotFixID='Q329048")
```

(Remember to replace the hotfix Q329048 with a hotfix that you know you have installed on the machine on which you run this to get a result.)

The previous script is broken into two lines because of the limitations of the printed book's page width. In VBScript, denote the continuation of a line using an underscore character (_). Because the break occurred within a string (between quotation marks), we must terminate the string, break the line, and then concatenate the remainder of the line. This page-break line is not needed in JScript because JScript uses a semicolon (;) as a required line termination character.

Tables 2.4 and 2.5 list the built-in functions used as the mortar and glue for putting together your own script for JScript and VBScript, respectively.

TABLE 2.4   Built-in JScript methods and functions

Methods			
abs	getFullYear	moveFirst	splice
acos	getHours	moveNext	split
anchor	getItem	parse	sqrt
apply	getMilliseconds	parseFloat	strike
asin	getMinutes	parseInt	sub
atan	getMonth	pop	substr
atan2	getSeconds	pow	substring
attend	getTime	push	sup
big	getTimezoneOffset	random	tan
blink	getUTCDate	replace	test
bold	getUTCDay	reverse	toArray
call	getUTCFullYear	round	toDateString
ceil	getUTCHours	search	toExponential
charAt	getUTCMilliseconds	setDate	toFixed
charCodeAt	getUTCMinutes	setFullYear	toGMTString
compile	getUTCMonth	setHours	toLocaleDateString
concat (Array)	getUTCSeconds	setMilliseconds	toLocaleLowerCase
concat (String)	getVarDate	setMinutes	toLocaleString
cos	getYear	setMonth	toLocaleTimeString
decodeURI	hasOwnProperty	setSeconds	toLocaleUpperCase
decodeURIComponent	indexOf	setTime	toLowerCase
dimensions	isFinite	setUTCDate	toPrecision
encodeURI	isNaN	setUTCFullYear	toString
encodeURIComponent	isPrototypeOf	setUTCHours	toTimeString
escape	italics	setUTCMilliseconds	toUpperCase
eval	item	setUTCMinutes	toUTCString
exec	join	setUTCMonth	ubound
exp	lastIndexOf	setUTCSeconds	unescape
fixed	lbound	setYear	unshift
floor	link	shift	UTC
fontcolor	localeCompare	sin	valueOf
fontsize	log	slice (Array)	write
fromCharCode	match	slice (String)	writeln
getDate	max	small	
getDay	min	sort	

TABLE 2.4 Built-in JScript methods and functions *(continued)*

Functions

GetObject	ScriptEngineBuildVersion	ScriptEngineMajorVersion	ScriptEngineMinorVersion
ScriptEngine			

TABLE 2.5 Built-in VBScript methods and functions

Methods

Clear	Raise	Test	WriteLine
Execute	Replace	Write	

Functions

Abs	Escape	LCase	Sgn
Array	Eval	Left	Sin
Asc	Exp	Len	Space
Atn	Filter	LoadPicture	Split
CBool	FormatCurrency	Log	Sqr
CByte	FormatDateTime	LTrim; RTrim; and Trim	StrComp
CCur	FormatNumber	Maths	String
CDate	FormatPercent	Mid	StrReverse
CDbl	GetLocale	Minute	Tan
Chr	GetObject	Month	Time
CInt	GetRef	MonthName	Timer
CLng	Hex	MsgBox	TimeSerial
Conversions	Hour	Now	TimeValue
Cos	InputBox	Oct	TypeName
CreateObject	InStr	RGB	UBound
CSng	InStrRev	Right	UCase
CStr	Int, Fix	Rnd	Unescape
Date	IsArray	Round	VarType
DateAdd	IsDate	ScriptEngine	Weekday
DateDiff	IsEmpty	ScriptEngineBuildVersion	WeekdayName
DatePart	IsNull	ScriptEngineMajorVersion	Write
DateSerial	IsNumeric	ScriptEngineMinorVersion	WriteLine
DateValue	IsObject	Second	Year
Day	Join	SetLocale	
Derived Math	LBound		

Properties

Where methods are the actions associated with an object, properties are the characteristics of the object that can be read (and sometimes set). If my dog were an object, he might have the properties of breed and name with corresponding values of Yellow Labrador and Trooper, respectively. My dog might have the methods of Play, Stay, Eat, or Sleep.

Again, using our previous example, Figure 2.6, we display the properties of the collection of objects returned from the class Win32_QuickFixEngineering. This class has no methods, but it does have a number of read-only properties. (You will find that most WMI properties are read-only. This characteristic is because many properties represent the physical state of an object, such as a disk having some amount of megabytes free. You can *get* how much space is left, but you can't *set* how much space is left. Make sense?) Look at the properties supported by classWin32_QuickFixEngineering: Caption, CSName, Description, FixComments, HotFixID, InstallDate, InstalledBy, InstalledOn, Name, ServicePackInEffect, and Status.

From just looking at this one class (there are over 180 Win32 WMI classes alone), you may be starting to get a sense of just how robust WMI can be and how easy it is to extract customized information from any computer that you manage.

Tables 2.6 and 2.7 show some of the native properties for JScript and VBScript, respectively.

TABLE 2.6 Native JScript properties

$1...$9	index	LOG10E	PI
0...n	Infinity	LOG2E	POSITIVE_INFINITY
arguments	input	MAX_VALUE	propertyIsEnumerable
callee	lastIndex	message	prototype
caller	lastMatch	MIN_VALUE	rightContext
constructor	lastParen	multiline	source
description	leftContext	name	SQRT1_2
E	length	NaN	SQRT2
global	LN10	NEGATIVE_INFINITY	undefined
ignoreCase	LN2	number	

TABLE 2.7   Native VBScript properties

VBScript	Global	IgnoreCase	Pattern
Description	HelpContext	Length	Source
FirstIndex	HelpFile	Number	Value

Events

Events are one of the primary signaling mechanisms for the Windows operating system. Events can be set by objects, applications, and even users. Scripting event-aware code can be simple, such as a Web page that changes a menu font or color by watching for the onmouseover event, or much more complex, such as watching and acting on a new event log or process.

ON THE CD

Figure 2.7 diagrams a short script that contains some fairly complicated code (included on the CD-ROM as Listing 2.8). Run this script from a command line to watch and report the creation of new processes, as well as the command line (if any) that was used to start them.

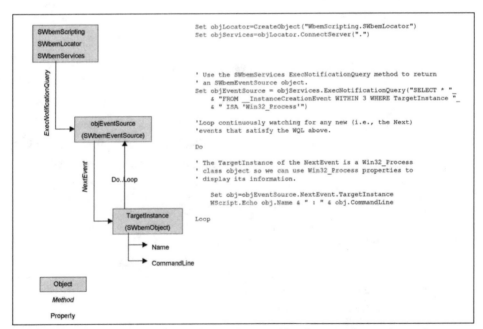

FIGURE 2.7   Tap into events to monitor action outside of your script, such as when a new process is created.

This script demonstrates one approach of harnessing WMI to access system events. You can modify the script to report on new additions to the event log or changes to the registry, in addition to processes.

In an earlier example, we used the method ExecQuery to return a subset of data from a particular WMI class. In this example, we use the method ExecNotificationQuery, which receives events and also supports WQL.

Notice, too, that the WQL query has gotten a bit longer. Let's break it down to see how this helps us get at the events. This statement should be on a single line.

```
SELECT * FROM __InstanceCreationEvent WITHIN 3 WHERE
    TargetInstance ISA 'Win32_Process'"
```

The statement basically says poll the system class __InstanceCreationEvent every three seconds and return to me any newly created Win32_Process objects.

The WITHIN 3 statement specifies the polling interval of the WMI query to check every three seconds. Specifying a short polling interval increases system resource utilization; however, for some events that may come and go (such as creating a process and then quickly deleting it), you may miss certain events. Other events that persist (such as new events added to the event log or a change in disk capacity) are better suited to a longer polling interval.

The ISA operator of the WHERE clause instructs the query to return instances of all classes within the class hierarchy.

```
Do
    Set obj=objEventSource.NextEvent.TargetInstance
    WScript.Echo obj.Name & " : " & obj.CommandLine
Loop
```

The last part of the script is a basic loop, which keeps the program running after every new event. The script calls the NextEvent method, which waits for a new event that satisfies the previous criteria. The script will wait (forever, if necessary) until an event is returned, polling the system at an interval as specified in the WITHIN statement. When a new event or events are detected, it assigns them to an object and displays the Name and Command Line properties of the object.

The previous example represents a *synchronous query*, which means the system waits to return with the complete result. (You can specify a Time Out value in the NextEvent method, but the query still runs synchronously.) However, most of the WBemServices methods offer an *asynchronous* counterpart (for example, ExecQueryAsync, ExecNotificationQueryAsync). These methods asynchronously execute the query and then return control of your script to do other useful things, such as monitoring multiple event providers or providing status. Running processes asynchronously can be handy.

Listing 2.9 shows the code to monitor the simple creation and deletion of process events. The code is more complex, and you may want to return to it after you have learned a bit more about scripting to understand its nuances; however, it does give a good idea for how to manage events asynchronously. You will need to run this script using *CScript.exe*, not *WScript.exe*, because it uses StdOut text stream object, which is not compatible with the *WScript.exe* engine.

LISTING 2.9   Monitor events asynchronously (run using *CScript.exe* only)

```
Set objLocator=CreateObject("WbemScripting.SWbemLocator")
Set objServices=objLocator.ConnectServer(".")

Set sink = WScript.CreateObject("WbemScripting.SWbemSink","SINK_")

Set nvsSink=WScript.CreateObject("WbemScripting.SWbemNamedValueSet")
nvsSink.Add "Created: ",1
nvsSink.Add "Deleted: ",0

objServices.ExecNotificationQueryAsync sink,"SELECT * FROM "_
    & "__InstanceCreationEvent WITHIN 1 WHERE TargetInstance ISA "_
    & "'Win32_Process'",,,,nvsSink("Created: ")

objServices.ExecNotificationQueryAsync sink,"SELECT * FROM "_
    & "__InstanceDeletionEvent WITHIN 1 WHERE TargetInstance ISA "_
    & "'Win32_Process'",,,,nvsSink("Deleted: ")

Do
   WScript.StdOut.Write(".")
   WScript.Sleep(1000)
Loop

Sub SINK_OnObjectReady(objObject, objAsyncContext)
    set obj=objObject.TargetInstance
    WScript.Echo
    WScript.Echo Hour(Now) & ":" & Minute(Now) & ":" & Second(Now)_
    & " " & objAsyncContext.Name & obj.Name & " : " & obj.CommandLine
End Sub
```

To use an asynchronous method you must create an SWbemSink object and pass that object to the method. Next, create status functions, such as OnCompleted, OnObjectPut, OnObjectReady, and OnProgress, to monitor the progress of that particular sink. You can also pass into the sink a context parameter that identifies the calling function if you wish to reuse the sink. The script then receives notifications from the sink of the query progress and returns to the data when it is ready.

In addition to listing processes, use events to track event logs or registry changes or to interface with other Microsoft APIs that support a scripting API interface to their event handling. Windows is event driven, and a lot of good information and status is available from tapping into these information beacons.

Statements

Statements are among the most fundamental programming techniques with which you should familiarize yourself. Statements provide body to a particular language and are the base elements for a language's syntax and structure. The common statements for JScript and VBScript are shown in Tables 2.8 and 2.9, respectively. The syntax for any given statement is often language dependent, but the concepts are the same across any language. The following sections demonstrate some of the more common statements and caveats that you may encounter during your scripting.

TABLE 2.8 Built-in JScript statements

/*..*/	catch	function	throw
//	continue	if...else	try
@cc_on	debugger	Labeled	var
@if	do...while	return	while
@set	for	switch	with
break	for...in	this	

TABLE 2.9 Built-in VBScript statements

Call	Exit	Property Get	Set
Class	For Each...Next	Property Let	Stop
Const	For...Next	Property Set	Sub
Dim	Function	Public	While...Wend
Do...Loop	If...Then...Else	Randomize	With
Erase	On Error	ReDim	
Execute	Option Explicit	Rem	
ExecuteGlobal	Private	Select Case	

Loops

Loops repetitively perform a set of operations a specified number of times. Loops often are nested (one loop inside another) to iterate in multiple dimensions. Take, for example, a pair of nested loops used to output a matrix. Also notice the use of the method WScript.StdOut.Write. This method provides a great alternative to the more basic WScript.Echo when you don't want to terminate every statement with a new line. However, if you decide to use the StdOut or StdIn text stream objects, you must use them with the *CScript.exe* (not *WScript.exe*) WSH engine. They will not work with WScript. You may have noticed this method was also used in Listing 2.9 to output the line of dots in the asynchronous events example.

LISTING 2.10   VBScript loop (CScript only)

```
WScript.Echo ("This nested loop will output a 9x9 matrix")
For i = 1 to 9
    For j = 1 to 9
    WScript.StdOut.Write ("(" & i & "," & j & ")")
    Next
    WScript.Echo
Next
```

Enumerators

Use enumerators to iterate through a *collection* of objects. A collection is an object typically returned from a WMI or ADSI query; for example, all the users in an Organizational Unit (OU) or all the members of a group.

An earlier example demonstrated querying WMI for a property of a specific item.

```
Set colQFE = objServices.ExecQuery("select * from_
    Win32_QuickFixEngineering where HotFixID='Q329048'")
```

This query returns a collection of all objects where the HotFixID equals Q329048. Alternatively, we could return all of the Win32_QuickFixEngineering items:

```
Set colQFE = objServices.ExecQuery("select * from_
    Win32_QuickFixEngineering")
```

and then loop through each item looking for whether the HotFixID value matches. Iterating through the results in this way is not as efficient as querying for the exact result we prefer, but this practice is useful if you aren't sure what you are looking for to begin with and want to start with a larger (or the entire) list.

Let's further look at this by listing the shares and disk space for a computer, shown in Listing 2.11.

ON THE CD

LISTING 2.11   Iterating through a collection using VBScript

```
Const LOCAL_DRIVE = 3
Const NETWORK_DRIVE = 4
Dim strServer
Dim iFreeSpaceGb
Dim iTemp
strServer="panzer"

Set objLocator=CreateObject("WbemScripting.SWbemLocator")
Set objServices=objLocator.ConnectServer(strServer)

strQuery="SELECT * FROM Win32_LogicalDisk"
Set colLogicalDisk = objServices.ExecQuery(strQuery)

WScript.Echo "Displaying Local and Network Drive free space "_
    & "for server : " & strServer

For Each objLogicalDisk in colLogicalDisk

Select Case objLogicalDisk.DriveType
    Case LOCAL_DRIVE, NETWORK_DRIVE
        if isNull(objLogicalDisk.FreeSpace) then
            iFreeSpaceGb = 0
        Else
            iTemp=CSng(objLogicalDisk.FreeSpace/1024^3)
            iFreeSpaceGb=FormatNumber(iTemp,1)
        end if
        WScript.Echo objLogicalDisk.Name & " : " & iFreeSpaceGb & " GB"

    Case Else
End Select

Next
```

DriveType is one of the properties of the Win32_LogicalDisk object and is referenced numerically per one of the values in Table 2.10.

TABLE 2.10 `Win32_LogicalDisk.DriveType` property table

Value	Type
0	unknown
1	no root directory
2	removable disk
3	local disk
4	network drive
5	compact disc
6	RAM disk

For documentation purposes, it is clearer to declare the types we are interested in as constants representing the `DriveType` property values, as follows:

```
Const LOCAL_DRIVE = 3
Const NETWORK_DRIVE = 4
```

This declaration allows us to substitute the word *LOCAL_DRIVE* instead of the constant value of 3, which helps the readability of the code.

Next, we declare the variable `strServer` and set it to a string value of a machine name; in this case, the machine name EUROPA.

```
Dim strServer
strServer="EUROPA"
```

TIP

You may have guessed, but this example also demonstrates how easy it is to use WMI on remote computers. Try changing the value of `strServer` to other computers on your network to enumerate their drives. (You must have appropriate permissions, and the remote computer must support WMI.)

Next, we must create the appropriate WMI objects to connect to the computer designated in `strServer`. Notice, too, that we are returning all records from the `ExecQuery`, and we are not filtering criteria, as we did in previous examples.

```
Set objLocator=CreateObject("WbemScripting.SWbemLocator")
Set objServices=objLocator.ConnectServer(strServer)

strQuery="SELECT * FROM Win32_LogicalDisk"
Set colLogicalDisk = objServices.ExecQuery(strQuery)
```

The next portion of script demonstrates a few statements in action. First, enumerate through every object in the collection using the VBScript For...Each statement. JScript requires the use of a special enumerator object, which we'll get to in a bit. The script then walks through every object in the collection and tests whether the DriveType is that of a Local Disk or Network Drive. Because we declared the constants earlier, we can refer to the constant name and not the value. You can see that referring to constants instead of values (Case LOCAL_DRIVE, NETWORK_DRIVE instead of Case 3,4, respectively) improves readability of the code. The Select...Case statement is preferable to multiple If...Then statements for testing more than one condition. In this example, we can perform multiple tasks, depending on the DriveType.

```
WScript.Echo "Displaying Local and Network Drive free space "_
    & "for server : " & strServer

For Each objLogicalDisk in colLogicalDisk

Select Case objLogicalDisk.DriveType
    Case LOCAL_DRIVE, NETWORK_DRIVE
```

Use the Select...Case Case Else to handle defaults (even if it is empty), and terminate the statement with End Select. This conditional could also have been handled with a slightly more complex If...Then statement.

```
If objLogicalDisk.DriveType = LOCAL_DRIVE_
   or objLogicalDisk.DriveType = NETWORK_DRIVE Then
...
Else (optional)
...
End If (mandatory)
```

Use either method—whichever seems more straightforward to you. Several mechanisms are available to solve the problem, and whereas one may be slightly more efficient or cleaner, ultimately, if you are scripting "quick and dirty," you should use what is more efficient and easier for you. As you become more proficient, you will naturally lean toward the more elegant scripting methodologies.

In the next bit of code, we do some rudimentary exception (error) handling and manipulate the free disk space for pleasant output.

When you are enumerating a collection of different objects and want to manipulate the data, try to insert code to make sure that your data values are expected and proper. For example, if you were to include all drive types (including floppy and CD-ROM drives), and a disk was not present, the Win32_LogicalDisk

FreeSpace property would return a null (not a zero). If you try to pass this null value to a function that wants a number, you will get an error.

In this example, we test for a null value, and, based on the result, set the variable iFreeSpaceGb to 0. (We could just as easily set it to a string "Not Available" and then convert the rest of our output to strings.)

```
if isNull(objLogicalDisk.FreeSpace) then
    iFreeSpaceGb = 0
Else
```

Next, we use the FormatNumber and CSng functions to convert the FreeSpace property from bytes to gigabytes. The function CSng converts the calculation to a *single* data subtype. (A single data subtype supports single-precision, floating-point numbers ranging from +/– 1.4E-45 to 3.4E38.) Although not necessary for this script, this function demonstrates forcing data type conversion to make absolutely sure of the variable's data subtype being passed to another function.) FormatNumber inserts the group delineator and sets the precision of an expression (for example, FormatNumber(2344.2321,1) outputs 2,3344.2). Lastly, we output the results and end the Select and Loop statements.

```
        temp=CSng(objLogicalDisk.FreeSpace/1024^3)
            iFreeSpaceGb=FormatNumber(temp,1)
        end if
        WScript.Echo objLogicalDisk.Name & " : " & iFreeSpaceGb & " GB"
    Case Else
End Select

Next
```

You might still be asking why this section was called "Enumerators." Why?

Whereas VBScript uses the syntax For...Each statement to iterate a collection, JScript directly uses the Enumerator object to walk through a collection. Listing 2.12 presents the same example coded for JScript to show not only the syntactical differences between VBScript and JScript but also to show that fundamentally, regardless of what language you choose, the script flow remains the same.

LISTING 2.12 Enumerating through a collection using JScript

ON THE CD

```
var LOCAL_DRIVE = 3;
var NETWORK_DRIVE =  4;
var iFreeSpaceGb;
```

```
var strServer="panzer";
var strOutput;

objLocator = new ActiveXObject("WbemScripting.SWbemLocator");
objServices =  objLocator.ConnectServer(strServer);
strQuery="SELECT * FROM Win32_LogicalDisk";
colLogicalDisk = objServices.ExecQuery(strQuery);

strOutput="Displaying Local and Network Drive free space " +
 "for server : " + strServer;

WScript.Echo(strOutput);


eLogicalDisk = new Enumerator(colLogicalDisk);
for (;!eLogicalDisk.atEnd();eLogicalDisk.moveNext()) {
    x=eLogicalDisk.item()
    switch (x.DriveType) {
        case LOCAL_DRIVE:
        case NETWORK_DRIVE:
            if(!x.FreeSpace){
                iFreeSpaceGb = 0 }
            else {
            iFreeSpaceGb=(x.FreeSpace/Math.pow(1024,3)).toFixed(1);
            }
            WScript.Echo(x.Name + " : " + iFreeSpaceGb + " GB");
        break;
        default:
        break;
        }
    }
```

With JScript, you instantiate a new ActiveXObject object instead of the VBScript CreateObject. Also, notice that each line is terminated with a semicolon and that the loops and conditionals are contained within braces ({...}) instead of ending with a type of End statement. JScript does not have a line-continuation character like VBScript (_), so to ensure that the code lines fit on a page, the example breaks the longer output string into a variable strOutput and then displays the variable.

```
var LOCAL_DRIVE = 3;
var NETWORK_DRIVE =  4;
var iFreeSpaceGb;
var temp;
```

```
var strServer="panzer";
var strOutput;

objLocator = new ActiveXObject("WbemScripting.SWbemLocator");
objServices =  objLocator.ConnectServer(strServer);
strQuery="SELECT * FROM Win32_LogicalDisk";
colLogicalDisk = objServices.ExecQuery(strQuery);

strOutput="Displaying Local and Network Drive free space ";
strOutput+="for server : " + strServer;

WScript.Echo(strOutput);
```

The `Enumerator` object contains the collection object from the `ExecQuery`. It also contains its own methods: `atEnd`, `item`, `moveFirst`, and `moveNext`. Use these methods in a loop to iterate through the enumerator object.

```
eLogicalDisk = new Enumerator(colLogicalDisk);
for (;!eLogicalDisk.atEnd();eLogicalDisk.moveNext()) {
```

Using JScript, you can use the `for...in` loop statement to iterate through an object's properties. However, to iterate through a collection, you must use an Enumerator object, because objects are built into JScript and have special support.

If you are new to JScript, you may notice an oddity of the `for…loop` syntax—it apparently begins with a semicolon. Actually, this character just signifies that the first argument is blank. Also, the JScript `for` loop is a bit more flexible than the VB-Script `For…Next` or `For…Each` loops, because it can contain any function in the arguments. Notice that in our statement, we have:

```
for ( ; !eLogicalDisk.atEnd() ; eLogicalDisk.moveNext())
```

Broken down, we can say that starting now, execute the `moveNext` method until the last enumerator object is reached. But whereas this function is similar to the VBScript `For…Each` object in collection, we could put any function in the JScript loop. For example:

```
for (foo ; bar ; f(foobar))
```

Of course, your function `f(foobar)` should get you from `foo` to `bar` or else you will be stuck in an infinite loop; but you can see how the function is more flexible than simply iterating arithmetically or consecutively.

Also, to simplify subsequent references, we have set the variable x to be the current enumerator item. We could have just as easily used the reference eLogical-Disk.item() to reference the properties instead.

```
x=eLogicalDisk.item()
```

JScript uses a switch...case, which is similar to the VBScript Select Case. Note the break statement. This statement causes the program to exit the switch statement. Without it, a later case could satisfy an earlier case, which could cause unforeseen problems. Also notice the default statement. This statement processes any criteria that have not been satisfied by an earlier case.

Also notice that in place of the VBScript isNull function, we use the inequality operator !. This example uses the JScript Math.Pow function to calculate 1024^3 and the JScript toFixed function instead of FormatNumber.

```
switch (x.DriveType) {
    case LOCAL_DRIVE:
    case NETWORK_DRIVE:
        if(!x.FreeSpace){
            iFreeSpaceGb = 0
            }
        else {
            iFreeSpaceGb=(x.FreeSpace/Math.pow(1024,3)).toFixed(1);
            }
        WScript.Echo(x.Name + " : " + iFreeSpaceGb + " GB");
    break;
    default:
    break;
    }
}
```

As you can see, the structure of the code remained intact between the VBScript and JScript versions, but the syntax differed substantially, both in the statements and functions or methods used. Some people feel that VBScript is a bit more natural, or easier, to use and pick up than JScript. However, JScript can be learned, and if you see yourself advancing to C# or another advanced language, you will find more similarities between these advanced languages and JScript.

Operators

Operators compare or operate on two expressions. Table 2.11 lists the operators used in both JScript and VBScript, respectively. JScript supports assignment operators, which both evaluate and operate an expression. For example, i++ in JScript is the same as i=i+1 in VBScript.

TABLE 2.11   Built-in JScript and VBScript operators

	JScript	*VBScript*
addition assignment	+=	
addition	+	+
concatenate	+	&
assignment	=	=
bitwise AND	&	
bitwise OR	|	
bitwise XOR	^	
comma	,	
conditional ternary	?:	
decrement	--	
division assignment	/=	
division	/	/
equality	==	=
exclusion		xor
exponential (x^y)	Math.pow(x,y)	^
greater than	>	>
greater than or equal to	>=	>=
identity *	===	=
implication		Imp
increment	++	
inequality	!=	<>
integer division		\
less than	<	<
less than or equal to	<=	<=
logical AND	&&	And
logical NOT	!	Not
logical OR	||	or
modulus assignment	%=	
modulus	%	Mod
multiplication assignment	*=	
multiplication	*	*
nonidentity *	!==	
subtraction assignment	-=	
subtraction	-	-
variable type	type of operator	TypeName function

Excluded: unsigned shifts and bitwise operators.

* Identity and nonidentity operators in JScript are similar to the equality and inequality operators, except they do not attempt type conversion.

Also note that the assignment operator for JScript is very different from the equality operator. The JScript expression `i=1` sets the variable `i` to the value `1`, whereas the equality operator of `i==1` tests `i` to see if it equals the value `1`. This difference is a common source of bugs in a JScript program and is demonstrated in the following code:

```
i=0;
if(i=1) {
    WScript.Echo("Oops, used an assignment operator.")
    }
WScript.Echo(i); //returns 1
```

The following script is probably what the script writer intended:

```
i=0;
if(i==1) {
    WScript.Echo("Used the equality operator.")
    }
WScript.Echo(i); // returns 0
```

VBScript does not differentiate between the assignment and equality operators, and = is used for both.

DEBUGGING

Inevitably you will spend quite a bit of time debugging your program of errors or mistakes. All programmers, no matter their skill and experience, spend time debugging their programs. Debugging takes on many forms, from simply correcting a misspelled variable name to entirely rewriting a bit of code. The most obvious bugs occur at runtime with an error message that halts the execution of your script; however, be careful of deeper, more subtle bugs, such as those that cause errors in the calculation of data. In these cases, the script may run from beginning to end without any sign of a problem, but you can end up with the wrong answer. In either of these examples, stepping through the program one line at a time is helpful in finding and correcting the mistakes in real-time as they are occur. It is also helpful to find and determine the answer using a method independent from your script and then comparing your findings to the results of your script. For example, if your script lists all users in a domain group, use Active Directory Users and Computers to manually verify that the script output matches that of the independent tool.

Also, to expose bugs, run your script in different environments under a variety of loads. For example, suppose your script inspects all of the computers in your specified domain. The results from running a script in a test environment of 10 computers compared to a production environment of 500 or 1,000 computers can expose bugs in your script not present in the smaller environment. These bugs might be related to the number of computers (such as if a variable is not sized to handle the larger count) or just expose a bug due to the possibly larger diversity in targets that your script must run against.

Syntactical mistakes are the easiest to correct, especially when using a programming environment such as the Microsoft Office VBA Visual Basic Editor. This editor flags and alerts you to certain mistakes as they occur when you type them (such as omitting a parenthesis). The Visual Basic Editor also includes a robust debugger that lets you step through your script line-by-line and observe its behavior as it runs. You can inspect the value of your programs variables using the editor's various Watch windows.

Windows Script Host also supports a debugger. To activate this debugger, execute your WSH script through either the CScript or WScript engine with the //D parameter to enable active debugging, as the following command demonstrates:

```
CScript //D myscript.vbs
```

You will then be prompted to choose which debugger to use (if you have multiple debuggers installed), as seen in Figure 2.8.

After you have selected a debugger, your script is loaded into the editor, but it does not begin to run. At this point you can choose how you wish your program to run.

Using the Microsoft Script Editor, from the Debug menu, you can choose many options of how you wish to run your program in the editor. Most common are *Continue*, which runs the script until the next breakpoint (we'll cover breakpoints in a bit) or to the end of the script if there are no errors; *Step Into*, which advances the program to the next line; and *Step Over*. Step Over does not enter a called procedure, whereas Step Into steps into any procedure, wherever the program takes it. For example, suppose you are stepping through your main branch of code and reach a command that calls a subroutine that loops through many entries. Step Over processes the procedure, but it does so without walking you through it line by line. It executes the code in the procedure automatically and then stops at the line immediately following the call to the procedure and returns manual control to you. Step Into, on the other hand, advances you into the procedure and stops and waits at the procedure's first line.

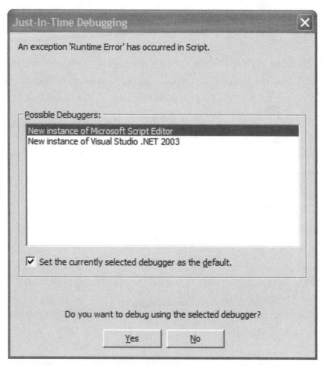

FIGURE 2.8 Debug your console-based applications using the built-in Microsoft Script Editor.

Figure 2.9 shows an example of trying to run a script with an obvious error in the code using the Microsoft Script Editor. Notice that the script editor alerts you of errors and highlights the line of the error.

Also in Figure 2.9, notice the Watch window and the Command Window - Immediate. The Watch window shows the values of current variables. You must add them by right-clicking the variable and then selecting Add Watch. In Figure 2.9, notice that the value of the variable strName is Topaz, but it was initially set to Trooper. This discrepancy is because we changed the value of strName in the Command Window - Immediate. You can set or display variables and run program commands directly in this window. For example, if you type WScript.Echo ("Something") in this window, the text Something will occur in the console that you had initially called your program.

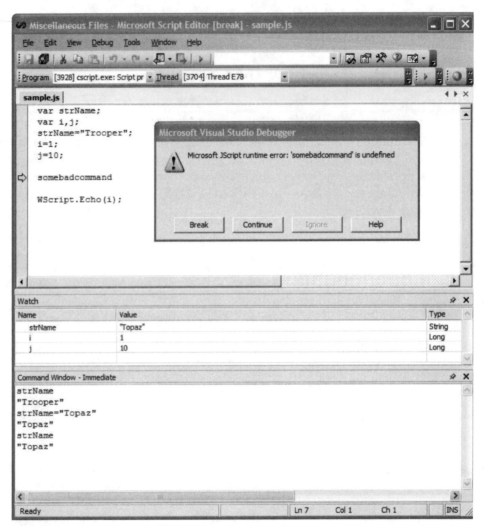

FIGURE 2.9   The Microsoft Script Editor provides a debugger for WSH console-based scripts.

Stepping through a program line by line gives you a good inspection of exactly what your script is doing and when. However, you can use a breakpoint to run to a specific point in your script and then stop. A breakpoint is a stop sign in your script. The script runs until the line that contains the breakpoint and then halts the debugger. From this point, you can step through the program, inspect variables, and continue to debug your program.

Install and play with the debuggers using a simple program, and familiarize yourself how they work—you will be pleased to add this tool to your scripting arsenal.

SUMMARY

The examples of this chapter were meant to give you a bit of flavor of both VBScript and JScript, as well a taste of the framework and architecture of a basic Windows script. Many of the objects, methods, and properties that you will use to create your scripts are the same as (or very similar to) those entities used by professional software developers.

KEY POINTS

- Scripting provides a completely customizable, reusable toolset from which you can build solutions to reduce repetitive work or automate a large or lengthy task.
- Both VBScript and JScript provide a similar set of scripting tools, and your choice of one over the other will more likely depend on your target audience, application, preference, or comfort level.
- Windows Script Host comes in two flavors: the console-based CScript and the window-based WScript. Some methods and properties work only with a specific console.
- Use comments to document your code and help others understand your thought process.
- Make a habit of declaring your variables to avoid errors as well as exhibit good programming practice. Also, declaring your variables lets a debugger know to show the variables in the Locals window.
- Objects represent a programming entity that includes both the data and functionality for the entity; for example, the WMI connection to a remote computer is referenced using an object.
- Methods provide the actions that an object can initiate; for example, the method ExecQuery returns a set of WMI data from a remote computer.
- Functions provide actions that can be called from a script or program, independent of any object.
- Properties are the characteristics of the object that can be read (and sometimes set); for example the Win32_LogicalDisk class has a property named FreeSpace, which represents the remaining space on a particular drive.

- Events, one of the primary signaling mechanisms for the Windows operating system, can be sent by objects, applications, and even users.
- Statements provide the vernacular of a particular language and are the base elements for a language's syntax and structure. Loops, conditionals, and declarations are all examples of statements.
- Operators compare or operate on two expressions; for example, =, >, *.

Chapter 3, "Windows Scriptable Technologies," is introduces some of the more popular Windows scripting technologies: Windows Script Host (WSH), Windows Management Instrumentation (WMI), and Active Directory Services Interface (ADSI).

3

Windows Scriptable Technologies

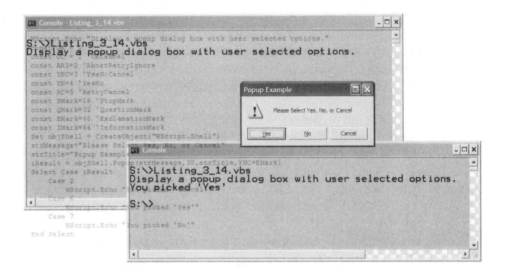

In Chapter 2, we covered programming basics demonstrated by the two most popular Microsoft scripting languages, VBScript and JScript. With these languages you can write scripts and tap into many different Microsoft application programming interfaces (APIs). Many of the components of the Windows operating system, as well as some enterprise software such as Microsoft® Exchange, support script access to manage their behavior.

In this chapter, we discuss a few of the more popular Windows scriptable technologies: *Windows Script Host (WSH), Windows Management Instrumentation (WMI),* and *Active Directory Services Interface (ADSI).* This trio of technologies provides a huge opportunity for systems administrators and engineers to automate routine, large, or distributed tasks. Many of the scripts contained in this book repeatedly leverage these technologies. After working with these tools, you will be able to transfer knowledge learned to pick up fairly quickly new scripting technologies as they are released.

WINDOWS SCRIPT HOST

The Windows Script Host (WSH) provides a language-independent mechanism for managing and accessing the Windows platform through scripts. Microsoft provides support for both VBScript and JScript, and WSH recognizes the scripting language by its file extension (.js or .vbs). WSH processes scripts using the engine called directly (for example, *cscript script.vbs*) or by using the registered WSH script engine (for example, typing a script name into the command line, such as *script.vbs*, followed by pressing the Enter key). CScript provides a command-line interface, whereas WScript provides a Windows-based interface.

In addition to the VBScript and JScript language interpreters, WSH provides access to a number of Windows platform resources, including the file system, network, shell, environment variables, and executable wrapper from which you can launch other tools from within your script. Additionally, WSH supports access to any ActiveX COM-supported automation interface, such as Microsoft Active Directory (AD) or WMI. These objects, methods, and properties offer access and write scripts to provide Windows management or maintenance functionality. Access the WSH infrastructure through its supported objects.

WScript

The WScript object provides basic programming functionality within WSH. In Chapter 2, we used the WScript object extensively to display output in the console or a dialog box, pause the script, and capture events using the asynchronous event handler. The WScript object performs other useful tasks as well.

WScript.Argument

The WScript object provides a mechanism for capturing command-line arguments in your script. Arguments are parameters passed to a program externally, such as from the command line. Let's look at an example using the common tool ping.

```
ping -n 10 127.0.0.1
```

In this example, `ping` is the command, and the rest of the command is arguments. `-n 10` is actually two arguments; `-n` specifies the number of echo requests to send, and `10` is the actual number sent. The address `127.0.0.1` is the target address for what you want to ping. You may take this for granted, but consider the logic. `ping 127.0.0.1 -n 10` also works. And if you use `-n` without a number following it, `ping` responds with an error specific to the number argument.

In your scripts, too, you will likely desire the use of arguments, and implementation is quick and easy using WSH. Get the arguments by accessing the property `WScript.Arguments`. The returned collection is a `WshArguments` object. In Listing 3.1, we capture and output the first argument sent to the script.

LISTING 3.1   JScript example of WScript.Arguments

```
var colArgs = WScript.Arguments;
var eArgs = new Enumerator(colArgs);
WScript.Echo(eArgs.item(0));
```

Preferably, we can shorten this to simply:

```
WScript.Echo(WScript.Arguments(0));
```

Let's expand this script a bit to show how to capture all of the arguments and then decide what to do based on what the arguments received. Argument-handling routines are highly reusable, and you may find your own argument routines to be essential components in your scripting toolbox. Listing 3.2 shows a basic argument-handling routine. It is argument-location independent (for example, `-a -f filename`, or `-f filename -a`) and also supports dependent arguments (such as `-f filename`). Also notice that we have iterated through the `WshArguments` collection using the `length` and `item` properties of the `Argument` object.

LISTING 3.2   Basic argument-handling routine

```
var colArgs = WScript.Arguments;
for(i=0;i<colArgs.length;i++)
    {
    switch(colArgs.item(i))
        {
        case "-a":
            WScript.Echo("flag -a received");
            break;
```

```
        case "-f":
                if(i+1<colArgs.length) {
                strText=colArgs.item(i+1);
                WScript.Echo("Dependent parameter: " + strText);
                i++;}
                else {
                    WScript.Echo("flag -f needs additional argument");
                    WScript.Quit(1);}
            break;

        case "-?":
        default :
            WScript.Echo("Flag not recognized. Display Help");
            WScript.Quit(1);
            break;
        }
    }
    WScript.Echo( colArgs.length + " Arguments Received");
```

The previous code sample loops through every argument that is passed and can also take specific action based on any (or all) arguments set. Also, the code supports dependent arguments, meaning if it sees the argument -f, it will look for and accept a following argument as some text (such as a filename). Error handling consists of ensuring that a dependent parameter is present. However, it does not check the validity of the arguments, nor does it include logic to error if the logic -f -a is entered. In this case, the script treats -a as the dependent parameter to -f. When run, we see output like that shown in Figure 3.1.

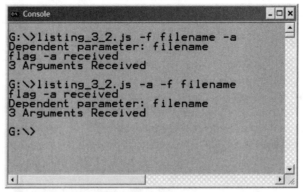

FIGURE 3.1 Argument-handling routines offer an easy means for configuring script-run parameters.

The previous script is a basic example that could definitely benefit from error handling (for example, validating a proper filename or searching for erroneous parameters before displaying anything to the user). You can see not only how easy it is to collect the parameters, but also how flexible subsequent argument manipulation can be.

WScript.Shell

The `objShell` object provides scripting access to the Windows shell. Use the shell object's methods to read or write to the registry, execute applications, and log events to the event log.

Instantiate the shell object with the following code:

VBScript
```
Set objShell = WScript.CreateObject("WScript.Shell")
```

JScript
```
var objShell = WScript.CreateObject("WScript.Shell");
```

With this object, you can access `WshShell` objects, methods, and properties, such as reading or writing to the registry.

WshShell—*Registry*

The `WshShell` object provides three methods for accessing the registry on a local machine: `RegRead`, `RegWrite`, and `RegDelete`. Before jumping into the shell object's registry methods, let's review the parts of the registry.

Use the WMI registry provider to access the registry on a remote machine.

TIP

Whenever working with the registry—especially when creating or deleting data— be sure to back up your registry first.

NOTE

The Windows registry editor's (*regedit.exe*) standard display shows two panes, similar to the left and right panes shown in Figure 3.2. The left pane contains *keys* and *subkeys*. The right pane contains *values*. Values are comprised of three parts— the *Name*, the *Type*, and the *Data*. The top-level keys (called *root keys*) can be abbreviated in your script as shown in Table 3.1.

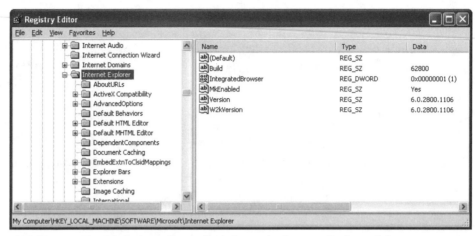

FIGURE 3.2   The Windows registry editor shows the keys and values of the system registry.

TABLE 3.1   Root key abbreviations

Root Key	Abbreviation
HKEY_CLASSES_ROOT	HKCR
HKEY_CURRENT_USER	HKCU
HKEY_LOCAL_MACHINE	HKLM
HKEY_USERS	HKEY_USERS
HKEY_CURRENT_CONFIG	HKEY_CURRENT_CONFIG

The WshShell object's registry manipulation method requires both the name of the key *and* the specific value name that you wish to read or write. This requirement is a bit different from that of the command-line *reg.exe* binary, which dumps keys as well as values.

For example, the key HKLM\SOFTWARE\Microsoft\Internet Explorer contains several values, including Build, Integrated Browser, Version, MkEnabled, W2kVersion, and possibly others, depending on the software installed on the machine. To read or write to a value, you must specify both the key and the value name, such as HKLM\SOFTWARE\Microsoft\Internet Explorer\Version.

Each value is defined as a specific data type. When creating a value using the RegWrite method, you can specify one of the data types shown in Table 3.2.

TABLE 3.2 WshShell registry data types

Data	Type	Example
string value	REG_SZ	Q123456 or C:\WINDOWS\
binary value	REG_BINARY	01 02 01 00 05 01 01 01
DWORD value	REG_DWORD	0x00000002 or 21
expandable string value	REG_EXPAND_SZ	%systemroot%\system32\

Note: The multi-string value data type REG_MULTI_SZ is not supported by the method RegWrite.

The following JScript example in Listing 3.3 displays the version of Internet Explorer on the local machine.

ON THE CD

LISTING 3.3 Reading the local machine registry

```
var strRootKey="HKLM";
var strRegPath="\\SOFTWARE\\Microsoft\\Internet Explorer\\version";
var objShell = WScript.CreateObject("WScript.Shell");
strValue=objShell.RegRead(strRootKey+strRegPath);
WScript.Echo(strValue);
```

TIP

In JScript, the backslash denotes a special character (such as the newline character, \n), so when you want to use an actual backslash, you must escape it with a second backslash, like so: \\.

Listing 3.4 shows how to insert a new value in the registry. Specify the entire key name, including the value name, the value, and, optionally, the value type. An unspecified value type defaults to a REG_SZ type. (Note that running this script adds the entry to your registry.)

ON THE CD

LISTING 3.4 Adding or updating a registry key and value

```
var strRootKey="HKLM";
var strRegPath="\\SOFTWARE\\Gadget\\ValueName";
var objShell = WScript.CreateObject("WScript.Shell");
strValue=objShell.
RegWrite(strRootKey+strRegPath,"ValueData","REG_SZ");
```

Deleting a registry key or value is also a straightforward procedure. Specify the key and value name to delete.

```
strValue=objShell.RegDelete("HKLM\\SOFTWARE\\Gadget\\ValueName");
```

Delete an entire key by ending your string with a key name followed by a backslash.

```
strValue=objShell.RegDelete("HKLM\\SOFTWARE\\Gadget\\");
```

WshShell—*Execute Applications*

The WshShell object also lets you execute applications from within your script. This functionality is tremendously useful because it allows you to call existing applications from your script and use the returned data.

Think of your script as a harness or wrapper around the executable that you wish to run. A few examples follow:

- A login script that pings a remote machine to verify network connectivity before running a program that requires a LAN connection.
- A script that specifies the run parameters of a Windows resource kit tool such as RoboCopy. Your script might contain a source and destination array of network names, which you loop through and spawn RoboCopy instances that each run under custom and automated parameters.

The two shell methods to execute applications are Exec and Run.

The Exec method is supported in WSH version 5.6. (You must use the Run method on machines running earlier versions of WSH.) The Exec method returns a WshScriptExec object, which supports the standard *stream* objects StdIn, StdOut, and StdErr. This support enables the script to read the output directly from the application without first having to save the output to a file. The following script in Listing 3.5 shows an example of how to use the Exec method to execute an external command and then process the results using the standard stream properties.

ON THE CD

LISTING 3.5 The Exec method supports standard streams, which makes it easy to capture output from an external command

```
WScript.Echo("Check Network using command line Ping")

Set objShell = CreateObject("WScript.Shell")
Set objWshScriptExec = objShell.Exec("%comspec% /c ping 127.0.0.1")
```

```
Set objStdOut = objWshScriptExec.StdOut
While Not objStdOut.AtEndOfStream
    strLine = objStdOut.ReadLine
    if instr(strLine,"Re") then
        strOutput=strOutput & strLine & vbcrlf
    end if
Wend


WScript.Echo(strOutput)
```

The premise of this script is to execute the shell object's Exec method and capture, format, and display the output. The output from the Exec method is kept in a new WshScriptExec object. (In the example script, this object is stored in a variable called objWshScriptExec.) StdOut, StdIn, and Stderr, all properties of the Wsh-ScriptExec object, themselves have additional methods and properties. This example script iterates through the StdOut property of the ObjWshScriptExec object.

The TextStream object is covered a bit later in a discussion of the FileSystem-Object. However, many of the methods and properties that apply to the TextStream object apply to the Exec method's standard stream as well.

NOTE

WSH standard stream processing does not have the flexible cursor (file pointer) positioning found in some other languages. You can work around this limitation by reading one line at a time into a temporary string and then parsing that line independently for the data that you seek.

TIP

```
While Not objStdOut.AtEndOfStream
    strLine = objStdOut.ReadLine
    if instr(strLine,"Re") then
        WScript.Echo strLine & vbcrlf
    end if
Wend
```

This code loops through each line of the output and looks for the string "Re", which corresponds to the ping command's return of *Request timed out* or *Reply from*. If the script finds a match, it concatenates the line to the previous result, which is displayed at the end. The VB constant vbcrlf represents a new line (in JScript, the newline character is \n) and instructs the code to insert a carriage return or line feed. Use these characters to format output that you do not want to run on\in a single line.

The %comspec% environment variable represents the command interpreter for the system. Windows 9x uses command.exe, and Windows NT and later use cmd.exe. Using the %comspec% environment variable ensures compatibility across a wider range of platforms that may be running different versions of WSH.

Older versions of WSH do not support the Exec object, but you can achieve similar functionality using the Run method. The Run method does not support access to the standard output, but you can work around this by using text files. Simply redirect the application's output to a text file (using >), and read the results from the output text file. Later in this chapter, we cover the FileSystemObject and show an example of how to use the TextStream object together with the Run method to do just this.

WshShell–*Log Events*

Use the shell object to log events to the Application Event Log of either a local or remote machine. The syntax to log an event is shown in Listing 3.6.

LISTING 3.6   Log an event to the Application Event Log using the shell object

```
WScript.Echo("Log an Event")
  strMessage="Success message logged to the Application Event Log"
Set objShell = CreateObject("WScript.Shell")
objShell.LogEvent 0, strMessage
```

Running this script logs an event to the Application Event Log of the local computer, as shown in Figure 3.3.

The LogEvent method supports three parameters: type, message, and computer (optional).

```
objShell.LogEvent type, message, computer
```

The message is a string, and computer is optional. If you specify the name of a computer, method logs that message on the remote computer's event log. Table 3.3 lists the different types of events that can be logged using LogEvent.

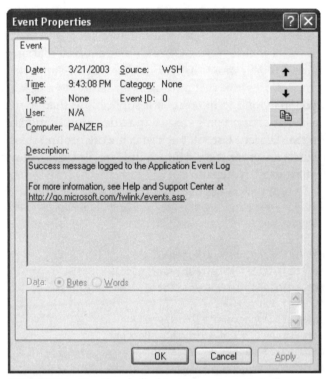

FIGURE 3.3 Use the shell object to post events to the Application Event Log.

TABLE 3.3 LogEvent type parameters

Value	Event Type
0	success
1	error
2	warning
4	information
8	audit success
16	audit failure

WshShell—SendKeys

WSH supports COM and can interoperate directly with applications such as Microsoft Office that support COM automation. Automation is the process by which an Object Model exposes its API to other COM-supported clients. For example, you can create a script that outputs information to a new Word document. However, you may want to interact with applications that do not support automation. The SendKeys method provides a way to drive applications that do not expose any automation objects. Use SendKeys with the Exec or Run and AppActivate methods to manipulate an application. The example in Listing 3.7 shows how to create a simple text file using Notepad.

LISTING 3.7   SendKeys enables you to manipulate applications that do not support
ON THE CD   automation.

```
WScript.Echo("Drive Notepad")
Set objShell = CreateObject("WScript.Shell")
Set objWshScriptExec=objShell.Exec("notepad.exe")
WScript.Sleep 1000
objShell.AppActivate "Notepad"
strMessage="Enter this into Notepad"
objShell.SendKeys strMessage
objShell.SendKeys "%fac:\test.txt%s"
WScript.Sleep 1000
objShell.SendKeys "%{F4}"
WScript.Echo("Used Notepad to create and save c:\test.txt")
```

Sending keys to a program can be a risky procedure because controlling the timing and predicting the outcome of the application can be difficult. For example, try running the above script twice. Notice that because the file already exists from the first run, Notepad prompts you to replace the file. Because we did not code for this additional keystroke, Notepad effectively breaks out of the script. Notice, too, the use of the Sleep function. This delay slows down the script to ensure that any preceding tasks have finished. For example, without the first delay, the script may try to activate Notepad before it has finished loading.

For the previously mentioned reasons, refrain from using SendKeys except for the most basic of applications and, even then, only when you know absolutely how the application will respond, every time.

TIP

The shell object method AppActivate activates an application by its name or process ID and changes focus to that application. If you have multiple applications of the same name open (and you choose not to use the process ID), AppActivate changes focus to one of them arbitrarily.

The SendKeys method parameters include the string of characters that you wish to send. Table 3.4 analyzes the string %fac:\test.txt%s that is sent to Notepad.

```
objShell.SendKeys "%fac:\test.txt%s"
```

TABLE 3.4 Analysis of sample SendKeys method parameters

%f	a	C:\test.txt	%s
Alt-F	Save As	C:\test.txt	Save
Accesses the File menu	Selects the shortcut key A for Save As	The resulting dialog box defaults the cursor into the Filename field, so simply send it the filename	Use Alt-S to save the file

WshShell—*Environment Variables*

WSH provides access to a system's environment variables through the shell object's Environment property. This property returns a collection of WshEnvironment objects, which can be enumerated using the VBScript For...Each loop or the JScript Enumerator object.

The environment is broken into four distinct groups: SYSTEM, USER, VOLATILE, and PROCESS. The script in Listing 3.8 displays the current SYSTEM and USER environment variables.

ON THE CD

LISTING 3.8 Using the shell object Environment property to display the system and user environment variables

```
WScript.Echo("Display Environment Variables- SYSTEM" & vbcrlf)
Set objShell = CreateObject("WScript.Shell")
Set colEnvironment = objShell.Environment("SYSTEM")
For Each objVariable In colEnvironment
    WScript.Echo(objVariable)
Next
WScript.Echo(vbcrlf & "Display Environment Variables- USER" & vbcrlf)
Set colEnvironment = objShell.Environment("USER")
```

```
For Each objVariable In colEnvironment
    WScript.Echo(objVariable)
Next
```

Notice the difference between this collection and some of the aforementioned collections. You do not need the `item` method to access the object; you can access it directly. To access a specific environment variable, specify it explicitly in the collection object as follows:

```
Set objShell = CreateObject("WScript.Shell")
Set colEnvironment = objShell.Environment("USER")
WScript.Echo(colEnvironment("Path"))
```

Likewise, you can change the writable environment variables by directly assigning new values to the variable, as shown in the following example:

```
Set objShell = CreateObject("WScript.Shell")
Set colEnvironment = objShell.Environment("USER")
colEnvironment("TEMP") = "c:\"
```

File System Object

WSH supports access to the file system through the use of the `FileSystemObject` (FSO) object model. The objects, methods, and properties of this model let you access files, directories, and drives on a system. For example, you can read, write, create, or delete files or query a drive for available space or size.

The objects of the FSO model include the following:

- `FileSystemObject`
- `Drive`
- `Drives`
- `File`
- `Files`
- `Folder`
- `Folders`
- `TextStream`

The example in Listing 3.9 shows how to use the FSO object model to return a set of folders and files from a path (in this example, the root of c:). By now, you should be familiar with the process of creating an object and using that object to access other objects, methods, and properties.

LISTING 3.9   Using the `FileSystemObject` to list the subfolders and files in a folder

```
Set objFSO = CreateObject("Scripting.FileSystemObject")
Set objFolder= objFSO.GetFolder("c:")
Set colSubFolder =objFolder.SubFolders
For Each objFile in colSubFolder
    WScript.Echo (objFile.type & " : " & objFile.name)
Next
WScript.Echo
For Each objFile in objFolder.Files
    WScript.Echo (objFile.name & "  (" & objFile.type & ")")
Next
```

In this example, we first create the object `Scripting.FileSystemObject`. Next, we get the folder object of a specific path and then iterate through the `SubFolders` and `Files` collection of the `Folder` object. Through this iteration we display the `name` and `type` of each of the folder and file objects returned from the `SubFolder` collection.

Currently, the FSO model supports only the manipulation of text files through the use of the `TextStream` object. Text files are great for temporarily storing the output of a command or script, which you can subsequently parse.

WSH versions earlier than version 5.6 do not support the `Exec` method and its standard I/O streams. A workaround is to use the shell object `Run` method and then use text files as intermediary storage for this data. The `TextStream` object supports the standard streams.

Listing 3.10 presents a script harness around the command `ipconfig`. This script launches the command so that it redirects its output to a text file. When the command is complete, the script reads the output and displays it. (Compare this listing with Listing 3.5, which uses the `Exec` method to harness the `ping` command.)

ON THE CD

LISTING 3.10   Use the `TextStream` object for reading and writing to a text file

```
WScript.Echo("Script harness for ipconfig.")
Set objShell = CreateObject("WScript.Shell")
strTemp=objShell.ExpandEnvironmentStrings("%temp%")
WScript.Echo("Writing to temp dir: " & strTemp)
strCommand="%comspec% /C ipconfig /all > " & strTemp & "\out.txt"
Return=objShell.Run(strCommand,0,true)
strOutput=ReadResults(strTemp & "\out.txt")
WScript.Echo(strOutput)
```

```
Function ReadResults(strFileName)
    Set objFSO = CreateObject("Scripting.FileSystemObject")
    If (objFSO.FileExists(strFileName)) Then
       Set objTextFile = objFSO.OpenTextFile(strFileName,1,False)
       strOutput=""
        Do While objTextFile.AtEndOfStream <> True
           strReadLine=objTextFile.ReadLine
           strOutput=strOutput & strReadLine & vbcrlf
          Loop
       objTextFile.Close
    Else
        strOutput="Error reading file."
    End If
    ReadResults=strOutput
End function
```

Let's break down the script and discuss how it works.

```
WScript.Echo("Script harness for ipconfig.")
Set objShell = CreateObject("WScript.Shell")
strTemp=objShell.ExpandEnvironmentStrings("%temp%")
WScript.Echo("Writing to temp dir: " & strTemp)
```

Notice in this example the usage of the Environment variables %temp% and, again, %comspec%. Regardless of end-user permission, %temp% directory is usually open for read and write access; its actual location may change from machine to machine. The %temp% directory is a safe and appropriate location for a temporary file such as this (as opposed to writing to the root or system directory, which may be restricted).

```
strCommand="%comspec% /C ipconfig /all > " & strTemp & "\out.txt"
Return=objShell.Run(strCommand,0,true)
```

The Run method is similar to the Exec method. It takes the following form:

```
Run(command, window style, wait on return)
```

The command is the name of the executable and its command-line properties that you wish to run. In our example, with the expanded %temp% environment variable fully expanded, the command executed within the command window is:

```
ipconfig /all > C:\DOCUME~1\temp\LOCALS~1\Temp\out.txt
```

The > symbol redirects the output to a new text file. The symbol >> similarly redirects the output, but it appends data if the file already exists.

Set the Run method window style to determine whether or not you see a command window. Setting this option to 0 (zero) means the command runs in the background. The parameter supports configurations from 0 to 10 and instructs the program how to display the window (for example, no window, minimized, or maximized). Table 3.5 shows the values for window display parameters.

TABLE 3.5 Run method window display values

Value	Description
0	Hides the window and activates another window.
1	Use when displaying a window for the first time. Activates and displays the window in its original size and position.
2	Activates window in minimized form.
3	Activates window in maximized form.
4	Displays window in current size and position, but keeps active window active.
5	Activates and displays window in current size and position.
6	Minimizes specified window and activates the next window down (in z-order).
7	Minimizes window, but keeps active window active.
8	Displays window in current state, but keeps active window active.
9	Restores a minimized window to its original size and position.
10	Displays the window based on the state of the calling program's window.

The parameter wait on return defines what the Run method should do after the command has finished executing. Setting the value to True suspends your script's execution until the command has completed, at which time it resumes. For example, if you change the command in Listing 3.10 to tracert or some other command that typically takes longer to complete and set wait on return to False, the script does not wait for tracert to complete; instead, the script displays only what tracert has begun to report. Note, however, that tracert continues to run in the background. Check back later and see that tracert has finished writing its output to the text file in the %temp% directory.

Setting the Run method window style to 0 and the wait on return value to True, as we did in the example, sets the program to not display any additional console windows. The program waits until the command has completed execution before proceeding.

After the command has run, we read its output back into the script for parsing. This practice presents a good opportunity to outsource work to a function. The function named ReadResults has one input parameter, strFileName, which represents the name of the file that we wish to read. The function returns to the calling program a string containing the contents of the file.

Functions in a script run only if they are called. If your entire program consisted of the code within a function, the program would not do anything when run. You must insert code (in the main body or another function) that calls the function. The following code calls the function ReadResults and passes the path and filename of the file that we wish to read.

```
strOutput=ReadResults(strTemp & "\out.txt")
WScript.Echo(strOutput)
```

The ReadResults function creates a new FileSystemObject and verifies that the file exists. If it does exist, the function opens the text file, loops through and reads each line of data, and stores the results to a variable. This variable is ultimately returned to the calling program. vbcrlf is a built-in constant for VBScript to designate a carriage return and line feed. The JScript equivalent is the newline symbol \n.

```
Function ReadResults(strFileName)
    Set objFSO = CreateObject("Scripting.FileSystemObject")
    If (objFSO.FileExists(strFileName)) Then
        Set objTextFile = objFSO.OpenTextFile(strFileName,1,False)
        strOutput=""
        Do While objTextFile.AtEndOfStream <> True
            strReadLine=objTextFile.ReadLine
            strOutput=strOutput & strReadLine & vbcrlf
        Loop
```

The OpenTextFile method requires a filename (such as strFileName) as an argument. The method supports three optional arguments. The first optional argument, iomode, lets you specify whether the text files should be opened for reading (1), writing (2), or appending (8). The second optional parameter, create, is a Boolean setting designating whether the file should be created if it does not exist. Because we are reading an already existing file (and have tested for this state), we do not want to create a new file and, therefore, set the create parameter to False. The last optional argument (which was not used in the previous code example) is the format value, which specifies whether the file should be opened in Unicode, ASCII, or the system default character set. (The Unicode character set is a superset of ASCII and provides multilingual character support.)

Unlike the standard streams that close automatically when the process ends, the TextStream object must be explicitly closed. We do this with the Close method. Lastly, we set the string containing the contents of the file we read to the function name and end the function.

```
     objTextFile.Close
Else
     strOutput="Error reading file."
End If
ReadResults=strOutput
End function
```

Echo, StdOut.Write, and PopUp

Displaying information to the user as your script runs is a useful and oftentimes necessary function. WSH supports several methods to display information to the user, some of which have been demonstrated previously. The most common methods are Echo, StdOut.Write, and PopUp.

Basic Echo

WScript.Echo is the simplest display method. It works with both *cscript.exe* and *wscript.exe* and displays a line with a newline character or text in a message box, respectively.

Figures 3.4 and 3.5 show sample output of WScript.Echo using both the *cscript.exe* and *wscript.exe* engines when the code from Listing 3.11 is run.

ON THE CD

LISTING 3.11 WScript.Echo

```
strLine3="This is Line3"
WScript.Echo "Line 1" , "Line 2", strLine3
strLine4="This is Line4"
WScript.Echo "Take 2- Line 1" , "Take 2- Line 2", strLine4
```

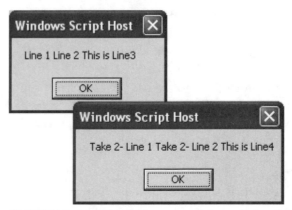

FIGURE 3.4 `WScript.Echo` processed through the console-based *cscript.exe* WSH engine.

FIGURE 3.5 Running `WScript.Echo` through the GUI *wscript.exe* WSH engine.

Each call of the `Echo` method using *wscript.exe* creates a new dialog box. To output multiple results to a single dialog box, consider storing the results of each individual call into a temporary string and then displaying that string when the script is complete, as in Listing 3.12.

ON THE CD

LISTING 3.12 Concatenating strings into a single variable to display in a single Echo dialog box

```
strLine3="This is Line3"
strTemp =  "Line 1" & " " & "Line 2" & " " & strLine3 & vbcrlf
strLine4="This is Line4"
strTemp = strTemp &  "Take 2- Line 1" & " " & "Take 2- Line 2" & " " &
strLine4
WScript.Echo strTemp
```

To mimic the output of the example in Figure 3.4, the previous example inserts spaces and a carriage return after the first line. Figure 3.6 shows the *wscript.exe* dialog box that more closely matches the *cscript.exe* output but in a dialog box.

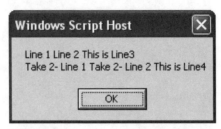

FIGURE 3.6 Concatenating strings into an output variable permits output to one dialog box.

A Finer Degree of Output Control

Earlier in the asynchronous event example in Chapter 2, we introduced the `StdOut.Write` method, which provides character-by-character control of the output. As seen in Figure 3.4, the `Echo` method inserts a newline character each time it is called. Use the `StdOut.Write` method to display output continuously to the same line. Listing 3.13 demonstrates `Echo` versus `StdOut.Write`, and Figure 3.7 shows the resulting output.

ON THE CD

LISTING 3.13 Comparison of `WScript.Echo` and `StdOut.Write`

```
For i =  33 To 43
    WScript.Echo Chr(i)
Next

For i =  33 To 43
    WScript.StdOut.Write Chr(i)
Next
```

A Deluxe Pop Up

The `PopUp` method provides a pop-up dialog box that is displayed when your script is running in the `cscript.exe` or `wscript.exe` engine. This method provides the most robust controls, including the display of buttons, icons, button response, and even a timed auto-close.

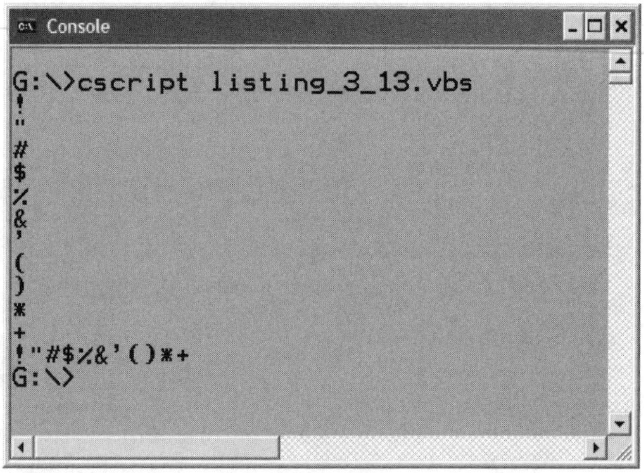

```
G:\>cscript listing_3_13.vbs
!
"
#
$
%
&
'
(
)
*
+
!"#$%&'()*+
G:\>
```

FIGURE 3.7 WScript.Echo inserts newlines, but StdOut.Write is displayed on one line.

Calling the PopUp method is straightforward. The code in Listing 3.14 creates a dialog box with an Exclamation Point and Yes, No, and Cancel buttons. The script then takes action based on the button selected.

ON THE CD

LISTING 3.14 The PopUp method

```
WScript.Echo "Display a popup dialog box with user selected options."
const OK = 0
const OC = 1 'OKCancel
const ARI=2 'AbortRetryIgnore
const YNC=3 'YesNoCancel
const YN=4 'YesNo
const RC=5 'RetryCancel

const SMark=16 'StopMark
const QMark=32 'QuestionMark
const EMark=48 'ExclamationMark
const IMark=64 'InformationMark

Set objShell = CreateObject("WScript.Shell")
strMessage="Please Select Yes, No, or Cancel"
strTitle="Popup Example"
iResult = objShell.Popup(strMessage,30,strTitle,YNC+EMark)
Select Case iResult
    Case 2
```

```
          WScript.Echo "You picked 'Cancel'"
    Case 6
          WScript.Echo "You picked 'Yes'"
    Case 7
          WScript.Echo "You picked 'No'"
End Select
```

Figure 3.8 shows the output of the PopUp method and the acknowledgement of user selection.

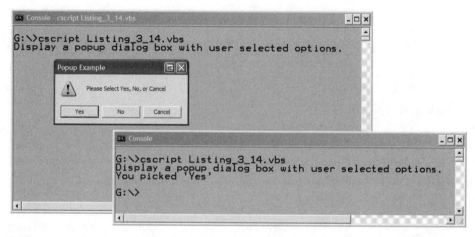

FIGURE 3.8   The PopUp method provides notification graphics and basic user interaction.

Network Object

WSH provides access to network shares, network printers, and network user information using the Network object. This functionality is similar in scope to the net command as opposed to the other network commands such as ping or ipconfig, which we explored earlier in this chapter.

An example of this object is its use in a WSH login script, which is detailed in Chapter 7. In this example, the login script enumerates current drive mappings and then maps only those that do not already exist (for example, if they were created persistent prior to the script execution). The login script also uses the Network object properties UserDomain, UserName, and ComputerName to welcome the user onto the network. Listing 3.15 demonstrates how to use the Network object to show all installed printers.

LISTING 3.15   Using the network object to enumerate installed printers

```
WScript.Echo("Display Installed Printers")
Set objNetwork = CreateObject("WScript.Network")
Set colPrinters = objNetwork.EnumPrinterConnections
for i = 0 to colPrinters.Count-1 Step 2
    WScript.Echo("Printer " & colPrinters.Item(i+1)_
   & " is installed on " & colPrinters.Item(i))
Next
```

Running this script displays the installed printers as shown in Figure 3.9.

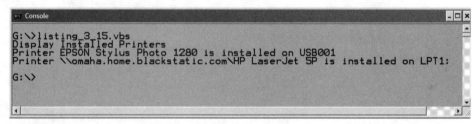

FIGURE 3.9   The `Network` object provides script access to printers, shares, and other network information.

Notice that the `EnumPrinterConnections` method returns a collection of printers, but the subsequent loop to iterate the collection is a `for…next` loop and not a `for…each` loop. This discrepancy is because the collection of objects returned is not an object itself. Instead, it is an array of values, alternating between the printer connection and the printer name.

The actual array is shown in the following table.

Array Element	Value
0	USB001
1	EPSON Stylus Photo 1280
2	LPT1:
3	\\omaha\HP LaserJet 5P

Therefore, we step by two through the array and print out on a single loop iteration both `colPrinter.Item(i)` and `colPrinterItem(i+1)`. Enumerating drives with the method `EnumNetworkDrives` returns a similar array-based collection.

TIP

Knowing whether an object returns a collection as an object or as an array is detailed in that object's SDK or documentation. No matter what reference material you use, when beginning programming, look to the SDK to learn about the nuances of that particular object. Appendix A lists many sources for the SDK and documentation useful in your scripting. SDKs do contain errors, however. If you have difficulty using a feature documented in the SDK, by all means, look to other resources on the Internet (such as newsgroups) or Microsoft Technical Support to confirm or correct the feature's proper operation.

WINDOWS MANAGEMENT INSTRUMENTATION

Windows Management Instrumentation (WMI) is the Microsoft Windows implementation of Web-based Enterprise Management (WBEM), which defines a cross-platform standard for managing enterprise computing objects. For example, use WMI to find the date of the basic input/output system (BIOS) of a particular machine or to see how much hard disk is free on a server. Hardware and software vendors create *providers*, which in turn provide access to *classes* that contain descriptions of objects or services.

For example, the Win_32 provider provides access to many classes—such as `Win32_OperatingSystem`. Other examples of providers that can be installed and used with WMI include registry, SNMP, DFS, disk quota, event log, IP route, ping, security, and terminal server.

WBEM defines the Common Information Model (CIM), which defines the object hierarchy. For example, you can use the same CIM functions to query the fan speed from a Hewlett Packard® server or Dell® server. Vendors provide WBEM- and CIM-compliant modules for their equipment that can be accessed using a common set of APIs. These objects and classes make up the WMI object model.

WMI comes installed with Windows 2000 and later For legacy support, download and install the WMI Core from Microsoft. Additionally, Microsoft provides GUI tools for examining the WMI *namespace* for a given computer. CIM is broken up into multiple namespaces, each of which contains one or more classes. The class is a logical grouping of objects; for example, the WMI Win32_Process class includes objects, methods, and properties that pertain to processes running on a computer. These tools provide a good opportunity to explore all of the different classes and objects available to a WMI programmer. (The WMI Core, WMI Tools, and other useful WMI information can be found at *http://msdn.microsoft.com/library/default.asp?url=/downloads/list/wmi.asp.*)

To get a sense of the different namespaces, run the WMI Tool *CIM Studio*. In the Connect to Namespace dialog box, click the Browse for Namespace icon, and then click Connect to list the namespace hierarchy for that machine, as shown in Figure 3.10.

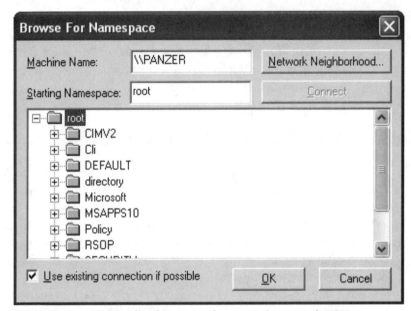

FIGURE 3.10   CIM Studio shows you the expansiveness of WMI.

Most of the administrative WMI classes that we use in the scripts in this text (and the default namespace for the WMI objects) are contained in the namespace `root/CIMV2`.

Connecting to a Machine with WMI

Connecting to a computer using WMI requires only a few steps, a few that you have already seen in some of the examples in Chapter 2. These steps can be further refined into a simple piece of code called a *moniker*. But first, let's look at each of the steps individually using the code in Listing 3.16 as our first example.

ON THE CD

LISTING 3.16   Use WMI to connect to a machine and display all installed updates

```
Set objLocator=CreateObject("WbemScripting.SWbemLocator")
Set objServices=objLocator.ConnectServer("localhost","root/cimv2")
objServices.Security_.ImpersonationLevel = 3

Set colQFE = objServices.InstancesOf ("Win32_QuickFixEngineering")

For Each objQFE in colQFE
    WScript.Echo objQFE.HotFixID & " : " & objQFE.Description
Next
```

The scripting API for WMI is called WbemScripting. To access an object on a given computer, you must instantiate a SWbemLocator object, which returns a SWbemServices object. The method ConnectServer allows you to specify the name of a specific server. The ConnectServer method, as used in Listing 3.16, connects the localhost machine and the root/cimv2 namespace. You can optionally specify a username and password under which to connect.

The next line sets the Security_.ImpersonationLevel to 3, which corresponds to wbemImpersonationLevelImpersonate. This level instructs the WMI objects to run under the context of the caller and is the default level in WMI Core Components 1.5 (Windows 2000 and later). In earlier versions of WMI, the default level was set to ImpersonationLevel=Identify, which may cause your script to fail. If you plan to run your scripts remotely under earlier versions of WMI Core Components, be sure to set the security to impersonate.

After the connection is made, the remaining script queries WMI for the desired data class. As briefly covered in Chapter 2, a few ways of querying the data are available. Use either the InstancesOf or ExecQuery method to return a collection of data.

```
Set colQFE = objServices.InstancesOf ("Win32_QuickFixEngineering")

Set colQFE = objServices.ExecQuery ("Select * from_
Win32_QuickFixEngineering")
```

The ExecQuery method supports the *Windows Management Instrumentation Query Language (WQL)*. This language offers additional flexibility than the simpler InstancesOf method, which is limited to a basic query of the specified class.

Using a WMI Moniker

A WMI moniker string is a shortcut of sorts to accessing the WMI data. The WMI moniker string is constructed with the prefix WinMgmts:, optional security settings, and the object path. The following single line of code accomplishes the same results as the first four lines of code in Listing 3.16. It combines the impersonation level, server name, namespace, and class into one moniker.

```
Set colQFE=GetObject("WinMgmts:{impersonationLevel=impersonate}"_
& "!//localhost/root/cimv2:Win32_QuickFixEngineering").Instances_
```

In the line of code above, the first underscore (_) denotes line continuation as in earlier examples—the code should be on a single line. The second underscore, however, is part of the name of the Instances_ method and is required. It differentiates the method from methods of the underlying object.

This moniker returns a SWBemObject (not a collection), so we can then use the SWBemObject.Instances_ method to return a collection that we can iterate as we did in the previous example.

The code in Listing 3.17 returns a similar object using the InstancesOf method.

ON THE CD

LISTING 3.17 A WMI moniker provides a shorter means of connecting to a WMI class

```
Set colQFE=GetObject("WinMgmts:{impersonationLevel=impersonate}"_
& "!//localhost/root/cimv2").InstancesOf("Win32_QuickFixEngineering")
For Each objQFE in colQFE
    WScript.Echo objQFE.HotFixID & " : " & objQFE.Description
Next
```

Let's look at another example. Listing 3.18 returns a specific instance of whether or not the World Wide Web publishing service is running on a machine named Europa.

ON THE CD

LISTING 3.18 Returns the status of the W3Svc service using WMI

```
strWMI="WinMgmts://localhost/root/cimv2:Win32_Service='w3svc'"
Set objService=GetObject(strWMI)
WScript.Echo objService.name & "  Status: " &objService.status
```

Understanding What Is Available: Dump the WMI Classes

WMI is very large and may seem overwhelming. When beginning, you may find yourself wondering what the name of the class or property is for any given object. Fortunately, a number of tools enumerate these names for you. Use the CIM Studio, which comes with the WMI Tools, to walk through and search for class and object details. Microsoft also offers a clever tool called the Scriptomatic (search for it on *msdn.microsoft.com*), which lists all classes and their properties and even generates WMI sample code for you to use.

Listing 3.19 presents a snippet of code that lists these classes. The InstancesOf class argument meta_class defines a schema query, which returns all classes within the current namespace. This query returns a collection of classes that can then be iterated and displayed using the SWBemObject Path_ property.

LISTING 3.19   Use the meta_class to display all WMI classes in a namespace

ON THE CD

```
Set colClass=GetObject("WinMgmts://localhost")_
.InstancesOf("meta_class")

For Each objClass in colClass
    WScript.Echo objClass.Path_
Next
```

In the same spirit of dumping information, the code in Listing 3.20 outputs all data for a given class instance. This code is particularly useful if you are unsure of the name or availability of a particular property, or if you simply want to display specific data from a remote machine.

LISTING 3.20   Use GetObjectText to display all WMI properties of a class

ON THE CD

```
strClass=WScript.Arguments(0)
Set colClass=GetObject("WinMgmts://localhost").InstancesOf(strClass)
For Each objClass in colClass
    WScript.Echo objClass.GetObjectText_
Next
```

Figure 3.11 shows an example of this script with the input argument Win32_Processor.

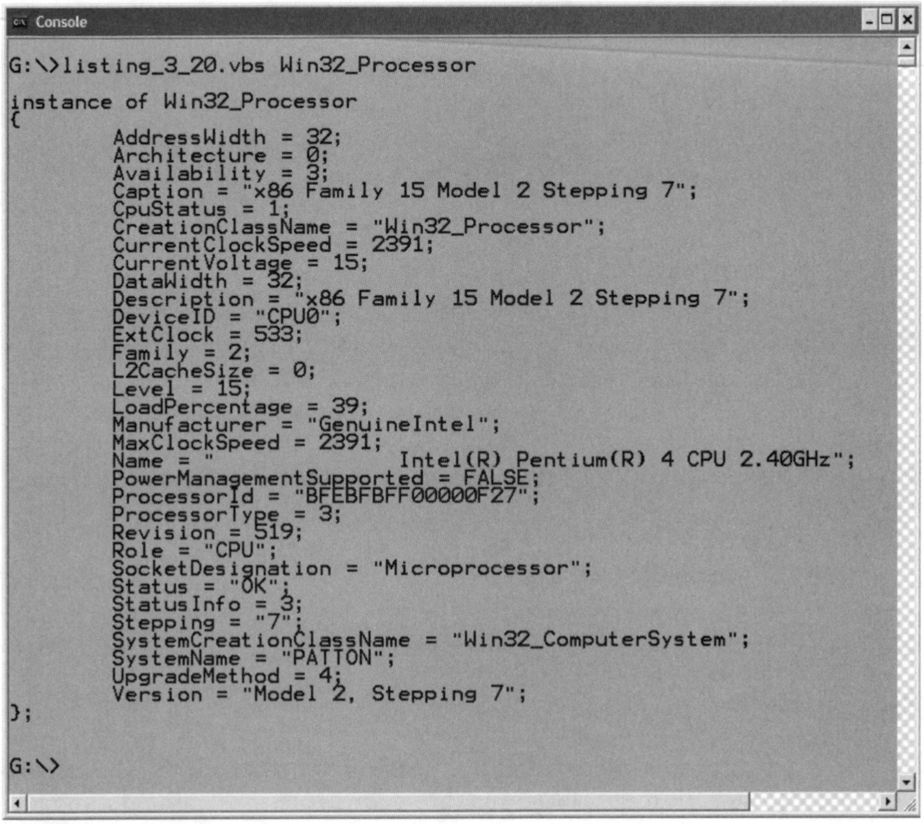

```
Console                                                              _ □ x
G:\>listing_3_20.vbs Win32_Processor

instance of Win32_Processor
{
        AddressWidth = 32;
        Architecture = 0;
        Availability = 3;
        Caption = "x86 Family 15 Model 2 Stepping 7";
        CpuStatus = 1;
        CreationClassName = "Win32_Processor";
        CurrentClockSpeed = 2391;
        CurrentVoltage = 15;
        DataWidth = 32;
        Description = "x86 Family 15 Model 2 Stepping 7";
        DeviceID = "CPU0";
        ExtClock = 533;
        Family = 2;
        L2CacheSize = 0;
        Level = 15;
        LoadPercentage = 39;
        Manufacturer = "GenuineIntel";
        MaxClockSpeed = 2391;
        Name = "            Intel(R) Pentium(R) 4 CPU 2.40GHz";
        PowerManagementSupported = FALSE;
        ProcessorId = "BFEBFBFF00000F27";
        ProcessorType = 3;
        Revision = 519;
        Role = "CPU";
        SocketDesignation = "Microprocessor";
        Status = "OK";
        StatusInfo = 3;
        Stepping = "7";
        SystemCreationClassName = "Win32_ComputerSystem";
        SystemName = "PATTON";
        UpgradeMethod = 4;
        Version = "Model 2, Stepping 7";
};

G:\>
```

FIGURE 3.11   WMI provides a large amount of information from a local or remote system.

These examples of returning data using WMI are fairly simple, and they are in-tentionally broad to give you an idea of all the different types of data that are easily obtainable. Refer to the WMI SDK for all of the WMI classes, objects, methods, and properties available. For example, try modifying the previous code to dump the available namespaces or all of the methods and properties of a class.

How to Find Something: Querying WMI

A few of the examples in Chapter 2 demonstrate functions and methods using WQL—also known as Structured Query Language (SQL) for WMI. WQL provides a robust method for extracting specific data from the huge WMI repository. In ad-dition, a well-constructed WQL query executes faster than iterating through an entire data set looking for a particular piece of information. WQL lets you perform *data*, *event*, and *schema* queries. The next examples focus on data queries because that type of query is probably what the bulk of your scripts will include. Refer to earlier examples of event queries (Chapter 2, under Events) and schema queries.

WQL is a subset of SQL. However, learning the fundamentals to return basic WMI data is not at all complicated. The following examples walk through querying and displaying information about the processes running on a system. Listing 3.21 shows the basic framework of a WQL query using the ExecQuery method. This example lists all of the processes on the local host system.

ON THE CD

LISTING 3.21   A basic WQL query

```
strWQL="SELECT * from Win32_Process"
Set colProcess=GetObject("WinMgmts://localhost").ExecQuery(strWQL)
For Each objProcess In colProcess
    WScript.Echo objProcess.GetObjectText_
Next
```

We mostly work with and modify the WQL line:

```
strWQL="SELECT * from Win32_Process"
```

A basic WQL query takes the following form:

```
SELECT [something] FROM [somewhere] WHERE [something] [is compared to]
[something else]
```

The following sample queries illustrate this. To use these, just substitute the new values of strWQL into Listing 3.21.

```
strWQL="SELECT * FROM Win32_Service WHERE State='Running' "_
& "and StartMode='Manual'"
```

```
strWQL="SELECT HotFixID FROM Win32_QuickFixEngineering"
```

```
strWQL="SELECT Name,Size FROM Win32_LogicalDisk WHERE MediaType=12"
```

The first example returns a collection of all services running from a manual mode. The second example returns only the HotFixID of any patches installed on the target system, and the third example returns the name and size of any fixed hard disk media (corresponding to MediaType of 12).

Remember the SQL slammer worm that infected unpatched versions of SQL Server? Imagine how easy it would be to troll Win32_Services for running instances of SQL Server and then enumerate Win32_QuickFixEngineering to see if the machines are patched.

A number of keywords make up a WQL query, and the main ones are covered here. More complex examples are demonstrated in some of the scripts in the following chapters. All keywords are defined in the WMI SDK.

The SELECT clause defines the query. You select what properties you wish to examine from the class. SELECT also supports the wildcard character *. If you are unsure of what properties you want to use (or you want to use most of them), using the wildcard is useful. However, when accessing remote computers across slower links (or where network bandwidth or time is important), specify the specific properties to reduce the amount of data sent and time required to execute.

Specify the WMI class you wish to query in the FROM clause. Further refine the query results set by specifying required matching criteria in the WHERE clause. The WHERE clause criteria support the matching of property names and values. This clause is optional and is used to narrow the result set. Do not enclose the property names in quotations, but do enclose the property values in quotations as shown in the next example:

```
strWQL="SELECT * FROM Win32_LogicalDisk WHERE Name='c:'"
```

In this example, the property name is Name, and the value you wish to match is c:. Name does not need quotes; the string value c: does.

You can also insert the WQL string directly into the ExecQuery parameter, as follows:

```
Set colProcess=GetObject("WinMgmts://localhost")._
ExecQuery("SELECT * FROM Win32_LogicalDisk WHERE Name='c:'")
```

In this example in Listing 3.22, the query only returns the instances where the memory used by the process is greater than 10MB. The results are shown in Figure 3.12.

ON THE CD

LISTING 3.22 WQL supports variable comparisons, allowing more complex result sets

```
iSize=10 ' in MB

strWQL="SELECT * from Win32_Process WHERE WorkingSetSize > "_
& iSize *1024 *1024

Set colProcess=GetObject("WinMgmts://localhost").ExecQuery(strWQL)
For Each objProcess In colProcess
    WScript.Echo objProcess.Caption & " " & objProcess.WorkingSetSize
Next
```

FIGURE 3.12   Use WQL to alert on process memory consumption.

WMI provides access to both local and remote machines to query their configuration and aid in their management. Many of the scripts in the following chapters rely on WMI and use many of the basic principles covered previously as key ingredients in how they work.

Another major scripting technology that you will likely rely on in your IT systems administration script toolkit is the *Active Directory Services Interface*, or *ADSI*.

ACTIVE DIRECTORY SERVICES INTERFACE

Microsoft Active Directory (AD) is the directory service introduced with Windows 2000 as a replacement of the Windows NT account and group management database. AD is based on the Lightweight Directory Access Protocol (LDAP). Microsoft Active Directory provides rich directory services that centrally contain detailed network, user, and group data. Microsoft provides many tools to enumerate, diagnose, and manage AD, and more and more commercial tools are available that can help as well. However, even with this increasing amount of support options, you may find yourself yearning for a direct interface to the directory. Perhaps you want to code a Web-based application to create user accounts, or you want to audit the membership of a specific OU or group on a regular basis. Or maybe you just want to create a simple tool that lists group membership in a specific format. Fortunately, Microsoft published a script-accessible interface called Active Directory Services Interface (ADSI), which allows you to write programs like these quickly and efficiently. With ADSI you can write scripts in VBScript or JScript to help manage network resources, such as working with network printers, setting network permissions, and querying, adding, or removing users and groups.

ADSI is integrated with Windows 2000 and later Windows operating systems. Additionally, you can download client extensions for Windows 9.x and Windows NT to allow these legacy operating systems to use many (not all) of the features of AD. Although these extensions do not currently support some AD features such as Kerberos, Group Policy, IPSec, SPN, or Mutual Authentication, the extensions do support ADSI, which is what we are discussing.

A directory is an organizational system that stores relevant information about specific objects. In the past (and even today), information was stored in disparate systems such as network logon, e-mail, remote access, security systems, human resource database systems, finance systems, and others. Each of these directories likely contained duplicate information and required time and money to keep the directories updated. AD is an example of a robust, full-featured directory service that can be a clearinghouse for business and technical objects—from contact phone numbers to firewall configuration information. AD provides the cornerstone for the Windows 2000 domain structure. Migration to and support of AD is slow but coming. Microsoft Exchange 2000 is an example of a groupware product that is fully AD integrated, and many third-party companies have begun to recognize and interoperate with AD as well.

TIP

Recognize the level of integration with AD. For example, some vendors may say they support AD, which may lead you to think that they actually store information in AD. However, in actuality, they may set up a conduit to synchronize AD with their own legacy directory. Although this setup is neither good nor bad, it is important to recognize this because it may affect how you architect your scripts to interact with either directory.

ADSI provides a single, well-defined set of objects, methods, and properties to interact with supported directories. Abstracted access to an independent directory resource is via an ADSI *provider*. Supported providers included in Windows 2000 and later releases include LDAP, Windows NT 4.0 directory (WinNT), Internet Information Services (IIS), Novell NetWare Directory Services (NDS), and NetWare 3 bindery (NWCOMPAT).

The scripts in this text focus mostly on querying AD and local machines for user and group membership information. These correspond to the LDAP and WinNT providers, respectively.

ADSI VS. AD

Remember, AD is an example of one type of directory services, whereas ADSI is an interface that can access not only AD but also other directory services. We work with ADSI and its LDAP and WinNT providers to access AD and Windows NT Server 4.0 directories, respectively.

TIP

LDAP accesses the directory service on a Windows 2000 domain controller. Although the WinNT provider provides access to the legacy Windows NT 4.0 directory, use the WinNT provider to access the local user and group accounts even on machines running Windows 2000 and later operating systems.

WHAT'S IN A NAME?

To work with any directory object, you must know what it is called. AD supports several types of naming convention.

Relative Distinguished Name (RDN)

The *relative distinguished name (RDN)* is the simple name of the object and its attribute type; for example, OU=Finance, or CN=jeff, or CN=Server1. Some RDN attributes and examples are shown in Table 3.6.

TABLE 3.6   Relative distinguished name attributes

Attribute	String	Example
domain component	DC	DC=Some, DC=Domain, DC=com
common name	CN	CN=Jeff
organizational unit	OU	OU=Users
organization	O	O=
street name	STREET	STREET=
locality	L	L=
state or province	ST	ST=
country	C	C=
user ID	UID	UID=

Distinguished Name

A *distinguished name (DN)* comprises a series of relative distinguished names, separated by commas, which fully describe the location of the object within the hierarchy of the directory. For example, if the user Jeff was in the OU=Home Users in the domain home.blackstatic.com, then the DN would be:

```
CN=Jeff, OU=Home Users, DC=blackstatic, DC=com
```

Once you have a good idea of how the different objects are named, you can reference them directly. Depending on what provider you use, your syntax will vary as will some of the attribute values.

For example to connect, or *bind*, to a computer named Europa, you can use either the WinNT or LDAP provider. In the LDAP provider binding string, the server name is optional, and serverless binding is preferred. Under these cases, ADSI finds the best domain controller within the default domain.

```
WinNT://europa, computer

LDAP://CN=europa, OU=computer, DC=blackstatic, dc=com

GC://CN=europa, OU=computer, DC=blackstatic, dc=com
```

When querying AD objects or ADSI data from Windows 2000 domain controllers, use the LDAP provider. The WinNT provider allows enumeration of local accounts on member and standalone servers. However, when querying AD, always use the LDAP provider whenever possible because it is much better supported.

The GC notation uses the LDAP provider to bind to the global catalog for faster searches. The global catalog contains a subset of AD data for rapid retrieval.

Specifying a server name is shown in the following code, but you should avoid doing this if possible, because ADSI tries to choose the best domain controller.

```
LDAP://dc1.blackstatic.com/CN=europa, OU=computer, DC=blackstatic,
dc=com
```

Use the following code to bind to user objects using similar nomenclature:

```
WinNT://europa/jeff , user

LDAP://CN=jeff, OU=users, DC=home, DC=blackstatic, dc=com
```

A SHORT TECHNICAL OVERVIEW OF ADSI

ADSI objects are component object model (COM) objects, which we introduced in Chapter 2. Accessing and using these objects is easy. First, you get a reference to the object. Then you request the attributes you wish to examine (for example, using GetInfo or Get methods). You can then view or manipulate the information. The name of the interface that you use to access ADSI objects is called IADs, and all ADSI objects support the methods and properties of this interface. The IADs interface defines how an ADSI object is named, what its parent is, and how to access it.

Many directory services objects inherit the methods and properties from the IADs object and can be used in a similar fashion as IADs. Therefore, for example, every user object supports the IADsUser interface in addition to the IADs interface. Recognize these object-specific interfaces by the IADs prefix in their name. In another example, IADsOU is an object specifically designed for managing OU and includes appropriate properties and methods for accessing those properties. Table 3.7 shows the IADsInterfaces available for both the LDAP and WinNT providers.

TABLE 3.7   LDAP and WinNT provider IADs objects

ADSI Objects of LDAP	ADSI Objects of WinNT
IADs	IADs
IADsClass	IADsClass
	IADsCollection
	IADsComputer
	IADsComputerOperations
IADsContainer	IADsContainer
IADsDeleteOps	
	IADsDomain
	IADsFileService
	IADsFileServiceOperations
	IADsFileShare
IADsGroup	IADsGroup
IADsLocality	
IADsMembers	IADsMembers
IADsNameTranslate	
IADsO	

TABLE 3.7   LDAP and WinNT provider IADs objects *(continued)*

ADSI Objects of LDAP	ADSI Objects of WinNT
IADsObjectOptions	
IADsOpenDSObject	IADsOpenDSObject
IADsOU	
IADsPathname	
	IADsPrintJob
	IADsPrintJobOperations
IADsPrintQueue	IADsPrintQueue
IADsPrintQueueOperations	IADsPrintQueueOperations
IADsProperty	IADsProperty
IADsPropertyList	IADsPropertyList
	IADsResource
	IADsService
	IADsServiceOperations
	IADsSession
IADsSyntax	IADsSyntax
IADsUser	IADsUser
IDirectoryObject	
IDirectorySearch	

Each of these interfaces supports many object-specific methods and properties, and the Microsoft ADSI SDK provides a good reference for looking up the details of each. You can also enumerate the properties of a given object through script, although it is not quite as easy as it was for enumerating the properties of a WMI class. We'll cover that script in a bit after we first talk more about accessing ADSI objects.

Directory Services Hierarchy

The AD is a hierarchy of container objects and leaf objects. Container objects can contain other container and leaf objects. Leaf objects cannot contain objects. All container objects support the IADsContainer interface. The methods of IADs Container include creating and deleting new objects and accessing information about the contained object or objects.

The Schema

The AD *schema* defines of all of the types of data that the directory can store. If you have installed Exchange 2000 or Microsoft Internet Security and Accelerator® (ISA) Server you may have remembered "extending" the schema. What this means is that these products added information to AD that enhanced specific objects or resources. For example, Exchange 2000 adds the Exchange container to the AD.

Every object in the AD must adhere to the schema definition. For example, an OU contains the name, address, multiple phone numbers, and over 60 other properties. However, if you want to add, say, an employee number for every one of your users, then you could either try to fit that number onto an existing attribute, or you could extend the schema and add a new attribute called something like EmployeeNumber.

BINDING TO AN OBJECT USING SCRIPT

When scripting with the ADSI interface, the first step is to bind to a desired object. The example in Listing 3.23 shows how to bind to the AD user object Trooper using its DN. From this we can list a number of iADS attributes from the object.

ON THE CD

LISTING 3.23   Using the ADSI LDAP provider to bind to a user object

```
strPath="LDAP://CN=Trooper, OU=Home Users, DC=home, "&_
 DC=bladestatic, DC=com"
Set objUser = GetObject(strPath)

'IADs Property Methods
WScript.Echo "Uniquely Identifies the object: " & objUser.ADsPath
WScript.Echo "The GUID: " & objUser.GUID
WScript.Echo "Relative Name: " & objUser.Name
WScript.Echo "The Parent: " & objUser.Parent
WScript.Echo "The Schema: " & objUser.Schema
WScript.Echo "The Class: " & objUser.Class
```

The first two lines in this example define the DN and bind the object objUser to the adsPath of the directory service object. Next, the example accesses and displays user attribute information from the directory using the six iADS properties. The output from this example is shown in Figure 3.13.

FIGURE 3.13   Retrieve user information from Active Directory using ADSI.

Note that if the DN is changed to another attribute, such as an organizational unit, the script still runs, as shown in Figure 3.14. This script only accesses the iADS attributes that are available to all ADSI objects. To witness the AD hierarchy, change the DN in Listing 3.23 to represent the OU home users.

```
strPath="LDAP://OU=Home Users, DC=home, DC=blackstatic, DC=com"
```

FIGURE 3.14   High-level attributes of an OU within the Active Directory hierarchy.

Figures 3.13 and 3.14 show how the schema of both the user Trooper and the OU Home Users are related. These examples provide sense to the hierarchical organization of the AD and how an object's location can be uniquely identified by the schema attribute. Notice also the value of the class attribute. The first example pointed to a user object that followed the User class. The second example pointed to an OU and followed that class. The classes are important to recognize because they help identify other iADS interfaces available to tap to extract additional class-specific attributes from your directory object (although a 1:1 correspondence between the class name and the IADs interface may not always be available).

For example, now that we know that the class is a user, we can extract any of the IADsUser properties, as shown in Listing 3.24 through the use of a Select Case statement.

ON THE CD

LISTING 3.24   Taking action based on the ADSI class of the object

```
strPath="LDAP:// CN=Trooper, OU=Home Users, DC=home, "_
& "DC=blackstatic, DC=com"

set objUser = GetObject(strPath)

'IADs Property Attributes
WScript.Echo "ADsPath: " & objUser.ADsPath
WScript.Echo "The GUID: " & objUser.GUID
WScript.Echo "Relative Name: " & objUser.Name
WScript.Echo "The Parent: " & objUser.Parent
WScript.Echo "The Schema: " & objUser.Schema
WScript.Echo "The Class: " & objUser.Class & vbcrlf

Select Case objUser.Class
    'IADs User Attributes
    Case "user"
        WScript.Echo "FullName: " & objUser.FullName
        WScript.Echo "Password Last Changed: "_
          & objUser.PasswordLastChanged

    'IADsOU Attributes
    Case "organizationalUnit"
        WScript.Echo "Description: " & objUser.Description

    Case Else
    WScript.Echo "Not a user or an organizational unit."

End Select
```

Figure 3.15 shows the running of this example with both a user and organizational unit object.

Listing the Properties of an Object

It's often useful to dump all of the properties of a particular object. We did this earlier in Listing 3.20 to enumerate an entire WMI class. The code in Listing 3.25 shows how to do this, too, with an ADSI object, although it's not quite as simple.

```
Console                                                                    _ □ ×
G:\>Listing_3_24.vbs
ADsPath: LDAP://CN=Trooper,OU=Home Users, DC=home, DC=blackstatic, DC=com
The GUID: 99c0aff219319b4296652a15b8cc3217
Relative Name: CN=Trooper
The Parent: LDAP:// OU=Home Users, DC=home, DC=blackstatic, DC=com
The Schema: LDAP://schema/user
The Class: user

FullName: Trooper the Dog
Password Last Changed: 4/16/2003 10:22:49 PM

G:\>Listing_3_24.vbs
ADsPath: LDAP://OU=Home Users,DC=home, DC=blackstatic, DC=com
The GUID: 9b49224e097e29468817b8ae46eb30f0
Relative Name: OU=Home Users
The Parent: LDAP:// DC=home, DC=blackstatic, DC=com
The Schema: LDAP://schema/organizationalUnit
The Class: organizationalUnit

Description: Container for Home Users

G:\>
```

FIGURE 3.15   Fork script logic based on attributes retrieved from the directory.

ON THE CD

LISTING 3.25   Enumerate properties of an AD object

```
strPath="LDAP://cn=Trooper, ou=Home Users, DC=Home, "_
 & "DC=Blackstatic, DC=com"

Set objUser = GetObject(strPath)
objUser.GetInfo
For i = 0 To objUser.PropertyCount-1
    strName= objUser.Item(i).Name
    iADsType= objUser.Item(i).ADsType
    WScript.Echo i & ") " & strName & " (" & iADsType & ")"
    Set objProperty = objUser.GetPropertyItem(strName, iADsType)
    For Each objEntry In objProperty.Values
        On Error Resume Next
        WScript.Echo objEntry.GetObjectProperty(objEntry.ADsType)
        On Error Goto 0
    Next
Next
```

The output of this script is shown in Figure 3.16.

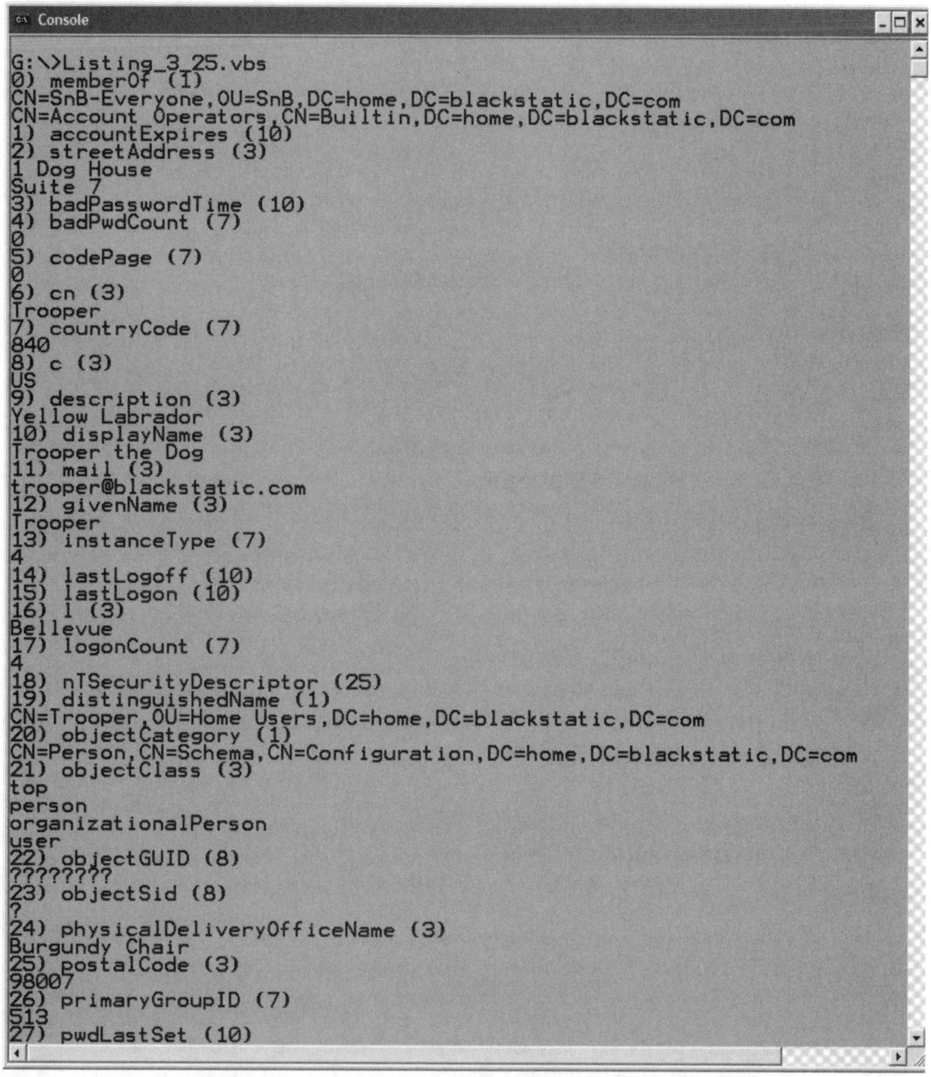

```
G:\>Listing_3_25.vbs
0) memberOf (1)
CN=SnB-Everyone,OU=SnB,DC=home,DC=blackstatic,DC=com
CN=Account_Operators,CN=Builtin,DC=home,DC=blackstatic,DC=com
1) accountExpires (10)
2) streetAddress (3)
1 Dog House
Suite 7
3) badPasswordTime (10)
4) badPwdCount (7)
0
5) codePage (7)
0
6) cn (3)
Trooper
7) countryCode (7)
840
8) c (3)
US
9) description (3)
Yellow Labrador
10) displayName (3)
Trooper the Dog
11) mail (3)
trooper@blackstatic.com
12) givenName (3)
Trooper
13) instanceType (7)
4
14) lastLogoff (10)
15) lastLogon (10)
16) l (3)
Bellevue
17) logonCount (7)
4
18) nTSecurityDescriptor (25)
19) distinguishedName (1)
CN=Trooper,OU=Home Users,DC=home,DC=blackstatic,DC=com
20) objectCategory (1)
CN=Person,CN=Schema,CN=Configuration,DC=home,DC=blackstatic,DC=com
21) objectClass (3)
top
person
organizationalPerson
user
22) objectGUID (8)
????????
23) objectSid (8)
?
24) physicalDeliveryOfficeName (3)
Burgundy Chair
25) postalCode (3)
98007
26) primaryGroupID (7)
513
27) pwdLastSet (10)
```

FIGURE 3.16 A script that enumerates all objects is sometimes useful to learn about the object.

Running this code binds to the object specified in the variable `objUser` and iterates through two loops to display the attribute names and values.

```
strPath="LDAP://cn=Trooper, ou=Home Users, DC=Home, "_
 & "DC=Blackstatic, DC=com"

Set objUser = GetObject(strPath)
```

The method `GetInfo` (also known as `iADs::GetInfo` in the ADSI SDK) loads all of the properties for a given object from the directory into a local cache. (You can use the method `GetInfoEx` to request only specific attributes.) The list of properties is zero based, and the first loop walks through them—from `0` to `objUser.PropertyCount  -1`. `PropertyCount` is a method of the `objUser` object returned from the `GetInfo` method. and it returns the number of properties in the list.

```
objUser.GetInfo
For i = 0 To objUser.PropertyCount-1
```

Properties can have single or multiple values. For example, the property `displayName` has a single value, whereas the property `memberOf` has multiple values and contains a list of each of the groups to which the user belongs. A second loop is needed to iterate through each of the possible multivalued properties. `On Error Resume Next` is a VBScript function that instructs the script to proceed to the next line if it encounters an error on the current line. This function is required because some properties may have no values and return an error when called.

Next, let's examine the ADSI methods used to access the properties and their values. We talked a bit about `GetInfo`, which returns a directory object's properties to the local property cache. The `iADsPropertyList` interface is used to access this cache. This interface supports the `Item` method, which provides access to a cached property, either by name or by index. Now that we have a specific item in the cache, we can begin to use it. First, the script displays the name and `ADsType` of the property.

```
strName= objUser.Item(i).Name
iADsType= objUser.Item(i).ADsType
WScript.Echo i & ") " & strName & " (" & iADsType & ")"
```

Every property has an `ADsType` that defines the format of the value (for example, is it a string or integer, case-sensitive or not, a security descriptor or 29 other possible types). The `ADsType` is important because some functions rely on knowing the type to properly set, display, or create the property. The `GetPropertyItem` method is one such function that requires this information. The `GetPropertyItem` method returns the property item from the cache, and the script iterates through the item's possible multiple values using the `Values` method. The method `Get-ObjectProperty` returns the actual values of the object, which we display.

```
Set objProperty = objUser.GetPropertyItem(strName, iADsType)
For Each objEntry In objProperty.Values
    On Error Resume Next
    WScript.Echo objEntry.GetObjectProperty(objEntry.ADsType)
    On Error Goto 0
Next
```

Figure 3.17 presents a schematic of the script to illustrate the flow of data from the directory, through the cache, through the successive loops, and ultimately to the display of the values.

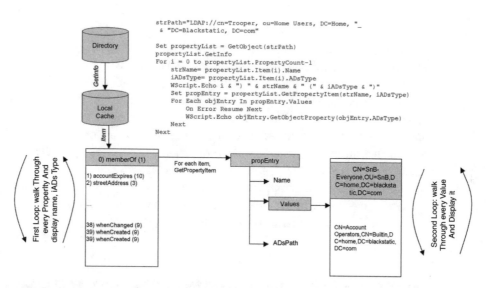

FIGURE 3.17   Schematic example of accessing and listing ADSI properties and corresponding values from directory to display.

Returning Specific Data from the Directory

The sample code in Listing 3.26 shows how to access specific properties directly from the directory.

ON THE CD

LISTING 3.26   Returning specific data from the directory

```
strPath="LDAP://cn=Trooper, ou=Home Users, DC=Home,"_
& "DC=Blackstatic, DC=com"

Set objUser = GetObject(strPath)
objUser.GetInfoEx Array("memberOf","displayName"), 0

WScript.Echo objUser.Get("displayName")

Set objProperty = objUser._
 GetPropertyItem("memberOf", ADSTYPE_DN_STRING)
```

```
For Each objEntry In objProperty.Values
    On Error Resume Next
    WScript.Echo "   " & objEntry.GetObjectProperty(objEntry.ADsType)
    On Error Goto 0
Next
```

We pass an array containing the names of the properties `memberOf` and `displayName` of the `GetInfoEx` method. (Remember to use `GetInfoEx` when returning specific properties from the directory.) For known single-valued properties (such as `displayName`), the script gets the value from cache and displays it. For multivalued properties, the script loops through the values as in the previous example. However, this time we know both the `name` of the property (`memberOf`) and its `ADsType` (`ADSTYPE_DN_STRING`). Use the previous script or look in the SDK for the `ADsType` for other properties. You can optionally use `ADSTYPE_UNKNOWN` if you don't know the `ADsTYPE` and are simply displaying an existing property value. (Note that using a type of unknown may cause some methods to report a error, especially if you are setting new values. In these cases, always send the exact `ADsType`.) Lastly, we iterate through the values and display them. The output of this example is shown in Figure 3.18.

FIGURE 3.18   Displaying the name and group membership of an object.

PROVIDERS RETURN DIFFERENT RESULTS

Because providers are ultimately connecting to different directory services, choosing different providers yields different results, even if the attributes are named the same. The code in Listing 3.27 shows how to bind to a user object using two different providers. The first example shows serverless binding to the user object cn=Jeff Fellinge (username=Jeff), and the second shows binding to Europa, the domain controller for the home domain, using the WinNT provider. (Remember that for a Windows 2000 domain controller, the local accounts are the domain accounts.) The output of this code, shown in Figure 3.19, shows some differences that you may see between the two providers (look at the GUID).

ON THE CD

LISTING 3.27   Choosing different providers may return different results

```
strPath="LDAP:// CN=Jeff Fellinge, OU=Home Users, "_
& "DC=home, DC=blackstatic, DC=com"

set objUser = GetObject(strPath)
WScript.Echo "Bind to user using LDAP provider:"
WScript.Echo "FullName: " & objUser.FullName
WScript.Echo "ADsPath: " & objUser.ADsPath
WScript.Echo "GUID: " & objUser.GUID
WScript.Echo "FullName: " & objUser.Schema & vbcrlf

objUser2="WinNT://europa/jeff, user"
set objUser2 = GetObject(objUser2)
WScript.Echo "Bind to user using WinNT provider:"
WScript.Echo "FullName: " & objUser2.FullName
WScript.Echo "ADsPath: " & objUser2.ADsPath
WScript.Echo "GUID: " & objUser2.GUID
WScript.Echo "FullName: " & objUser2.Schema
```

```
Console                                                                  _□x
G:\>Listing_3_27.vbs
Bind to user using LDAP provider:
FullName: Jeff Fellinge
ADsPath: LDAP://CN=Jeff Fellinge,OU=Home Users, DC=home, DC=blackstatic, DC=com
GUID: 4dd4d7a99e9f754da1b94cc6c99fbb67
FullName: LDAP://schema/user

Bind to user using WinNT provider:
FullName: Jeff Fellinge
ADsPath: WinNT://HOME/europa/jeff
GUID: {D83F1060-1E71-11CF-B1F3-02608C9E7553}
FullName: WinNT://HOME/Schema/User

G:\>
```

FIGURE 3.19   Choosing different providers yields different results.

LDAP or WinNT?

Some attributes, such as a user's last login date, are stored on each domain controller that that user has logged in to, and the property is not replicated (meaning that when one is updated, it is not copied to the other domain controllers). To find the latest login date, you must query each domain controller and use the latest value. Even when individually querying Windows 2000 domain controllers, remember to first use the LDAP provider (assuming it can provide the properties you seek). In some cases, such as when querying local user accounts and groups of member servers or standalone servers, you are required to use the WinNT provider. For other AD queries, such as group, users, group membership, and most other attributes, bind to the LDAP provider.

MODIFYING THE DIRECTORY

So far, all of our examples have revolved around getting and displaying values using ADSI. ADSI also lets you create objects and set their properties; for example, adding or removing members from a group, deleting an object, or creating a new user, OU, or group, to name a few. These processes are done by interface-specific methods (for example, iADSContainer::Create, iADSGroup::Add) as well as generic ones (iADs:Put).

ADSI uses a *property cache*. When you get information from the directory, ADSI retrieves it into a local cache on your computer. Similarly, when you create an object or modify an object's properties, these actions are done in your cache only. You must then use the iADs:SetInfo method to actually update the directory.

Creating a User

Creating a user is a straightforward procedure, as shown in Listing 3.28.

ON THE CD

LISTING 3.28 Creating a user with ADSI

```
strPath="LDAP://OU=Home Users, DC=home, DC=blackstatic, DC=com"
Set objOU = GetObject(strPath)
Set objNewUser=objOU.Create("User","cn=New User")
objNewUser.Put "sAMAccountName","NewUser"
objNewUser.SetInfo
```

However, this most basic of examples creates quite an empty shell of a user. The account is disabled by default, and the user does not have a logon name, password, or personal information set, as shown in Figure 3.20.

Let's walk through this example. First, you must bind to the location where you want to create the new user object and then, using iADSContainer::Create method, create that user. Notice that creating the object returns a reference to the new object (which the script calls objNewUser). Now we work exclusively with this new User object.

```
strPath="LDAP://OU=Home Users, DC=home, DC=blackstatic, DC=com"
Set objOU = GetObject(strPath)
Set objNewUser=objOU.Create("User","cn=New User")
```

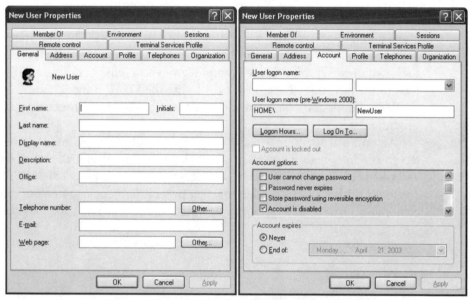

FIGURE 3.20   Creating a user account with the minimal options set is a barren account.

Next, specify the mandatory or optional attributes for that object. The ADSchema User object requires the following attributes: Common-Name, Instance-Type, NT-Security-Descriptor, Object-Category, Object-Class, Object-Sid, and SAM-Account-Name. Don't worry too much if that list seems long. Many of the required attributes are generated automatically when the object is created, but you must set the rest before you update the directory. You will receive an error if you do not have all of the mandatory attributes defined when you update the directory. In fact, we must set only the SAM-Account-Name to actually create the object. However, as Figure 3.20 shows, the shell is quite empty.

To update the SAM-Account-Name attribute, use the iADS::Put method, specifying both the attribute and the value. Lastly, use the iADS::SetInfo method to update the directory with this new user object.

```
objNewUser.Put "sAMAccountName","NewUser"
objNewUser.SetInfo
```

TIP

Many people find the older ADSI SDK to be a bit less friendly to script writers than the WMI SDK. To find all of the mandatory and optional attributes for any given class, switch to the SDK Index tab and look for ClassName [AD Schema]. For example, to find all of the details of the User class, including attributes, look for User [ADSchema].

Creating and Setting Other Objects

Use a similar approach to create other directory objects. The example in Listing 3.29 shows how to create a new OU, group, and user and then add that new user to the newly created group. These examples give you a sense of how to manipulate objects in the directory. They show how to use the LDAP provider, and you can use similar methods and objects to add local users to your machines using the WinNT provider.

An optional attribute when creating a group object is groupType. This attribute lets you define whether the group is a local, domain-local, global, or universal group and whether it is a distribution or security group. The script includes the group type constants, and you can use them together by summing their values. For example, to create a universal security group, you would set the groupType attribute to:

```
ADS_GROUP_TYPE_UNIVERSAL_GROUP + ADS_GROUP_TYPE_SECURITY_ENABLED
```

or

```
0x00000008 + 0x80000000
```

or

```
0x80000008.
```

Another way to represent this hex value in the script is &H80000008.

LISTING 3.29   Creating and assigning a user to a new OU and group

```
Const ADS_GROUP_TYPE_GLOBAL_GROUP = &H2
Const ADS_GROUP_TYPE_DOMAIN_LOCAL_GROUP = &H4
Const ADS_GROUP_TYPE_UNIVERSAL_GROUP = &H8
Const ADS_GROUP_TYPE_SECURITY_ENABLED = &H80000000

strPath="LDAP://DC=home, DC=blackstatic, DC=com"
Set objADs = GetObject(strPath)

Set objNewOU= objADs.Create("organizationalUnit","OU=New OU")
objNewOU.SetInfo
Set objADs =Nothing
Set objADs =GetObject("LDAP://" & objNewOU.distinguishedName)

Set objNewUser= objADs.Create("User","cn=New User")
objNewUser.Put "sAMAccountName","NewUser"
objNewUser.SetInfo
```

```
Set objNewGroup= objADs.Create("Group","cn=New Group")
objNewGroup.Put "sAMAccountName","NewGroup"
objNewGroup.Put "groupType",_
 ADS_GROUP_TYPE_GLOBAL_GROUP + ADS_GROUP_TYPE_SECURITY_ENABLED
objNewGroup.SetInfo

objNewGroup.Add("LDAP://" & objNewUser.distinguishedName)
```

These examples barely touch on all of the different possibilities of ADSI. The ADSI-centric scripts in the remaining chapters demonstrate many of the nuances of using ADSI.

SEARCHING ADSI

Thus far, the scripts have demonstrated a few of the basics for how to bind to an object and return its properties. From this, you could create a *recursive* function—one that dives into every container and enumerates its contents and then further dives into any new containers it finds and so forth. But you can use an easier and much faster way to search the directory. ADSI supports searching with *ActiveX Data Objects*, or *ADO*. Searching with ADO is a very efficient and powerful method to return data.

Limitations to using ADO are that you can search using only the LDAP provider, and you cannot update the directory using ADO. You are limited to searching and retrieving data. But, nonetheless, ADO is a great tool with which to crawl the directory looking for data.

You can use multiple avenues for constructing ADO-based queries. The following script shows one method using three basic ADO objects—Connection, Command, and RecordSet. We'll break down the script shortly, but first let's look at the entire script. The script in Listing 3.30 uses ADO to search the entire directory for users.

ON THE CD

LISTING 3.30 ADO provides a quick and efficient means of searching the directory

```
Set objConnection = CreateObject("ADODB.Connection")
objConnection.Provider = "ADsDSOObject"
objConnection.Open

Set command = CreateObject("ADODB.Command")
Set command.ActiveConnection = objConnection
```

```
command.CommandText = "<LDAP://DC=Home, DC=Blackstatic, DC=com>;"_
  & "(&objectCategory=user);"_
  & "Name,ADsPath;"_
  & "SubTree"
Set objRecordset=command.Execute

While Not objRecordset.EOF
    WScript.Echo objRecordset.Fields.Item("Name").Value & "   "_
     & objRecordset.Fields.Item("ADsPath").Value
    objRecordset.MoveNExt
Wend
```

Connection

The first object that we need to set up when creating an ADO ADSI query is the `Connection` object. This object is used to authenticate users. Create a new `Connection`, and specify the ADSI OLE DB provider. The OLE DB provider for ADSI is called `ADsDSOObject`.

```
Set objConnection = CreateObject("ADODB.Connection")
objConnection.Provider = "ADsDSOObject"
```

Optionally, set the authentication of the connection, such as the username and password. If you do not specify authentication information, the credentials of the currently logged-on user are used. The next code shows these optional connection properties, as well as the modified `Connection::Open` method passing the username and password. The first parameter of the `Open` method is an alternative string representing the DSN, username, and password, and is optional.

```
objConnection.Properties("User ID") = strUser
objConnection.Properties("Password") = strPassword
objConnection.Properties("Encrypt Password") = True
objConnection.Open "", strUser, strPassword
```

Our original example uses the credentials of the currently logged-on user, and we simply open the connection with the `Connection::Open` method without any parameters.

```
objConnection.Open
```

After the `Open` method, the connection is authenticated and established. If you choose to use specific authentication and the credentials are incorrect, your script returns an error when you attempt to `Open` the connection.

Command

The Command object is optional, but it includes additional parameters that are useful in regulating the query. Create a new Command object and set the ActiveConnection to the previously created Connection object.

```
Set objCommand = CreateObject("ADODB.Command")
Set objCommand.ActiveConnection = objConnection
```

Next, set the CommandText to define the query. (This code is all on one line, but it has been broken repeated times using the line continuation character (_) due to page-width constraints.)

```
objCommand.CommandText = "<LDAP://DC=Home, DC=Blackstatic, DC=com>;"_
                & "(objectCategory=user);"_
                & "Name,ADsPath;"_
                & "SubTree"
```

The ADSI ADO provider understands two query dialects: an LDAP dialect and a SQL dialect. Listing 3.30 uses the LDAP dialect. The differences between these dialects are covered in more detail in the next few sections.

The Command object supports a number of optional parameters. These parameters are called Properties of the Command object and include such things as Timeout, Time Limit, Sort on, Cache results, and so on.

Lastly, after setting the properties of the Command object, call the Command::Execute method to perform the query and return the results to a RecordSet object.

```
Set objRecordset=objCommand.Execute
```

RecordSet

The RecordSet object contains the results of the query and is used to iterate and display them. The script iterates through the collection using the EOF (end of file) and MoveNext methods. The script displays the name and value of the results using the Fields, Item, and Value methods and properties as shown in the next code:

```
While Not objRecordset.EOF
    WScript.Echo objRecordset.Fields.Item("Name").Value & "   "_
            & objRecordset.Fields.Item("ADsPath").Value
    objRecordset.MoveNExt
Wend
```

SQL vs. LDAP Query Dialect

ADSI queries support either the LDAP or SQL query dialect. Both query types include the same basic parameters of source, criteria, and returned fields; however, the syntax for each is quite different. Table 3.8 shows the basic query parameter map between the two dialects.

TABLE 3.8 Differences between LDAP and SQL query dialect

	LDAP	*SQL*
source	base distinguished name	FROM
criteria	search filters	WHERE
returned fields	attributes	SELECT attributes
directory search instructions		search scope

In code, these two dialects take the following form:

LDAP query dialect

```
objCommand.CommandText = "<LDAP://DC=Home, DC=Blackstatic, DC=com>;"_
                & "(objectCategory=user);"_
                & "Name,ADsPath;"_
                & "SubTree"
```

SQL query dialect

```
command.CommandText = "Select Name,ADsPath FROM "_
& "'LDAP://ou=Home Users, DC=Home, DC=Blackstatic, DC=com' "_
& "WHERE objectCategory='User'"
```

The SQL query is similar in syntax to the WQL queries discussed previously.

The LDAP dialect includes four parameters, separated by semicolons: base distinguished name, LDAP search filters, attributes, and search scope. Spaces are not allowed between the LDAP parameters, and if used, the script returns an error.

Base Distinguished Name

The base distinguished name identifies the start of the query and is in the form of an ADsPath. The base distinguished name is similar the FROM clause in SQL. Enclose the ADsPath in brackets.

Search Filters

The LDAP search filters operator defines the terms to restrict the query results. Nest the filters to achieve a robust and flexible method for defining the returned data set. A filter must be enclosed in parentheses.

The following code returns all objects where the `objectClass` is a user:

```
(objectClass=user)
```

When you run a query with the previous filter for the first time, it returns both user and computer objects. This is because the `objectClass` parameter is multivalued and contains all classes to which an object belongs, as seen in Table 3.9. A computer object is by default a member of the following object classes: `top`, `person`, `organizationalPerson`, `user`, and `computer`.

TABLE 3.9 Hierarchical in nature, ADSI objects may belong to multiple object classes

Computer Object	User Object
objectClass	objectClass
top	top
person	person
organizationalPerson	organizationalPerson
user	user
computer	

Use the single-value property `objectCategory` to find specific types of objects. `objectCategory` is indexed and provides a more efficient search query. Like the `objectClass` property, some objects share the same `objectCategory`, so construct your filters to ensure you get exactly what you are looking for. For example, both the `contact` and `user` object have an `objectCategory=person` class. To return just the actual users we must use a combination of filters. Filters support parenthetical nesting with logical operators, so pairing the two previous examples together, we obtain the filter that returns all objects with an `objectClass=user` and `objectCategory=person`. (This method effectively screens out both the computer and category objects.) Notice that we have used the & logical operator, which means "and," and that the location of that operator is outside the two individual filters, in a format such as the following:

```
(&(filter A)(filter B))
```

```
(&(objectClass=user)(objectCategory=person))
```

The filter can also include the wildcard character *. The following example returns any object with a common name cn= containing the string eff.

```
(cn=*eff*)
```

The filter syntax, including logical operators, is defined in RFC 1960, which defines the string representation of LDAP search filters. A partial listing is shown in Table 3.10.

TABLE 3.10   Partial listing of LDAP search filter syntax

and	&
or	|
not	!
equal	=
approx	~=
greater than	>=
less than	<=
present	=*
none	Null
any	*

Attributes

Specify what attributes you want to return from your query in the attributes section of LDAP dialect.

Search Scope

Lastly, specify how the query should handle hierarchical containers by setting the search scope. Specify whether to search in this location only or extend the search to child elements by specifying the search scope. The search scope can be set to a number of parameters, such as subtree, base, or one-level.

SUMMARY

Now you are armed with three of the most useful and powerful scripting technologies for the Windows platform—Windows Script Host, Windows Management Interfaces, and Active Directory Services Interfaces. Together, WSH, WMI, and ADSI can serve as a cornerstone to your scripting endeavors.

KEY POINTS

- Windows Script Host provides a variety of system-related objects, methods, and properties focused on helping automate Windows-based tasks, such as reading a text file, managing the registry, executing applications, or displaying output to the console.
- Windows Management Instrumentation (WMI) provides a detailed and robust set of management information about local and remote systems, such as software versions, hardware installed, running processes, and many other system configuration data.
- Active Directory Services Interface (ADSI) provides a mechanism to query and update several types of directories, including Active Directory, as well as local computer accounts.

Now the excitement really begins. The next 10 chapters walk through a variety of real-world problems and look at the opportunities and technologies to solve them via scripting.

Away we go!

4 Enumerating and Dumping the Users, Groups, and Computers of Active Directory

```
Console                                          _□x
G:\>adsiquery -gre admin
Administrators
DHCP Administrators
Schema Admins
Enterprise Admins
Domain Admins
DnsAdmins
intranet Admins


------------------------------------------------
Enumerating Objects
------------------------------------------------

Administrators (6 members)
    svcSMS
    svc_sql2000
    Jeff Fellinge
    Domain Admins
    Enterprise Admins
    Administrator

DHCP Administrators (0 members)
-empty-

Schema Admins (2 members)
    Jeff Fellinge
    Administrator

Enterprise Admins (2 members)
    Jeff Fellinge
    Administrator

Domain Admins (4 members)
    svcBackup
    svcSMS
    Jeff Fellinge
    Administrator

DnsAdmins (0 members)
-empty-

intranet Admins (0 members)
-empty-

G:\>_
```

OVERVIEW

Microsoft Active Directory (AD) provides rich directory services that centrally contain detailed network, user, and group data. AD defines the logical structure of your Windows 2000 and Windows 2003 network and provides the foundation for managing resource access. Active Directory Group Policy, Access Control Lists, Domain Name Services (DNS), sites, and domains exemplify services contained within AD.

Microsoft provides a number of tools to help manage AD, from graphical MMC plug-ins (such as AD Users and Computers, Sites and Services, and Domains and Trusts) to a number of command-line tools available in the resource kits. These tools allow you to add or edit users and groups, change group membership, define policy, configure domains, trusts, and otherwise manage AD for your enterprise.

You may find yourself yearning for a custom approach to extracting and presenting data from AD. In Chapter 3, we demonstrated how easy it is to query data from AD and create new users or groups. For example, you may want to audit the membership of a sensitive group on a reoccurring basis or list all users within a particular group or groups.

The scenario presented in this chapter outlines the requirements for a script to provide a command-line tool to audit the membership of specific groups within AD. Although existing tools to list and enumerate groups are available, developing your own script to accomplish this teaches you how to interact with AD and format output data in a manner of your choosing.

This tool uses WSH and ADSI to create a command-line script to enumerate and dump the membership of selected groups. The tool accepts command-line parameters to input criteria and uses regular expressions to find matching groups.

SCENARIO

You are a systems engineer of a medium-sized cellular service company. Your company provides cellular phone service to many rural areas not serviced by the larger companies. Three years ago, you deployed AD across the company and now manage all of the domains. The domain structure consists of a corporate domain and several child domains for service and test departments.

Several line managers have approached you to ask for help in managing access to their groups. For example, the Finance department wants a weekly report of all users' permissions to access customer data. Security access control lists (ACLs) set on AD groups govern access to customer data stored on Windows 2000 file servers. Finance administrators (the users) work on teams assigned to larger customers. The teams define the AD group membership that regulates access to specific customers' data. Group membership changes periodically as finance administrators rotate across customer teams. The finance manager wants to review a weekly list of who can access client data.

A second request for help comes from your own Information Technology group. The security manager wants a list of all domain computers with anti-virus software installed and running. The manager needs a simple list of computers in each domain (one computer per line), which can then be fed into a second command-line tool to check the status of the software.

ANALYSIS

You realize that these two requests, although different, each fundamentally consists of pulling data from the AD and then displaying it in a customized manner. You decide that a flexible console-based tool could be deployed and scheduled to run on a reoccurring basis for each request. The output could be viewed in real-time by the script runner and redirected to a text file or to a second program for later processing. This tool is used primarily by the IT administrators, who likely either run the application under their own credentials or schedule the script to run using a dedicated service account not used by other users. Based on the requirements, the script could consist of five primary functions and several helper functions:

The Main Program: Evaluates the arguments and calls the program's functions.

- Evaluates the input parameters, and sets the appropriate processing flags.
- Calls the processing functions.

ReadCriteria: Reads the group-matching criteria from a file.

- Uses FileSystemObject to read a list of criteria from a text file. The criteria will be used to filter results from the AD. The script permits criteria to be included on the command line as well as imported from a specified text file. Example criteria might include the names of the groups to which the finance administrators belong.

QueryObjects: Builds and executes the ADSI query.

- Initializes and executes an ADO query against the specified LDAP directory service. Uses the LDAP dialect filter to display user, computer, or group objects that match user-specified filter criteria.

Enumerate: Displays the results.

- Iterates through the returned data objects to list memberships. Group objects display their memberships, and computer and user objects display all groups of which they themselves are members.

ListGroupMemberships: Enumerates and displays all the groups of which a specified user is a member.

- Determines and displays group memberships of a specified user or computer object. ListGroupMemberships uses the Groups method to return a collection of the groups in which an object is a member.

ListMembersOfGroup: Enumerates and displays all of the members of a specified group.

- ListMembersOfGroup uses the property Members to return and display a collection of members of the specified group object.

The ancillary functions follow:

`CurrentDomain:` Discovers the current domain of the user context running the script.

■ Binds to the RootDSE (the root of the directory data tree on a directory server) to determine the domain of the account running the script. To query other domains, the current domain can be overridden by the `-D [domain]` parameter and a new domain specified.

`Heading:` Formats the output headers.

■ Provides a consistent look for displaying output headers.

`Usage:` Displays a help screen for how to use the script.

■ Displays help for how to use the script.

`ParamError:` Displays error text if a dependant script parameter is not included.

■ Displays an error if the parameter is not included. The parameters `-D [Domain]`, `-d [delimiter]`, and `-f [file]` each require additional dependent parameters.

`Tabulate:` Aligns text output.

■ Aligns columnar output at specified horizontal character positions. (Requires Windows Script version 5.6.)

Distribution and Installation

The script consists of a single JScript file and runs on Windows 2000 or later. Your company uses Microsoft Outlook®, which by default blocks access to .vbs and .js files. For this reason, you decide to post the script on your intranet, from which privileged users can download and run locally.

Scheduling the script to run and dump output to a text file on a weekly basis provides convenient and reliable group membership updates for the finance administrators. First, automate the reoccurring execution of the script using the Windows Scheduled Tasks application. Next, launch the Windows 2000 Scheduled Task Wizard and set up a new task. Copy the script to its run location and specify the program to launch the script. For example, choosing to launch the script with a batch file that contains the specified parameters makes future updates easy. An example batch file to launch the script might consist of one line:

```
%windir%\system32\cscript.exe c:\scripts\ADSIQuery.js -D home -gre
admin > c:\scripts\output.txt
```

This batch file calls the *cscript.exe* engine to process the *ADSIQuery.js* script. The parameters sent to the script are a specified domain (`-D home`) and `-gre admin`, which tells the script to list group objects (`g`), use regular expressions to evaluate the criteria (`r`), and enumerate the members of the groups (`e`) and includes the search criteria `admin`. Use the always handy redirect symbol > to send the console output to a specified text file. (Scripts that require user interaction or that display progress windows that must be clicked through represent poor candidates for scheduling or redirected output.)

Save this batch file, and schedule it to run weekly at 4:00 a.m. every Monday. Specify the username and password under which the batch file should run. Scheduling the script to run in this manner allows it to run independently of the currently logged-in user. Scheduled in this manner, this batch file will run weekly and output its results into the *c:\scripts* directory.

In addition to scheduling the task to run repeatedly using fixed parameters, you also want to provide it to other company system administrators to use ad hoc. Simply run the script from a command prompt with any of the following parameters:

```
Usage:
ADSIQuery -cugvrfe [filename] -D [Domain] -d [delimiter] [criteria]:
-c list computers
-u list users
-g list groups
-v verbose mode
-r use regular expressions for criteria matching
-f [filename] load criteria from a text file
-e Enumerate the results for group membership
-D [domain] specify domain other than default
-d [delimiter] specify delimiter to separate fields.
-a all non-RegExp criteria must match (default = any).
-? Help

Example:
ADSIQuery -ue jeff : Returns users that match 'jeff'
and will list all groups to which they belong
ADSIQuery -cr dc[12] : Returns computers that match
the regular expression dc[12] (e.g., dc1 or dc2)
```

NOTE

You must use the cscript.exe engine because the script uses the StdOut.Write method, which is not supported in the wscript.exe engine. As you'll see in the code, the script determines how it was called and runs or displays a warning to the user as appropriate. Building these safety nets and checks into your scripts might take a bit more time, but they greatly reduce possible frustration when other users use your programs.

TIP

To access the directories of untrusted domains, consider using the runas command to execute a command shell under a user from that domain. (Alternatively, you could open a Terminal Server, SSH, or Telnet session to a computer in that domain and run the script from that session.) The syntax for starting a command shell under another user in Windows XP is:

```
%windir%\System32\runas.exe /user:domain\user "cmd"
```

The runas application prompts you for a password for this user and then opens a command prompt running under the context of the alternate user. From this new command prompt, execute the script against the desired domain.

Output

The script's design centers on console output. Open a command shell, and use the console to drive the script and review the output. Using the console ensures that the program works either through other applications or through scheduled services.

Summary of Solution

To query AD group memberships, create a console-based script and schedule it to run in the domains that contain the finance administrator's data. Create a customized batch file for each domain that contains the names of the groups to be enumerated as well as the script parameters germane to that domain. Redirect the results to a central location accessible to the finance administrators who will review it on a weekly basis. Additionally, distribute the script to the security manager and other IT administrators so that they can use the tool to list AD objects and identify group memberships or user and computer memberships for any domain in your company.

OBJECTIVE

This script demonstrates using WSH and ADSI to query AD and several methods of displaying and formatting the results. The script uses the following objects and methods:

ADODB.Connection
- Provider
- Open

ADODB.Command

- ■ `ActiveConnection`
- ■ Properties
- ■ `CommandText`
- ■ `Execute`

ADSI

- ■ `GetObject`

Array Object

- ■ `push`

Enumerator

- ■ `atEnd`
- ■ `MoveNext`

LDAP Dialect Query

- ■ Base DN
- ■ Filter
- ■ Attributes
- ■ Scope

RecordSet

- ■ `EOF`
- ■ `Fields.Item(x).Value`

RegExp

- ■ `test`

RootDSE

- ■ `GetObject("LDAP://RootDSE")`
- ■ `Get("defaultNamingContext")`

Scripting.FileSystemObject

- ■ `OpenTextFileReadLine`
- ■ `FileExists`
- ■ `AtEndOfStream`
- ■ `Close`

String Functions

- ■ `split`

WScript.Arguments

- ■ `colArgs.length`
- ■ `colArgs.Item`

WScript.StdOut

- Write
- Column

THE SCRIPT

Figure 4.1 shows a sample output of the script *ADSIQuery*, which ran to enumerate and display the membership of the admin groups in the default domain. In this example, the first argument, -gre, actually contains three parameters. The parameters specify to list only groups (g), use regular expressions (r), and enumerate the membership (e). The second argument, admin, defines the actual criteria used in the filter. Figure 4.2 shows another example of the script, this time listing all computers in the domain. (The regular expression . (dot) represents the wildcard character.) This example illustrates the parameter input flexibility of the script. The parameters can be combined or input separately in any order. For example, the script processes the following parameters with the same effect:

```
ADSIQuery -grev admin
ADSIQuery admin -g -e -v -r
ADSIQuery -rv -ge admin
```

An exception is the delimiter and domain parameters, which must be input separately (for example, -grev -D domainname -d ~).

This example lists all computers in the domain as a simple list, suitable for import into other applications.

ON THE CD

The script is presented on the CD-ROM as /chapter 4/scripts/ADSIQuery.js. Before continuing in this next section, open the actual script from the CD-ROM and run it a few times in your test environment to get a feeling of how it works. Explore the entire code on your own and then continue with the walkthrough, which hopefully answers any questions you may have asked yourself. Plus, seeing the code in its entirety in your script editor gives you a good sense of its scope and also allows you to tweak it as we walk through its function.

The remainder of the chapter walks through *ADSIQuery* to describe how it works and why certain decisions were made concerning its design. The script is coded in JScript to demonstrate many of the features of that language.

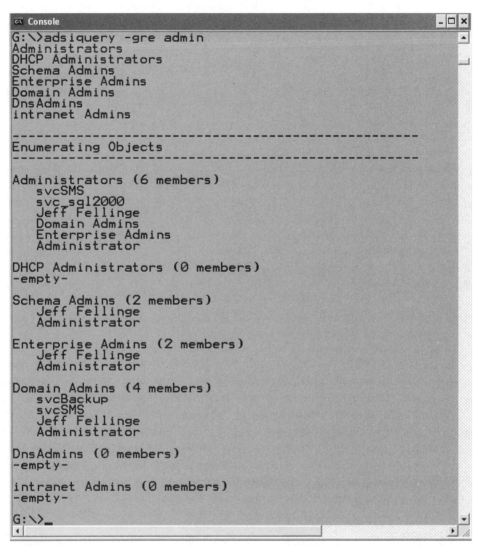

```
Console                                                    _ □ ×
G:\>adsiquery -gre admin
Administrators
DHCP Administrators
Schema Admins
Enterprise Admins
Domain Admins
DnsAdmins
intranet Admins


-----------------------------------------------------------
Enumerating Objects
-----------------------------------------------------------

Administrators (6 members)
    svcSMS
    svc_sql2000
    Jeff Fellinge
    Domain Admins
    Enterprise Admins
    Administrator

DHCP Administrators (0 members)
-empty-

Schema Admins (2 members)
    Jeff Fellinge
    Administrator

Enterprise Admins (2 members)
    Jeff Fellinge
    Administrator

Domain Admins (4 members)
    svcBackup
    svcSMS
    Jeff Fellinge
    Administrator

DnsAdmins (0 members)
-empty-

intranet Admins (0 members)
-empty-

G:\>_
```

FIGURE 4.1   The script *ADSIQuery.js* provides a flexible method for querying AD and enumerating memberships.

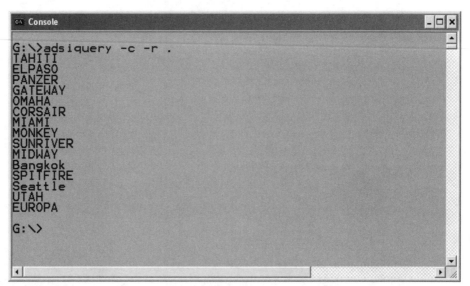

FIGURE 4.2 Parameters can be entered in any order and with varying levels of output detail.

Another benefit of a console-based tool such as this is that it is quick to add new functionality or adjust the output of the script. For example, if you wanted to return the phone numbers of any user objects, you could quickly add the AD attribute TelephoneNumber to the LDAP dialect query and then display this attribute when iterating through the RecordSet. As you'll see, some of the more generic functions are reused throughout the scripts in this text, such as reading text files, displaying help or errors, and other reusable code. Even if you have your own ADSI dump tool of choice, you may find the functions and methodologies useful for reuse in entirely different scripts.

Initialization

The beginning lines of the script declare the variables and set their initial values. In JScript, variables are considered local if they are defined within a function. Otherwise, the function accesses the variable if it is defined outside of any function but before that function is called. For example, all of the variables defined in a program are available to the functions called by that program as well, so long as they are defined before the function is executed. However, if a child function declares a variable with the same name as a previously used variable, then the local scoped variable is used instead.

Temper your usage of global variables and instead explicitly pass variables or objects to functions. Using global variables generally reduces the portability of your code and increases the risk of bugs or potential security holes. Although the programming community frowns on the widespread use of global variables, many agree that global variables can be valuable if used in restrained and specific situations.

This script takes advantage of regular expressions to validate proper filenames. Regular expressions provide powerful text-based filters that shine at pattern matching (among many other things), and this script demonstrates just a few of the utilities regular expressions offer. Regular expressions are well defined, and many Unix tools (for example, grep, sed, awk) support regular expression pattern matching extensively. Although pattern matching syntax generally persists across regular expression implementation, the methods to manipulate regular expressions vary by language. Search the Web for more information about how to define these filters and some of the more advanced features they support.

In JScript, regular expressions are defined between forward slashes, like so: /RegExp/. A bracket denotes optional characters within a position. For example, the regular expression /[DdHh]og/ would match the text Dog, dog, Hog, and hog. The first regular expression in *ADSIQuery* (reValidFile) defines invalid characters used in a filename. The script tests the filename supplied by the user at script run-time against reValidFile, and if the test returns true, the script rejects the filename. The regular expression /[*\?\[\]\{\}\|\+\(\)]/ simply looks for any of the characters *, ?, [,], {, }, |, +, (, or). The back slashes \ represent regular expression escape characters that define these otherwise special characters.

The Main Body

The script uses array objects to store the list of criteria (aCriteria) as well as the ADsPath of matching objects (aObjects). This script accesses the members of these arrays sequentially, which fits well into how an array accesses its members by index.

```
//------------------------------------------------------------------
// ADSIQuery.js
//------------------------------------------------------------------
var i=0,j=0;
var strCriteriaFile, strCriteria, iTotalCriteria;
var strQBDN, strQFilter, strQAttributes, strQScope;
var reValidFile = /[\*\?\[\]\{\}\|\+\(\)]/;
var strDelimit=" ";
var aCriteria = new Array();
var aObjects = new Array();
```

This script relies on a custom object (objState) to define process flags. JScript allows easy creation and use of custom objects that can be referenced by the property (object.elementname). Referencing by property provides an easy-to-read reference to a specified flag. For example, to determine the state of the group property, we can call it directly (objState.group). Alternatively, we could have defined individual flags for each value (such as, iGroup, iComputer), but containing them all in a single object as properties is a tidy design to keep all state variables together. Plus we can pass a single object to a function to determine state rather than multiple independent variables.

```
var objState = new Object;
objState.group=0;
objState.computer=0;
objState.user=0;
objState.reCustom=0;
objState.help=0;
objState.UseRegExp=0;
objState.verbose=0;
objState.enumerate=0;
objState.domain=0;
objState.all=0;
```

The script uses functionality not available through the *wscript.exe* engine. The script uses the JScript IndexOf method to examine the FullName property of the WScript object to determine if the script was executed using *wscript.exe*. If so, the script displays a warning advising the user to rerun the script using the *cscript.exe* engine.

```
if(WScript.FullName.indexOf("WScript")>=0) {
 WScript.Echo("Due to some of the functionality of this script, " +
   "Please run this program using the cscript.exe engine.\n" +
   "for example: cscript.exe userquery.js");
 WScript.Quit()
  }
```

The function CurrentDomain determines the current domain of the account running the script. This function allows the script to be run in any domain without having to hardcode the domain name in the script. To accommodate running the scripts in other domains, the script supports a parameter (-D domain_name) to specify a domain to search.

If an alternate domain is specified, the script binds to that domain and executes the directory service query using the security of the account running the script. This practice illustrates the utility of the aforementioned runas *command:* runas *allows you to run the script under a different user account. By specifying the domain and user account, you can query any domain from your computer, assuming the domain can communicate with your computer at a network level. Alternatively, add a parameter to the script to define a different username, request the password, and then execute the query using that alternative security context.*

```
strDomain=CurrentDomain();
```

Parameter enumeration consists of pulling the command-line arguments passed to the script, iterating through each of them, and setting state variables as appropriate. The following code loops through each of the WScript.Arguments and tests them against regular expressions that define the script parameters.

The first test checks for the presence of a hyphen (-) at the beginning of the argument string using the JScript indexOf method. IndexOf searches one string for the presence of another and, if it is found, returns the starting location of the substring. If we want to find out if a string begins with a hyphen, we can check to see if indexOf returns a 0 (denoting the beginning of the string). If the substring is not found, indexOf returns a -1.

```
var colArgs = WScript.Arguments;
for(i=0;i<colArgs.length;i++) {
    strArg=colArgs.Item(i)
    if(strArg.indexOf("-") == 0) {
```

The second nested set of tests looks for matching parameters and sets the appropriate state variables for later processing. Again, indexOf is used to test for the presence of the character in the argument string. If the method returns a value greater than or equal to 0, then the parameter is in the argument string. The script evaluates arguments that begin with a hyphen in one pass, and the tests are independent of order. This means that the script evaluates the arguments -cgue or -eugc each with the same result because indexOf only looks for the presence of the parameter character within the string. This design provides order independence and a compact way to evaluate script parameters.

```
if(strArg.indexOf("c") >=0) objState.computer=1;
if(strArg.indexOf("g") >=0) objState.group=1;
if(strArg.indexOf("u") >=0) objState.user=1;
if(strArg.indexOf("e") >=0) objState.enumerate=1;
```

```
if(strArg.indexOf("?") >=0) objState.help=1;
if(strArg.indexOf("r") >=0) objState.UseRegExp=1;
if(strArg.indexOf("v") >=0) objState.verbose=1;
if(strArg.indexOf("a") >=0) objState.all=1;
```

Some parameters expect dependent parameters. The parameter f denotes the use of a filename, and the script evaluates the next argument as the filename from which to import the search criteria. The script checks that the successive argument exists by checking the current argument number against the number of the arguments using the length property. The script throws an error if the dependent argument does not exist. Next, the script validates that the successive argument does not contain illegal characters in the filename.

The JScript function for testing a string against a regular expression is re.text(string). If the test is true, the script knows that the string satisfies the regular expression. In our case, the test looks for filenames containing illegal characters.

Assuming a well-constructed filename, the script sets the objState.file property and strCriteriaFile.

To skip over the argument containing the filename during the next iteration, the script increments the counter (i). Otherwise, the filename would be added as search criteria.

```
if(strArg.indexOf("f") >=0)  {
    if(i+1 > colArgs.length-1) ParamError();
    if(!reValidFile.test(colArgs.Item(i+1))) {
        strCriteriaFile=colArgs.Item(i+1);
        objState.file=1;
        i++;
        }
    else {
        WScript.Echo("Import filename is not valid.");
        WScript.Quit();
        }
    }
```

The script handles the delimiter (-d) and domain name (-D) parameters a bit differently than the filename parameter. The script uses a simple conditional to check whether the argument equals a -d or -D. If true, then the script looks for the next argument as the delimiter or domain name, respectively. Independently checking the parameters like this demonstrates a different method of interpreting parameters from using regular expressions. The delimiter defines the character that is used when displaying verbose data. A custom delimiter aids future parsing in applications such as Excel in which you can define a custom delimiter from the traditional comma or tab, which may be included within legitimate data. The domain name allows the user to query a different domain from that of the account running the script.

```
if(strArg=="-d") {
    if(i+1 > colArgs.length-1) ParamError();
    strDelimit=colArgs.Item(i+1);
    i++;
    }
if(strArg=="-D") {
    if(i+1 > colArgs.length-1) ParamError();
    strDomain=colArgs.Item(i+1);
    objState.domain=1;
    i++;
    }
}
```

Lastly, an `else` statement catches the remaining arguments that are not parameters (such as those that do not begin with hyphens). The script adds these arguments to the array `aCriteria` using the `push` method. Although we could have added a new member to the end of the array by index (for example, `aCriteria[aCriteria.length]="some criteria"`), using the `push` method is a bit simpler and more elegant. This method allows you to specify any number of criteria via the command line.

```
else {
    aCriteria.push(strArg);
    }
}
```

By this design, parameters must begin with a hyphen (but can be listed in any order), and multiple individual criteria may be entered at the command line. The following arguments demonstrate the flexibility of this argument design. The script interprets the following parameters as the same:

```
ADSIQuery -gru admin operator -f filename.txt -D home
ADSIQuery -gruf filename.txt admin operator -D home
ADSIQuery -D home operator -fgru filename.txt admin
```

Figure 4.3 shows how the script processes these arguments to construct the array `aCriteria`.

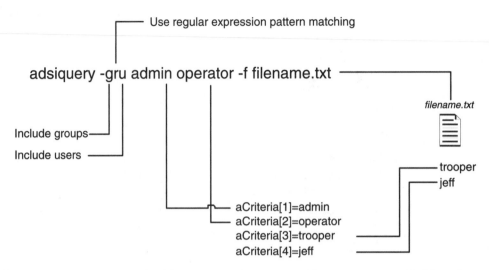

FIGURE 4.3 *ADSIQuery* flexibly processes the command-line arguments to build up the `aCriteria` array used to search Active Directory.

After processing the arguments, the script evaluates the state variables and takes appropriate action. The `-?` parameter sets the `objState.help` property, which in turn calls the `Usage` function to display a brief help message and then exit the script. Separating the help message into its own function allows other functions to call `usage` as well.

ADSIQuery frequently uses the state property verbose (`objState.verbose`) throughout the script to display more information. For example, it is used in the following code in conjunction with the file state property to display the specified filename back to the user. If the file state property (`objState.file`) is set, the script calls the `ReadFile` function to open the file specified by the user and read the criteria into the array `aCriteria`. Because any criteria entered at the command line is already defined within this array, it is passed to the `ReadFile` function, which then adds new members to it using the `push` method. The returned array then contains both the command-line criteria and the criteria imported from the file.

```
if(objState.help) Usage();
if(objState.verbose && objState.file){
    Heading("File Input");
    WScript.Echo("Importing criteria from file: " + strCriteriaFile);
    }
if(objState.file) aCriteria=ReadFile(strCriteriaFile, aCriteria);
iTotalCriteria=aCriteria.length;
```

If objState.Verbose is set, the script iterates through aCriteria and displays all of the discovered criteria.

Next, the script optionally displays the name of the domain that will be searched and then calls the QueryObjects function.

The QueryObjects function builds and executes the query and then displays the results. QueryObjects stores the results in a second array (aObjects) used in subsequent enumeration functions. The enumeration functions (specified by the parameter -e) list the memberships for each of the returned objects. The query and enumeration functions are described in more detail later in the script where the function code is listed.

```
if(objState.verbose){
    Heading("Search Criteria - " + iTotalCriteria + " found" );
    for(i=0;i<iTotalCriteria;i++) {
        WScript.Echo(aCriteria[i]);
        }
    WScript.StdOut.Write("\nRegular Expression Searching: ");
    WScript.StdOut.Write(objState.UseRegExp?"On\n":"Off\n")
    }
if(objState.verbose){
    Heading("Default Detected Domain");
    WScript.Echo(strDomain);
    }
aObjects=QueryObjects(aCriteria, strDomain, objState);
if(objState.enumerate){
    Heading("Enumerating Objects");
    Enumerate(aObjects);
    }
```

ReadFile

The function ReadFile reads a text file containing the criteria (one item per line) and inserts each criteria item into an array. In addition to the previous check validating the filename, ReadFile calls the method FileExists to check that the file exists before trying to open it. The function could be expanded to include error handling, such as verifying the file is a text file (and not binary), checking for proper formatting, or stopping processing if the file is too large.

The function uses the OpenTextFile method to open the file. The required arguments used by this method in this script are the filename of the file to open and the iomode. The iomode defines how the file should be opened, such as read only (1), write (2), or append (8). OpenTextFile supports additional optional requirements covered in Chapter 3 under the discussion of the FileSystemObject.

The script calls the `ReadLine` method to read each line of data from the text file into a string. This string is added to the criteria array using the array `push` method. Earlier in this chapter, we discussed global and local scope variables in JScript. Note that this function requires an array as an argument. However, `ReadLine` knows this object as `aContents`, regardless of the name of the array used in the calling function. At the end of the function, `aContents` is returned to the calling function using the `return` statement.

```
//-------------------------------------------------------------------
// Read a list of criteria from a file
//-------------------------------------------------------------------
function ReadFile(strFileName, aContents)
{
var objFSO = new ActiveXObject("Scripting.FileSystemObject");
var strReadLine;
if(objFSO.FileExists(strFileName)){
    objTextFile = objFSO.OpenTextFile(strFileName,1);
    while (!objTextFile.AtEndOfStream) {
    strReadLine=objTextFile.ReadLine();
    aContents.push(strReadLine);
    }
    objTextFile.close();
}
else {
    WScript.Echo("Error reading file.");
    WScript.Quit();
}

return aContents;
}
```

QueryObjects

The function `QueryObjects` constructs and executes the LDAP query. `QueryObjects` requires three parameters: the array containing a list of criteria, a string defining the domain directory server to search, and the array of state variables. The `objState` array contains information such as whether to search for user, computer, or group objects and whether verbose output should be displayed.

As discussed in Chapter 3, an ADSI ADO query is comprised of three parts: the `Connection` object, `Command` object, and `RecordSet` object. The `Connection` object opens the connection to the ADSI directory service.

Although we do not do so here, you can optionally set the username and password and whether or not to encrypt the password by setting these properties in the Con- nection *object. For example, an extension of this script might be to add a username argument to the script, which in turn prompts the user to supply a password. An easy method for capturing the password is using the* StdIn.Read *method. However, this method has its drawbacks in that the password is displayed in clear text in the console instead of hashing the password with ∗ characters. Also, if you do choose to prompt for a username and password, be sure to set the* Connection("Encrypt Password") *property to true. Otherwise, the username and password are clearly visible in the network traffic request to the server.*

```
//------------------------------------------------------------------
// Build and Execute the ADSI query
//------------------------------------------------------------------
function QueryObjects(aCriteria, strDomain, objState)
{
objConnection = new ActiveXObject ("ADODB.Connection");
objConnection.Provider = "ADsDSOObject";
objConnection.Open;
```

The Command object contains the instructions for how to perform the query. First, we set the ActiveConnection to that of the Connection object that was just created, and then set any optional command properties. For example, to sort the results by name, add the property objCommand.Properties("Sort On")= "Name".

To accommodate large data sets greater than 1,000 results, the script sets the command object's Page Size *property, which turns on support for paging and sets the page size. The page size is the number of objects returned from the server to the client (for example, a page size of 999 would require 11 pages for the server to return data from 10,000 objects). Multiple pages are delivered transparently for the most part. However, the user may see brief pausing between pages, depending on the speed of the network. Without paging, the query would return only the first 1,000 results and then quit.*

Next, we begin to build the LDAP dialect query. ADSI supports both LDAP and SQL dialect queries. This script uses the LDAP dialect because nesting multiple filter criteria is fairly straightforward when using the LDAP dialect. A drawback to using the LDAP dialect is that it does not support non-LDAP directory services such as the ADSI WinNT provider (for those few cases when you must use this other provider).

```
objCommand = new ActiveXObject("ADODB.Command");
objCommand.ActiveConnection = objConnection;
objCommand.Properties("Page Size") = 999;
var strQBDN;
var strQFilter;
var strQAttributes;
var strQScope;
var reCriteria=new RegExp(".");
```

The components of the LDAP dialect query and their represented variables are the base DN (`strQBDN`), filter(`strQFilter`), attributes (`strQAttributes`), and scope (`strQScope`). The base DN provides the ADO query with its starting point and scope. This script assumes a starting point at the top of the domain (for example, `LDAP://DC=domain, DC=com`); however, the base could be defined as a lower container as well (such as `LDAP://OU=users, DC=domain, DC=com`). The scope sets how deep the query should travel. *ADSIQuery* searches the base container and all subcontainers (`scope="SubTree"`), but you could optionally limit it to the base container only (`scope="Base"`) or only to the base and children of the base (`scope="One-Level"`). The query attributes include a list of all of the ADSI attributes to be returned from the query. The script uses the `objectClass` and `objectCategory` attributes to determine if an object is a user, computer, or group and the `ADsPath` attribute to uniquely identify the returned object. The script displays the object's `Name`.

```
strQBDN = "<LDAP://" + strDomain + ">;";
strQAttributes =  "Name,ADsPath;";
strQScope =  "SubTree";
```

The LDAP dialect filter defines the query search criteria. Chapter 3 introduced some examples of constructing a basic filter, and the following code builds on these examples to accommodate user selections. For example, if the user ran the script with the following syntax:

```
ADSIQuery -ucg trooper topaz
```

the resulting filter would resemble the following code:

```
(&(|(&(objectClass=user)(objectCategory=person))(objectCategory=compute
r)(objectCategory=group))(|(Name=*trooper*)(Name=*topaz*))));
```

which translates to: *get all objects that have an `objectClass` of user AND `objectCat-egory` of person or computer or group AND a name that contains `trooper` or a name that contains `topaz`.*

This logic may be slightly clearer by breaking out the filter into different lines:

```
(&
(|
(&(objectClass=user)(objectCategory=person))
(objectCategory=computer)
(objectCategory=group)
)
(|(Name=*trooper*)(Name=*topaz*))
);
```

A computer object is also a member of the `objectClass=user`, so the user objects must be further defined as all `objectClass=user`, *and* `objectCategory=person`. Using the `objectCategory` is preferred because it is an indexed single value, whereas the `objectClass` is a multivalued property and not as efficient when used for searching. The *Contact* AD object also has an `objectCategory=person`, and to return just user objects (not including computer or contact objects), we use the filter `objectClass=user` and `objectCategory=person`.

```
strQFilter="(&";
strQFilter+="(|";
if(objState.user) strQFilter +=
  "(&(objectClass=user)(objectCategory=person))";
if(objState.computer) strQFilter +=  "(objectCategory=computer)";
if(objState.group) strQFilter +=  "(objectCategory=group)";
strQFilter += ")";
```

The script also supports searches using regular expressions; however, the LDAP dialect does not, so custom code must be written. The script looks for the flag `objState.UseRegExp`, which indicates whether the user has specified to use regular expressions for the search criteria (by the script argument -r). If the flag is set, the script builds and executes the ADO query with the user, computer, or group filter and then tests the query results against the regular expression criteria in a separate procedure. As you might imagine, this additional processing dramatically slows the script in two ways. First, the ADO query returns a much larger set of data that must be manipulated (for example, all users). Next, the script must examine every object against each piece of criteria through a pair of nested loops.

If `objState.UseRegExp` is not set, then the script loops through each of the items contained in the criteria array and adds the criteria to the LDAP dialect filter using the or logical operator or | or &, depending on whether the user specified the all parameter (-a). The inclusion (and specification) of either of these operators increases the utility of the script because they let the user specify whether the output results should include all of the criteria or any of the criteria. If the verbose flag is set, the script displays the query filter that it built from the criteria and other user selections. This is often helpful in troubleshooting or pulling for other scripts or applications.

```
if(!objState.UseRegExp)
    {
    if(objState.all) {
        strQFilter+="(&";
      }
    else {
        strQFilter+="(|";
    }
    for(i=0;i<iTotalCriteria;i++) {
        strQFilter += "(Name=*" + aCriteria[i] + "*)";
        }
    strQFilter += ")";
    }
strQFilter += ");"
if(objState.verbose){
    Heading("Query Filter");
    WScript.Echo(strQFilter);
    }
```

The function then concatenates each of the LDAP dialect query components to define the Command object CommandText property. Once defined, the script calls the Execute method to perform the query and return the results to the specified RecordSet object (objRecordset).

```
objCommand.CommandText=strQBDN + strQFilter + strQAttributes +
 strQScope;
objRecordset=objCommand.Execute();
```

Displaying the results consists of looping through the RecordSet object that contains the query results. In addition to displaying the name, the script adds the object's ADsPath to a new array. Storing the results programmatically allows them to be later used without having to rerun the query. ADsPath is a unique identifier stored in the array, and additional information about the object can be easily referenced by this identifier.

The following code contains the loops that process the regular expression search criteria. For every object returned from the query, the script loops through each of the criteria and calls the regular expression test method to compare the object against the criteria. If the test is true, then the object is displayed and the ADsPath is added to the aforementioned array. Although regular expression searching offers more flexibility in criteria definition, the searches take longer and the returned RecordSets are larger than when using the LDAP filter search criteria.

```
if(objState.verbose){
    Heading("Results");
    }
while (!objRecordset.EOF) {
    if (objState.UseRegExp) {
        for(i=0;i<iTotalCriteria;i++) {
            reCriteria = new RegExp(aCriteria[i],"i");
            if(reCriteria.test(objRecordset.Fields.Item("Name")
             .Value)) {
                WScript.Echo(objRecordset.Fields.Item("Name")
                 .Value);
                aObjects.push(objRecordset.Fields.Item("ADsPath")
                 .Value);
            }
        }
    }
    else {
        WScript.Echo(objRecordset.Fields.Item("Name").Value);
        aObjects.push(objRecordset.Fields.Item("ADsPath").Value);
        }
    objRecordset.MoveNext;
    }
return aObjects;
}
```

Enumerate

In addition to listing the AD objects that satisfy a particular set of criteria, the script provides enumeration of the objects to display their relationships with other objects.

If the enumeration flag is set (objState.enumerate), the main program calls the Enumerate function. The Enumerate function inspects the Class of every object. If the Class is a user or computer, then the function sends the object ADsPath to the function ListGroupMemberships. This function determines and displays all of the groups of which a particular user or computer is a member. If the object Class is a group, then the Enumerate function calls the ListMembersOfGroup function to display all objects that are a member of that group.

The aObjects array contains the ADsPath of each of the objects to be inspected. The script calls the GetObject method to return a pointer to that object to the variable objADs. This pointer to the object is passed to the other functions to use for collecting this membership data.

```
//-----------------------------------------------------------------------
// Enumerate the Results
//-----------------------------------------------------------------------
function Enumerate(aObjects)
{
for(i=0;i< aObjects.length;i++) {
    objADs=GetObject(aObjects[i]);
    switch (objADs.Class) {
        case 'user':
        case 'computer':
            ListGroupMemberships(objADs);
        break;
        case 'group':
            ListMembersOfGroup(objADs);
        break;
    }
}
}
```

ListGroupMemberships

The `ListGroupMemberships` function lists all of the group memberships of a user or computer object. The input parameter to this function is a pointer to an AD object. Using this pointer, the `Groups` method can be called, which specifically returns all of the groups of which object is a member. The `Groups` method returns a pointer to an `IADsMembers` interface, which the script enumerates using an `Enumerator`. The `Groups` method `IADsMembers` interface `Count` property has known issues (it doesn't work), and so the script instead iterates through the objects to count their number.

```
//-----------------------------------------------------------------------
// List the group memberships of a user or computer
//-----------------------------------------------------------------------
function ListGroupMemberships(objADs)
{
groupList = objADs.Groups();
iCount=0;
eGroups = new Enumerator(groupList);
for (;!eGroups.atEnd();eGroups.moveNext()) {
    iCount++; }
```

Whereas the LDAP query returns a user's name as simply the name (such as `Trooper`), binding to the object using the `GetObject` method and calling the `IADs.Name` property returns the object's relative name (for example, `CN=Trooper`). We want the script to display the name alone without the relative name attribute.

To do this, the script uses the regular expression function split to search the relative name for the delimiter = and then display only the name behind it. The split method compares a string against a delimiter and returns an array that comprises the fields between the delimiter. In our previous example, if strUser="CN=Trooper", then strUser.split("=") returns a zero-based array of two members. (A zero-based array means that the first member is referenced as zero. For example, a[0]= "first member", a[1]= "second member".) The first element (referenced by a [0] is "CN" and the second element [1] is "Trooper". Putting this together, strUser. split("=")[1] = "Trooper".

```
WScript.Echo("\nName: " + objADs.Name.split("=")[1] +
   " (member of " + iCount + " groups)");
if(objState.verbose) {
   WScript.Echo("objADs: " +objADs.objADs );
   }
```

Next, the function displays all of the groups of which that computer or user object is a member. Because the IADsMembers pointer groupList still exists, we can enumerate it again and display the name, class, and ADsPath for each of the objects. The script calls the custom function Tabulate to handle displaying the entries in an aligned-left column format. Tabulate ensures that all of the data in each of the columns begin at the same horizontal character location. The Tabulate function simply positions and displays the data given the string to display, the column number, and the delimiter character or characters. Tabulate relies on the standard streams Column property, which requires Windows Script version 5.6.

If the membership is empty, then iCount will equal 0 and the text "-none-" is displayed.

```
if (iCount>0) {
    eGroups = new Enumerator(groupList);
    for (;!eGroups.atEnd();eGroups.moveNext()) {
        x=eGroups.item();
        WScript.StdOut.Write(x.Name.split("=")[1]);
        if(objState.verbose) {
            Tabulate(x.Class, 35, strDelimit);
            Tabulate(x.objADs, 45, strDelimit);
            }
        WScript.StdOut.Write("\n");
        }
    }
else {
    WScript.Echo("-none-");
    }
}
```

ListMembersOfGroup

The previous function displays all of the groups to which a user or computer object belongs. The next function displays all of the members (user, computer, or group) of a group object. Using a pointer to a group object as input, the ListMembersOfGroup function calls the Members method to return a collection of objects that are a member of that group. Similar to the previously mentioned function, Groups, the Members method returns a pointer to an IADsMembers interface, which the script then enumerates and displays. The IADsMembers interface returned with the Members method supports the Count property that returns the number of objects in the collection. This means a separate iteration is not necessary to calculate iCount, as was required in the previous function.

The script presents the results by enumerating the IADsMembers collection and calling the Tabulate function to display the data.

```
//---------------------------------------------------------------------
// List the individual members of a group
//---------------------------------------------------------------------
function ListMembersOfGroup(objADs)
{
memberList = objADs.Members();
iCount=memberList.Count;
WScript.Echo("\n"+ objADs.Name.split("=")[1] +
   " (" + iCount + " members)");
if(objState.verbose) {
    WScript.Echo("objADs: " +objADs.objADs );
    }
if (iCount>0) {
    eMembers = new Enumerator(memberList);
    for (;!eMembers.atEnd();eMembers.moveNext()) {
        x=eMembers.item();
        WScript.StdOut.Write("   " + x.Name.split("=")[1]);
        if(objState.verbose) {
            Tabulate(x.Class, 35, strDelimit);
            Tabulate(x.objADs, 45, strDelimit);
            }
            WScript.StdOut.Write("\n");
        }
    }
else {
    WScript.Echo("-empty-");
    }
}
```

CurrentDomain

The function CurrentDomain determines the domain of the account running the script. For example, if you run the script under an account somedomain\user, then the script determines that you are in the somedomain domain. The script determines this by serverless binding to the LDAP RootDSE object. RootDSE provides information about the directory service. To determine the domain of the account running the script, the script gets the property defaultNamingContext. This property is the distinguished name of the domain of which the directory server (in this case RootDSE) is a member.

```
//----------------------------------------------------------------
// Discover the domain of the user context running the script
//----------------------------------------------------------------
function CurrentDomain()
{
var objRootDSE = GetObject("LDAP://RootDSE");
return objRootDSE.Get("defaultNamingContext");
}
```

Heading

The following function, Heading, illustrates a programming technique of offloading repeated actions to a dedicated function. Offloading formatting to a dedicated function makes the code easier to read and ensures format consistency. Isolating repeated actions benefits your code in several ways. First, it makes the code easier to read because anytime a reader sees the code Heading(sometext), the reader knows what to expect. Also, using a dedicated function simplifies code maintenance because you only need to update the code in one location. For example, if you wanted to change the heading formatting, you simply update the code in the function and the changes are immediately reflected anywhere the function is called the next time the script is run.

```
//----------------------------------------------------------------
// Format headers in the output
//----------------------------------------------------------------
function Heading(strMessage)
{
WScript.Echo("\n--------------------------------------------------");
WScript.Echo(strMessage);
WScript.Echo("--------------------------------------------------");
}
```

Usage

The Usage function displays instructions for how to use the script as a courtesy to the user. It displays a brief explanation of the script, gives the usage, lists the supported parameters, and includes an example.

```
//------------------------------------------------------------------
// Display the help screen
//------------------------------------------------------------------
function Usage()
{
WScript.Echo("Tool to search Microsoft Active Directory");
WScript.Echo("and display and enumerate group memberships.");
WScript.Echo("Usage:");
WScript.Echo("ADSIQuery -cugvrfe [filename] -D [Domain] " +
   "-d [delimiter] [criteria]:");
WScript.Echo("-c list computers");
WScript.Echo("-u list users");
WScript.Echo("-g list groups");
WScript.Echo("-v verbose mode");
WScript.Echo("-r use regular expressions for criteria matching");
WScript.Echo("-f [filename] load criteria from a text file");
WScript.Echo("-e Enumerate the results for group membership");
WScript.Echo("-D [domain] specify domain other than default");
WScript.Echo("-d [delimiter] specify delimiter to separate fields.");
WScript.Echo("-a all specified criteria must match (default is any).");
WScript.Echo("-? Help");
WScript.Echo("");
WScript.Echo("Example:");
WScript.Echo("ADSIQuery -ue jeff : Returns users that match 'jeff'");
WScript.Echo("and will list all groups to which they belong");
WScript.Echo("ADSIQuery -cr dc[12] : Returns computers that match");
WScript.Echo("the regular expression dc[12] (e.g., dc1 or dc2)");
}
```

ParamError

ParamError is another example of a function that is repeatedly called in the script. Rather than maintain a number of instances of code, offloading this feature into its own function improves readability and streamlines the code.

```
//-------------------------------------------------------------------
// Display An error if a dependant script parameter is not included
//-------------------------------------------------------------------
function ParamError()
{
Heading("Parameter Error Received. Please check syntax.\n");
Usage();
WScript.Quit();
}
```

Tabulate

One problem of using WScript.Echo to display many rows of different fields is that the columns are often unaligned with a jagged appearance that is sometimes difficult to read. This script displays columnar data in organized columns using the Tabulate function. Tabulate displays a string at a specific text column and provides a clean method of presenting data in orderly columns. Tabulate uses the StdOut.Column property to get the current horizontal character position of the current StdOut line.

 The StdOut.Column *property is available with Windows Script version 5.6 and later. If you run this script on a default-build Windows 2000 computer, the* Column *property does not work correctly. Download and install Windows Script 5.6 to fix this.*

It inserts delimiter characters until the desired column is reached and then displays the string. For example, calling the function Tabulate("MyData", 30, " ") displays the string MyData on the 30th column. Spaces would be inserted until the spacing is aligned on the 30th character. This function also supports custom delimiter characters (by default, a space, but this character can be defined with the -d [delimiter] argument), which may aid in parsing of the output.

```
//-------------------------------------------------------------------
// Display output in left-aligned columns
//-------------------------------------------------------------------
function Tabulate(strData, iColumn, strDelimit)
{
while(WScript.StdOut.Column<iColumn)
    {WScript.StdOut.Write(strDelimit);}
WScript.StdOut.Write(strData);
}
```

SUMMARY

Quickly and easily pulling data from AD can prove a huge timesaver for small tasks or larger projects. *ADSIQuery* demonstrates how to use ADO to query AD and present the results in a flexible format suitable for input into other scripts or programs. Many of the techniques presented in this script can be applied to other directory services as well—for example, Exchange or a machine's local accounts (using the WinNT provider). Additionally, the tool can be trimmed down, say, to even provide a command-line phone tool to quickly look up phone numbers of employees. This script also becomes a basis for several of the other scripts in this text that rely on querying data from AD.

KEY POINTS

- A benefit of a strictly console-based script is that it can usually be scheduled using Windows Scheduled Tasks without a lot of special preparation
- Use the `runas` command to invoke a command prompt using administrator credentials or credentials from another domain to permit your scripts to run without changing your current logon session.
- Regular expressions bring extremely robust text parsing and filtering functionality to your scripts. Although they can be complicated to learn, their utility pays back the learning curve tenfold. To learn more about the power and usage of regular expressions, visit the RegEx documentation on the JScript reference MSDN site at *http://msdn.microsoft.com/library/en-us/script56/html/js56reconIntroductionToRegularExpressions.asp*.
- Use functions to offload repeated tasks. Doing so not only makes your code simpler and easier to read, it also allows you to reuse functions in other scripts.
- Custom objects and arrays provide storage containers for your data. Whereas an array is sequentially indexed, an object lets you assign custom properties and values.
- The WSH `FileSystemObject` provides methods to enable your scripts to read and write text files.
- Use either LDAP or SQL query syntax to define efficient ADSI ADO queries. An ADO query consists of a `Connection` object, `Command` Object, and `RecordSet` object.
- Use the `objRootDSE` object to get information about the default directory server of the user running the script using a serverless binding (for example, `LDAP://RootDSE`).

5

Dumping a Domain User's Detailed Account Information

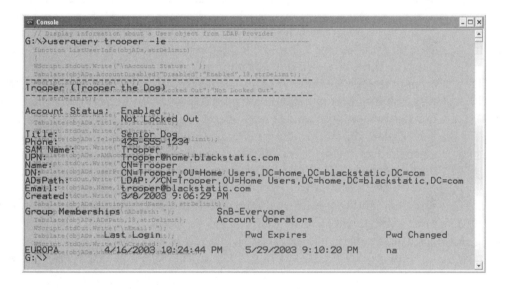

```
Console                                                              _ □ ×
     Display information about a User object from LDAP Provider
G:\>userquery trooper -le
  function ListUserInfo(objADs, strDelim)
  {
  WScript.StdOut.Write("\nAccount Status: ");
  Tabulate(objADs.AccountDisabled?"Disabled":"Enabled", 18, strDelim);
────────────────────────────────────────────────────────
Trooper (Trooper the Dog)         ocked Out":"Not Locked Out",

Account Status:  Enabled
                 Not Locked Out

Title:           Senior Dog
Phone:           425-555-1234
SAM Name:        Trooper
UPN:             Trooper@home.blackstatic.com
Name:            CN=Trooper
DN:              CN=Trooper,OU=Home Users,DC=home,DC=blackstatic,DC=com
ADsPath:         LDAP://CN=Trooper,OU=Home Users,DC=home,DC=blackstatic,DC=com
Email:           trooper@blackstatic.com
Created:         3/8/2003 9:06:29 PM
Group Memberships            SnB-Everyone
                             Account Operators
                Last Login            Pwd Expires            Pwd Changed
EUROPA          4/16/2003 10:24:44 PM  5/29/2003 9:10:20 PM   na
G:\>
```

OVERVIEW

The ADSI LDAP provider exposes a number of directory service objects that are available only on individual domain controllers. For example, the LastLogin property is stored only on the domain controller that authenticated the user and is not replicated to the other domain controllers. (This also means that in an environment with multiple domain controllers, not all domain controllers have information about the `LastLogin` date of all users.) However, the information is still reachable using the LDAP provider. Instead of performing a serverless binding, specify a specific domain controller to query. The *ADSIQuery* script described in Chapter 4 used ADSI LDAP queries to return and display a few properties from a broad collection. The *UserQuery* script presented in this chapter complements *ADSIQuery* by providing more detailed properties from fewer objects. You will notice that some of the techniques used in Chapter 4 have been carried over into this script, but this script also presents new alternative approaches to solving some of the tasks examined in Chapter 4.

SCENARIO

As the manager of the help desk, you regularly check the status of user accounts, such auditing disabled accounts or expired passwords, finding the full name or phone number of a user, or determining in what domain or organizational unit an account resides. The MMC plug-in, *Active Directory Users and Computers*, provides much of this information, but you prefer a command-line tool that you can use to quickly dump information about a specific user account. Plus, you prefer to obtain and display additional information not provided directly by this plug-in, such as when the account expires or when the account was last used. Your company also recently migrated to the Active Directory-integrated Exchange 2000, and you definitely want to extend your script to pull Exchange-specific data as well. You also want to be able to specify what information to collect and display, such as enumerating group membership.

You decide that a basic script is the way to go because you can customize the input to quickly specify any domain, and you can format the output for a variety of uses.

ANALYSIS

This console-based script uses both the ADSI LDAP and WinNT providers. In some cases, you may find the need to tap into the WinNT provider to fill in some of the gaps left open, such as obtaining some of the few properties not available from the LDAP provider or enumerating the local user accounts from a non-Windows 2000 domain controller. The script uses the LDAP provider to query the last login information stored on each individual domain controller to which the user connects. The business objectives outlined in the scenario can be broken down into the technical requirements presented in Table 5.1.

TABLE 5.1   Business objectives mapped into technical requirements

Business objectives	Technical requirements
Check the status of individual user accounts.	Create an ADO ADSI query and select user accounts that match the specified name pattern and `objectCategory='person'`.
Check disabled accounts, expired passwords.	Check for ADSI `AccountDisabled` and `IsAccountLocked` properties.
Find information about the user, including full name, phone number, domain, and OU information.	Get ADSI properties, such as `Title`, `TelephoneNumber`, `sAMAccountName`, `uUPN`, `Name`, `DN`, `ADsPath`, and `create` date.
Display when account was last used.	Need to determine `LastLogon` property, which is stored on each domain controller. Find all domain controllers by looking for computers with a `PrimaryGroupID=516`. Use ADSI WinNT provider to Query each domain controller for the `LastLogon` for that specified user. Trap errors in case a user has not logged on to the network using that domain.
Display Exchange 2000 mailbox information.	List ADSI `mail` property.
List group membership.	Iterate through the ADSI `Groups` property. This property lists only domain groups of which the user is a member and not members of local computer groups.

The script reuses some of the basic operational functions from the *ADSIQuery* script discussed in Chapter 4, including the argument-handling code and the enumeration code. The ADO directory service query procedures have been rewritten to use SQL dialect instead of LDAP dialect, and the ADO code has been offloaded to its own function to increase reusability. By isolating the query to its own function, the same function can be used to perform the script's two main queries: to search for criteria matching user accounts and to search for domain controllers.

The Main Program: Evaluates the arguments and builds the query syntax to send to the `QueryDS` function.

- Evaluates the input parameters and sets the appropriate processing flags.
- Constructs the directory service queries.
- Calls the processing functions.

`ListUserInfo:`

■ Displays ADSI LDAP properties about a user object stored in the directory service, such as name, e-mail address, telephone number, whether the account is locked out or disabled, and the different account naming structures.

`ListLastLogon:` Displays date information using LDAP and WinNT providers.

■ Uses the LDAP provider to bind to each domain controller of the specified (or default) domain and returns the date and time of last login, password expiration, and password last changed date.

■ Uses the WinNT provider to bind to the individual domain controllers to return the `LastLogon` date.

`ListGroupMemberships:` Lists the groups of which the object is a member.

■ Represents a truncated version of the group enumeration function presented in Chapter 4.

`QueryDS:` Returns an array of all the objects that satisfy a given `CommandText`.

■ Represents a leaner version of the LDAP query engine presented in Chapter 4. This version accepts the `Command` object `CommandText` property as input, executes the specified query, and returns an array containing the `ADsPath` of each object in the result set.

`GetDomainInfo:` Gets the name of the current domain and the domain controllers.

■ Determines the name of the current domain and the domain controllers. The ancillary functions have been reused from Chapter 4:

`CurrentDomain:` Discovers the current domain of the user context running the script.

`Heading:` Formats the output headers.

`Usage:` Displays a help screen for how to use the script.

`ParamError:` Displays error text if a dependant script parameter is not included.

`Tabulate:` Aligns text output.

Distribution and Installation

This script, *UserQuery.js*, reuses a number of functions presented in Chapter 4 and is written in JScript. The script will be distributed to help desk staff, who will run the script from a command prompt. The script uses the `WScript.StdOut` methods and must be run with *cscript.exe* and not *wscript.exe*.

Output

The script displays user account information in block format and optionally lists the groups in which the user is a member and last login information stored in each domain controller. The example shown in Figure 5.1 shows the output of the script when run with these arguments:

```
UserQuery -le trooper
```

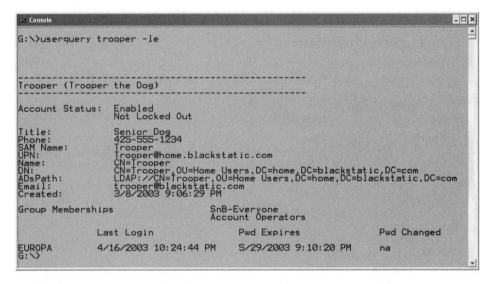

FIGURE 5.1 *UserQuery.js* presents detailed user information queried from ADSI.

Summary of Solution

UserQuery provides an example of how to update an existing script to perform a different function. You will see similarities to the *ADSIQuery* code of Chapter 4; however, note that the script is leaner in some places and adds functionality in others. The script intentionally uses different methodologies to solve similar problems tackled in Chapter 4. Examining multiple, alternate approaches may bring additional clarity as you learn these different techniques. Plus, this multipronged approach may help you answer lingering questions from Chapter 4.

OBJECTIVE

Design a command-line tool that queries and displays user account status, creation date, password expiration dates, group membership, and distribution list membership. This script demonstrates WSH arguments, ADSI, ADO, and LDAP and WinNT providers.

ADODB.Connection

- `Provider`
- `Open`

ADODB.Command

- `ActiveConnection`
- `Properties`
- `CommandText`
- `Execute`

ADSI—Connect to both LDAP and WinNT providers

- `GetObject`
- Groups

Array

- `push`

Enumerator

- `atEnd`
- `MoveNext`

JScript error handling

- `Try…Catch`

SQL dialect query

- SELECT, FROM, WHERE

RecordSet

- `EOF`
- `Fields.Item(x).Value`

RootDSE

- `GetObject("LDAP://RootDSE")`
- `Get("defaultNamingContext")`

String functions

■ `split`

WScript.Arguments

■ `colArgs.length`
■ `colArgs.Item`

WScript.StdOut

■ `Write`
■ `Column`

THE SCRIPT

UserQuery is a command-line tool that displays user account information as shown in Figure 5.1. The properties presented in this script demonstrate only a few of the more than 50 IADsUser interface properties that you could add to tailor the script to your own environment or needs.

Running the script queries the default LDAP directory service (because `-D [domain name]` is not specified) on the well-known LDAP port (TCP port 389).

The `-l` argument instructs the script to list the login dates of the user account. When a user logs in to a Windows 2000 network, the login date and time is stored only on the domain controller that authenticated the user. This means that to accurately display the last login date, the script must query all of the domain controllers in the domain for the last login date. The computer running the script to call these ADSI queries using the WinNT provider uses TCP port 445 to make these direct machine queries. The WinNT provider uses a different port than the LDAP provider because Windows computers do not run LDAP services to provide local account information. However, Windows 2000 domain controllers support both WinNT and LDAP provider queries, and these two services that listen on two separate ports provide similar but not identical information about the same directory of objects.

ON THE CD

The script is presented on the CD-ROM as /chapter 5/scripts/UserQuery.js. Before continuing in this next section, open the actual script from the CD-ROM and run it a few times in your test environment to get a feeling of how it works. Explore the entire code on your own and then continue with the walkthrough, which hopefully answers any questions you may have. Plus, seeing the code in its entirety in your script editor gives you a good sense of its scope and also allows you to tweak it as we walk through its function.

The Main Body

Now, let's explore the script *UserQuery.js*. The script opens with the traditional declaration of variables. The variables `i` and `j` represent generic counter variables used throughout the script. The variable `iTotalCriteria` represents the total number of criteria detected. The object `objState` includes various state flag parameters used to instruct the script what actions to take. The states defined in this script follow:

`objState.help`	Display the help to the user.
`objState.verbose`	Display more detailed information beyond the default.
`objState.domain`	Use the alternate domain name specified by the user.
`objState.lastlogin`	Enumerate the detected domain controllers and query each for its recorded last login property.
`objState.enumerate`	Display the groups to which the specified user account belongs.

UserQuery tracks data using three arrays: `aCriteria`, `aObjects`, and `aDCs`. The first array, `aCriteria`, contains the list of names (or partial names) that the script uses to match user account names. The second array, `aObjects`, contains the `AdsPath` of each of the actual user accounts that match all of the criteria contained in `aCriteria`. (The `IADs` interface `AdsPath` property represents the entire unique LDAP name of an object contained in AD.) The last array, `aDCs`, contains the `AdsPath` of each of the domain controllers discovered in the target domain.

```
//------------------------------------------------------------------
// UserQuery.js
//------------------------------------------------------------------
var i=0,j=0;
var strCriteria, iTotalCriteria;
var strDelimit=" ";
var objState = new Object;
var aCriteria = new Array();
var aObjects = new Array();
var aDCs = new Array();
objState.help=0;
objState.verbose=0;
objState.domain=0;
objState.lastlogin=0;
objState.enumerate=0;
objState.info=0;
```

```
objState.all=0;
if(WScript.FullName.indexOf("WScript")>=0) {
    WScript.Echo("Due to some of the functionality of this script, " +
    "Please run this program using the cscript.exe engine.\n" +
    "for example: cscript.exe UserQuery.js");
    WScript.Quit()
    }
```

The function `strDomain` uses the same methodology detailed in Chapter 4 to discover the name of the domain account running the script. Also, the argument enumeration routines used in this script are the same as those used in *ADSIQuery*. If the user specifies the help parameter (`-?`), the script calls the `Usage` function and then exits.

```
strDomain=CurrentDomain();
var colArgs = WScript.Arguments;
for(i=0;i<colArgs.length;i++) {
    strArg=colArgs.Item(i)
    if(strArg.indexOf("-")==0)           {
        if(strArg.indexOf("i")>=0)  objState.info=1;
        if(strArg.indexOf("l")>=0)  objState.lastlogin=1;
        if(strArg.indexOf("e")>=0)  objState.enumerate=1;
        if(strArg.indexOf("?")>=0)  objState.help=1;
        if(strArg.indexOf("v")>=0)  objState.verbose=1;
        if(strArg.indexOf("a") >=0) objState.all=1;
        if(strArg=="-d") {
            if(i+1 > colArgs.length-1) ParamError();
            strDelimit=colArgs.Item(i+1);
            i++;
            }
        if(strArg=="-D") {
            if(i+1 > colArgs.length-1) ParamError();
            strDomain=colArgs.Item(i+1);
            objState.domain=1;
            i++;
            }
        }
    else {
        aCriteria.push(strArg);
        }
}
if(!objState.info && !objState.lastlogin && !objState.enumerate)
    { objState.info=1;
    }
if(objState.help) Usage();
```

UserQuery conducts ADSI queries slightly differently than *ADSIQuery* does. In *ADSIQuery*, a dedicated function handled the entire building and execution of the query. However, in the *UserQuery* script, the main program constructs the query string and then sends only the CommandText string to the query function, QueryDS. The CommandText property of the ADO Command object defines the query to be executed. QueryDS then builds the connection and command objects, executes the query, parses the RecordSet, and returns the results to the calling function in an array.

ADSIQuery used the query engine only once, so it made sense to house the entire query process in a single function. *UserQuery*, however, executes several different queries during each run, and separating a bulk of the query into its own function makes it reusable and reduces code duplication.

Also, compare the SQL dialect used to build the ADSI query to the LDAP dialect used in *ADSIQuery* in Chapter 4. Both query strings accomplish the same result, and the CommandText string used by QueryDS may consist of either dialect. The query searches for user objects that match any of a given set of criteria. The script appends all of the criteria to the query string, which means that that script only has to execute the query once, as opposed to looping through every criteria and executing a query for each.

The SQL dialect-based query begins with a SELECT statement. In the following code, the query string SELECT ADSPath,Name FROM LDAP://" + strDomain means that only two properties (Adspath, and Name) are returned from the LDAP server to the client running the script. As covered in Chapter 3, defining LDAP://domainname without specifying the name of a specific domain controller is called serverless binding. With serverless binding, Windows calculates the best domain controller to connect to and query.

Continuing on with the query string, the syntax WHERE objectClass = 'user' AND objectCategory = 'person' defines only user objects.

```
iTotalCriteria=aCriteria.length;
strCommandText="SELECT ADsPath, Name FROM 'LDAP://" +
  strDomain + "' WHERE objectClass = 'user' AND " +
  "objectCategory = 'person'";
```

Next, the code appends each piece of criteria to the query string. The query construction presented in the following code consists of the following base logic:

```
(user object) AND (Name= criteria 1 OR Name=criteria 2 …OR…
Name=criteria n)
```

Deconstruct this logic into the following pseudo-code:

user object
If Criteria > 1 then
  (AND Name=criteria 1
Loop through Criteria 2 through n:
  OR Name=criteria n
Follow with a closing parenthesis:
)

The wildcard characters (*) concatenated before and after each piece of criteria permit the matching of partial criteria to the user object's Name property. For example, *Tro* matches Trooper. Depending on the application, you may want to change the query to match the actual username, sAMAccountName. For example, if your organization designated contractor accounts with the prefix c- (for example, c-jefff), and you wanted to list details about all contractor accounts, you could simply change the query string to return sAMAccountName instead of Name and then match that property instead.

```
if(iTotalCriteria > 0){
    strCommandText+=" AND (Name='*" + aCriteria[0] + "*'";
    for(i=1;i<iTotalCriteria;i++) {
        strCommandText+=(objState.all ? " AND ":" OR ") + "Name='*" +
        aCriteria[i] + "*'";          }
        strCommandText+=")";
}
```

At this point, the CommandText for the query is completely defined. An actual example of the CommandText of a completed query with three criteria items follows:

```
SELECT ADsPath, Name FROM 'LDAP://DC=blackstatic,DC=com' WHERE
objectClass = 'user' AND objectCategory = 'person' AND (Name='*jeff*'
OR Name='*kellie*' OR Name='*trooper*')
```

Next, the script displays optional information about the query and then executes the query by sending the CommandText to the function QueryDS. The QueryDS function returns the results in the array aObjects. A detailed description of how QueryDS operates follows further in this script under the actual function.

```
if(objState.verbose){
    Heading("Query String");
    WScript.Echo(strCommandText);
```

```
        Heading("Using Domain");
        WScript.Echo(strDomain);
        }
    aObjects=QueryDS(strCommandText);
```

Earlier we mentioned that the query engine used in *ADSIQuery* was separated into two parts. The first part defines the unique characteristics of the query string, and the second part represents a more generic function reusable by different types of query. The next bit of code demonstrates the reuse of the QueryDS function and begins with a new definition of the CommandText property. This second query returns all of the domain controllers in the specified domain. Later, the script will bind each of these domain controllers to extract the last login date of the specified user. If the script is not run with the last login parameter (-1), then the script skips this entire query. Binding individually to multiple machines can take a bit of time, which is why the script offers this functionality optionally.

The query syntax uses the SQL dialect, although the QueryDS function accepts and processes CommandText using either SQL or LDAP dialect query strings. It begins using similar syntax to the previous user object criteria syntax. This query requires only the IADs interface properties AdsPath and PrimaryGroupID. The AdsPath defines the unique location of the domain controller. A domain controller in a Windows 2000 AD domain has a defined PrimaryGroupID of 516. (The PrimaryGroupID is the relative identifier (RID) for the primary group of the user.)

NOTE

Finding the specific, unique criteria that distinguishes your desired search parameter (such as finding domain controllers by a PrimaryGroupID of 516) may seem to you to be like looking for a needle in a haystack. First, try seeking the information on the Internet—you may be surprised at the wealth of detailed scripting information available. If you still need additional information, try enumerating all of the properties of a few different objects and see if you can spot unique characteristics that define your search parameter. (Chapter 3 presents an ADSI property enumeration script.)

Also, be wary of queries that may miss objects. For example, in looking for all domain controllers, you could query the contents of the built-in organizational unit "Domain Controllers"; however, some organizations may not always use that container and may move domain controllers into other containers. Searching for a RID of 516 returns the domain controllers no matter where they may have been moved in the AD hierarchy.

An example of a completed query for domain controllers looks something like:

```
SELECT ADsPath,PrimaryGroupID FROM 'LDAP://DC=blackstatic,DC=com'
WHERE PrimaryGroupID = '516'
```

This `CommandText` is sent to the `QueryDS` function, and the results are stored in the array `aDCs`. Optionally, the script also then displays the `AdsPath` of the domain controllers.

```
if(objState.lastlogin) {
    strCommandText="SELECT ADsPath,PrimaryGroupID FROM 'LDAP://" +
    strDomain + "' WHERE PrimaryGroupID = '516'";
    aDCs=QueryDS(strCommandText);
    if(objState.verbose){
        Heading("Detected Domain Controllers");
        for(i=0;i<aDCs.length;i++) {
            WScript.Echo(aDCs[i]);
            }
        }
    }
```

At this point, all of the user account objects and domain controllers have been identified, and now the script is ready to display the information about each of the objects.

The script loops through the array `aObjects` and binds to each user account object using the ADSI LDAP provider. This reference is then sent to three functions for final processing.

The first function, `ListUserInfo`, simply collects property information about the object and displays it to the console. The second function, `ListGroupMemberships`, lists all of the groups of which the user object is a member.

The last function, `ListLastLogon`, uses the LDAP `sAMAccountName` property of the object to bind to each of the domain controllers using the ADSI WinNT provider. User account date information is extracted from each domain controller and displayed in a matrix format.

These three functions close out the main program of the script.

```
for(i=0;i<aObjects.length;i++) {
    objADs=GetObject(aObjects[i]);
    WScript.Echo("\n\n");
    Heading(objADs.sAMAccountName + " (" + objADs.displayName + ")");
    if(objState.info) {
        ListUserInfo(objADs,strDelimit);
        }
```

```
        if(objState.enumerate) {
            WScript.StdOut.Write("\n\nGroup Memberships");
            ListGroupMemberships(objADs);
            }
        if(objState.lastlogin) {
            ListLastLogon(aDCs,objADs);
            }
        }
    WScript.Echo;
    WScript.Quit(0);
```

ListUserInfo

The ListUserInfo function displays the values of various ADSI LDAP properties of the specified user account object, as previously seen in Figure 5.1. The function accepts an established reference to the object as an input parameter. (The main program (the calling function) established this reference previously by calling the method GetObject(AdsPath).) ListUserInfo in turn calls the Tabulate function to left-align the LDAP properties to a specific column.

This function also uses the JScript conditional ternary operator ("x?y:z") to display one of two results, depending on the conditional result. Similar to the if(statement, true, false) function in Excel, the ternary operator provides a shortcut to evaluate conditional logic. The ternary operator evaluates the statement on the left side of the question mark, and, if true, then the code executes the first statement to the right of the question mark. Otherwise, the code executes the second statement to the right of the colon.

Take for example, the statement:

(scripting is fun ? of course it is : don't worry, it will get better!)

If "scripting is fun" evaluates as true, then the program returns "of course it is". Otherwise, if false, the program returns "don't worry, it will get better!" The ternary operator is an elegant avoidance of the longer if…then…else conditional statement.

```
//-------------------------------------------------------------------------
// Display information about a User object from LDAP Provider
//-------------------------------------------------------------------------
function ListUserInfo(objADs,strDelimit)
{
WScript.StdOut.Write("\nAccount Status: " );
Tabulate(objADs.AccountDisabled?"Disabled":"Enabled",18,strDelimit);
WScript.StdOut.Write("\n");
Tabulate(objADs.IsAccountLocked?"Locked Out":"Not Locked Out",
 18,strDelimit);
```

```
WScript.StdOut.Write("\n\nTitle: ");
Tabulate(objADs.Title,18,strDelimit);
WScript.StdOut.Write("\nPhone: ");
Tabulate(objADs.TelephoneNumber,18,strDelimit);
WScript.StdOut.Write("\nSAM Name: ");
Tabulate(objADs.sAMAccountName,18,strDelimit);
WScript.StdOut.Write("\nUPN: ");
Tabulate(objADs.userPrincipalName,18,strDelimit);
WScript.StdOut.Write("\nName: ");
Tabulate(objADs.Name,18,strDelimit);
WScript.StdOut.Write("\nDN: ");
Tabulate(objADs.distinguishedName,18,strDelimit);
WScript.StdOut.Write("\nADsPath: ");
Tabulate(objADs.ADsPath,18,strDelimit);
WScript.StdOut.Write("\nEmail: ");
Tabulate(objADs.mail,18,strDelimit);
WScript.StdOut.Write("\nCreated: " );
Tabulate(objADs.whenCreated,18,strDelimit);
}
```

Whereas the previous function `ListUserInfo` displayed properties from the AD LDAP directory service, the next function `ListLastLogon` binds to each domain controller to pull information about the specified user account from the WinNT provider.

Even though a Windows 2000 or Windows 2003 domain controller does not have local users such as a member or standalone server, the WinNT provider still provides additional information about domain users not available by the LDAP provider.

ListLastLogon

The function `ListLastLogon` queries each of the domain controllers discovered in the domain for user account date information, including the date and time the user last logged on, when his password expires, and when the password was last changed. You may be surprised that a user's actual last logon is not aggregated up to the LDAP server but instead stored locally on each domain controller that authenticated the user. To determine the actual last logon, you must query each of the domain controllers and find the most recent date. Figure 5.2 shows an example of the output of the script run with `ListLastLogon`.

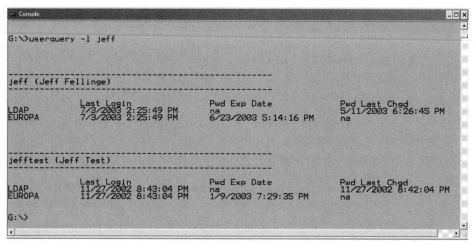

FIGURE 5.2 `ListLastLogon` shows the last logon date of the user accounts using both the ADSI LDAP and WinNT providers.

The variable i is used in this function—and it is also used in the main program that called this function. JScript uses a variable globally if it is not declared again in a subfunction, which might lead to some unexpected results. Avoid using global variables by declaring all of your variables explicitly in every function using them.

`ListLastLogon` loops through each of the members of the aDCs array and binds to the user object on each computer. The syntax for specifying a user object from a computer's WinNT provider follows:

```
WinNT://ComputerName/UserName
```

The aDCs array contains the ADsPath (for example, `LDAP://CN=dc1,OU=Domain Controllers,DC=blackstatic,DC=com`) of the domain controller and uses this property to bind to the ADSI LDAP object to get its `Name` property (for example, *CN=dc1*). The function takes this `Name` property and uses the string method `split` to return the data to the right of the regular expression (`"="`), which in this case is the simple name of the computer (such as *dc1*). (Remember, the `split` method returns an array containing the chopped up string, and, in this example, there is only one equal sign and so there are two parts—one to the left of the equal sign and one to the right. The first part would be denoted by the array element `array[0]` and the second element as `array[1]`.)

Lastly, the function concatenates the prefix `WinNT://` with the name of the computer followed by a slash (`/`) and the name of the user. Using this string, the function then binds to that user object on that computer.

After displaying the table headings, the function loops through and displays the name of the current domain controller. The function then tries to get the user account date properties from this domain controller. If a user has not logged on to a particular domain controller, then the value does not exist. This exemplifies the type of error that you simply must trap and handle because you don't know when or where it will occur. This function uses the JScript try…catch statement to protect the execution of the code while it attempts to get the property.

The JScript try…catch statement is a useful programming tool that provides JScript sophisticated error handling. The statement pair allows you to safely *try* risky operations. If they fail and result in an error, they route the program instead through secondary code. This secondary code only executes in an error situation. In addition to providing a safety net for your code, try…catch supports the error object from which you can get useful debugging information, including the error description and error code. As you can see in the following code, using try…catch almost resembles the structure of an if…then…else statement, which makes it readable even to new programmers who are perhaps not familiar with it.

If the property doesn't exist, an error occurs, and the catch routine intercepts the error (in the next example, the error is contained in the variable e) and runs an alternate set of statements. In the following code, the function assigns the variable dtTemp with the value of the date property, and, if it doesn't exist, the function assigns the variable dtTemp with the string na.

```
//----------------------------------------------------------------
// Query each DC to get date information from the WinNT provider
//----------------------------------------------------------------
function ListLastLogon(aDCs,objADs)
{
var dtTemp, i;
strUser=objADs.sAMAccountName;
WScript.StdOut.Write("\n");
Tabulate("Last Login",15,strDelimit);
Tabulate("Pwd Exp Date",40,strDelimit);
Tabulate("Pwd Last Chgd\n",65,strDelimit);

Tabulate("LDAP",0,strDelimit);
try {dtTemp=objADs.LastLogin;}
catch(e) {dtTemp="-none-";}
Tabulate(dtTemp,15,strDelimit);
Tabulate("na",40,strDelimit);
try {dtTemp=objADs.PasswordLastChanged;}
catch(e) {dtTemp="-none-";}
Tabulate(dtTemp,65,strDelimit);
```

```
WScript.StdOut.Write("\n");
for(i=0;i<aDCs.length;i++) {
    var strDC=GetObject(aDCs[i]).Name.split("=")[1];
    var objADs=GetObject("WinNT://" + strDC + "/" + strUser);
    WScript.StdOut.Write(strDC);
try {dtTemp= objADs.LastLogin;}
    catch(e) {dtTemp="-none-";}
    Tabulate(dtTemp,15,strDelimit);
    try {dtTemp= objADs.PasswordExpirationDate;}
    catch(e) {dtTemp="-none-";}
    Tabulate(dtTemp,40,strDelimit);
    Tabulate("na",65,strDelimit);
    WScript.StdOut.Write("\n");
    }
}
```

ListGroupMemberships

The ListGroupMemberships function is similar to that used in the *ADSIQuery* presented in Chapter 4; however, much abbreviated. Given an existing AdsPath reference to a user object, the function enumerates all of the groups of which the object is a member using the Groups method. The group Name property is represented as the relative distinguished name CN=groupname, and so the regular expression split method is used to truncate the name to only the group name. This name is displayed to the console. Figure 5.3 shows an example of the output of the script run with the -e argument, which exercises the ListGroupMemberships function.

```
//------------------------------------------------------------------
// List the group memberships of a user or computer
//------------------------------------------------------------------
function ListGroupMemberships(objADs)
{
eGroups = new Enumerator(objADs.Groups());
for (;!eGroups.atEnd();eGroups.moveNext()) {
    x=eGroups.item();
    Tabulate(x.Name.split("=")[1],35,strDelimit);
    WScript.StdOut.Write("\n");
    }
}
```

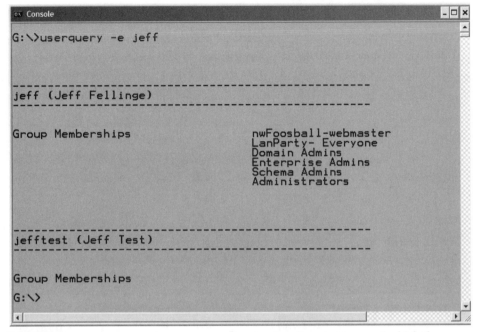

FIGURE 5.3 ListGroupMemberships enumerates the domain groups of which the user account is a member.

CurrentDomain

The CurrentDomain function returns the current domain name of the account running the script. More detail about this function can be found in Chapter 4.

```
//----------------------------------------------------------------
// Discover the domain of the user context running the script
//----------------------------------------------------------------
{
var objRootDSE = GetObject("LDAP://RootDSE");
return objRootDSE.Get("defaultNamingContext");
}
```

QueryDS

The QueryDS function represents a significant change from the LDAP query design used in the Chapter 4 script *ADSIQuery*. Here, the entire ADO query design is isolated in its own independent function. The input to the function is a string representing the Command object CommandText property that defines the query. The function returns an array containing the query results. The CommandText string contains the heart of the query and can be quite complex, as seen in the previous code descriptions. The rest of the ADO query mechanics are relatively straightforward, as seen in the next code and described in detail in Chapter 3. As with previous ADO queries, first the Connection object is created, then the Command object, and finally the RecordSet object is processed and the interesting data is added to the array and returned to the calling function.

```
//--------------------------------------------------------------------
// Query Active Directory given an LDAP query parameter
//--------------------------------------------------------------------
function QueryDS(strCommandText)
{
var aObjects = new Array();
objConnection = new ActiveXObject ("ADODB.Connection");
objConnection.Provider = "ADsDSOObject";
objConnection.Open;
objCommand = new ActiveXObject("ADODB.Command");
objCommand.ActiveConnection = objConnection;
objCommand.Properties("Page Size") = 999;
objCommand.Properties("Sort On") = "Name";
objCommand.CommandText=strCommandText;
objRecordset=objCommand.Execute();
while (!objRecordset.EOF) {
    aObjects.push(objRecordset.Fields.Item("ADsPath").Value);
    objRecordset.MoveNext;
}
return aObjects;
}
```

Heading

The Format headers function provides a common display for reused headers.

```
//--------------------------------------------------------------------
// Format headers in the output
//--------------------------------------------------------------------
function Heading(strMessage)
{
WScript.Echo("\n--------------------------------------------------");
WScript.Echo(strMessage);
WScript.Echo("--------------------------------------------------");
}
```

Usage

The Usage function displays instructions for how to use the script as a courtesy to the user. It displays a brief explanation of the script, gives the usage, lists the supported parameters, and includes an example.

```
//-------------------------------------------------------------------
// Display the help screen
//-------------------------------------------------------------------
function Usage()
{
WScript.Echo("Tool to list user information");
WScript.Echo("Usage:");
WScript.Echo("UserQuery -lev -D [Domain] " +
  "-d [delimiter] [criteria]:");
WScript.Echo("-i list account information");
WScript.Echo("-l list last login information");
WScript.Echo("-e Enumerate the results for group membership");
WScript.Echo("-v verbose mode");
WScript.Echo("-D [domain] specify domain other than default");
WScript.Echo("-d [delimiter] specify delimiter to separate fields.");
WScript.Echo("-a all specified criteria must match (default is any).");
WScript.Echo("-? Help");
WScript.Echo("");
WScript.Echo("Example:");
WScript.Echo("UserQuery -le jeff : Displays user information about");
WScript.Echo("user 'jeff' including login dates and groups");
WScript.Echo("to which they belong");
WScript.Echo("UserQuery jeff : Defaults to display user information");
WScript.Quit(0);
}
```

ParamError

ParamError is another example of a function that is repeatedly called in the script. Rather than maintain a number of instances of code, offloading this feature into its own function improves readability and streamlines the code.

```
//-------------------------------------------------------------------
// Display An error if a dependant script parameter is not included
//-------------------------------------------------------------------
function ParamError()
```

```
    {
    Heading("Parameter Error Received. Please check syntax.\n");
    Usage();
    WScript.Quit();
    }
```

Tabulate

`Tabulate` left-aligns a string on a designated text column. Refer to Chapter 4 for a detailed explanation of how this function operates.

```
//---------------------------------------------------------------
// Display output in left-aligned columns
//---------------------------------------------------------------
function Tabulate(strData, iColumn, strDelimit)
{
while(WScript.StdOut.Column<iColumn)
    {WScript.StdOut.Write(strDelimit);}
WScript.StdOut.Write(strData);
}
```

SUMMARY

Whereas *ADSIQuery* in Chapter 4 introduced ADSI and ADO techniques to return a broad range of data from AD, the *UserQuery* script in this chapter provides similar query capabilities but offers deeper inspection of user objects. The script also demonstrates how to combine LDAP and WinNT provider queries to collect many more data properties about a user object. *UserQuery* also presents a few alternate methods of tackling ADSI queries, such as using SQL dialect and offloading the actual query to a discrete function. After exploring the script in this chapter, review *ADSIQuery* in Chapter 4 and compare the different approaches to pulling data from ADSI.

KEY POINTS

- Use ADSI not only to search the entire directory for many objects but also to return detailed data about specific objects.
- The ADSI `IADsUser` interface provides over 50 viewable and mostly definable properties for user account objects.

- Whereas *ADSIQuery* used the LDAP dialect, *UserQuery* uses SQL dialect to compose the ADSI queries. Both offer benefits, depending on your needs.
- Use the ADSI LDAP provider whenever possible to fetch data from the directory server. However, you may find yourself needing to use the WinNT provider to connect directly to a computer to get some property information.
- Make your functions generic to increase reusability. For example, for an ADO query, by defining the command string in your calling function, you can define a reusable function that accepts any command string and then processes the `Connection`, `Command`, and `RecordSet` objects independently.
- The JScript ternary operator provides a great shortcut to simple conditionals (such as *scripting is fun ? of course it is : don't worry, it will get better!*).
- Use the JScript `try...catch` statement for robust error handling and debugging.

6 Listing Soon-to-Expire Domain User Accounts

```
Console                                                              _ |□| X|
G:\>datequery -c -w 14 -D BLVU
--------------------------------------------------------------------------
Accounts Created in the last 14 days.
--------------------------------------------------------------------------
Account Name                    PDE  Account Created
Administrator                    x   Sun May 11 01:50:19 PDT 2003
Guest                           xx   Sun May 11 01:50:19 PDT 2003
IUSR_PATION-VM-DC                x   Sun May 11 01:50:19 PDT 2003
IWAM_PATION-VM-DC                    Sun May 11 01:50:19 PDT 2003
jeff                             x   Sun May 11 02:55:02 PDT 2003
kellie                               Sun May 11 02:54:44 PDT 2003
krbtgt                           x   Sun May 11 01:57:25 PDT 2003
laptop                               Sun May 11 02:54:20 PDT 2003
topaz                                Sun May 11 02:54:02 PDT 2003
trooper                              Sun May 11 02:54:32 PDT 2003
TsInternetUser                   x   Sun May 11 01:50:19 PDT 2003

P:Password never expires, D:Account disabled, E:Password expired.

G:\>_
```

OVERVIEW

In the last two chapters we examined scripts that pull a variety of information from Active Directory using ADSI. As demonstrated in these scripts, querying and displaying ADSI properties can be a fairly straightforward process accomplished using varying techniques. This chapter offers more ADSI exposure by deconstructing a script that works with user account date properties, such as determining when accounts were created, when their passwords expire, or listing the individual properties of many accounts. The examples in this script demonstrate techniques to work with objects, dates, and new data types such as large integers and bitfields.

The code presented in this chapter builds from the library of functions presented in Chapters 4 and 5. Some functions have been rewritten to offer an alternate method of solving a problem or to provide additional capabilities. Compare and contrast the techniques in this chapter to the previous ADSI scripts; you can see how easy it is to tweak and extend a script once you understand its basic framework. Essentially, these scripts revolve around querying a directory for specific objects that satisfy a given criteria and displaying the data in a unique and useful format. This script is the last of the three console-based ADSI-focused scripts, although you will find that many of the other scripts in this book that require computer or user lists borrow many of the ADSI techniques from these chapters.

SCENARIO

Your company adopted a password policy that requires users to change their passwords every 90 days, and as the information security officer, you technically enforce this policy using AD Group Policy. As a reminder and courtesy to employees, the help desk manager wants to send an e-mail to all users whose passwords will expire within two weeks to encourage them to rotate their passwords before they expire and also to offer assistance, if needed. You have offered to help provide this data to the manager. You decide that you will write a console-based script to provide this information. Using a console script, you can easily schedule the script to regularly run and output this information for the manager. Additionally, you have been looking for a method to track account status across your domain, and you also want your script to display accounts with passwords set to never expire, accounts created in the previous week, and disabled or locked-out accounts. Although it may be burdensome to view accounts in real-time whenever they are created, you would like the ability to review account creation activity weekly.

The help desk manager uses a ticketing application that accepts a list of user accounts to automatically generate the password-rotation notification e-mails. To accommodate this functionality, a console-based script that outputs data to either the console or a text file is desired.

ANALYSIS

The script combines ADSI query and display methods with date manipulation procedures. Date manipulation can be tricky at times, but as we will see, both JScript and VBScript include date-handling methods that make the calculations manageable. The business objectives can be mapped into technical requirements, as seen in Table 6.1.

TABLE 6.1 Business objectives mapped into technical requirements for the *DateQuery* script

Business objectives	Technical requirements
List all accounts with passwords expiring in two weeks.	Compare today's date with every user account's password expiration date and display those expiring in two weeks. Calculate user account password expiration date as the date the password was last changed plus the maximum password age allowed.
Schedule the script to run regularly.	Develop a script to support Windows Script Host *cscript.exe* engine, which will run from the command line and will support running from a scheduled task using predefined launch arguments.
Display accounts with passwords set to never expire.	Check the UserAccountControl property to see if the account is set to never expire. (Note that this is different from Chapter 5, which listed if a password had *actually* expired.)
Display accounts created in the previous week.	Compare today's date with the date that each account is created, and display a list of only those created in the previous week.
Display disabled accounts.	List whether an account is disabled according to the UserAccountControl.
Display locked-out accounts.	List whether an account is locked out using the ADSI WinNT provider IsAccountLocked attribute.
Output the script results to the console.	Develop a script to support Windows Script Host *cscript.exe* engine.

This script audits the user-account password-rotation policy defined within your network's AD by scanning and generating lists that show account creation and password expiration dates. An effective password management policy includes specifying the use of strong passwords, locking out accounts if an incorrect password is attempted more than a few times, and specifying the maximum age of a password before it must be changed. Many of these policies can technically be enforced using AD Group Policy, and some can be accessed and queried using ADSI.

Creating a new group policy object (GPO) provides a centralized enforcement of your security policy. Within the group policy, you define both the password policy and the account lockout policy. The password policy, shown in Figure 6.1, lets you specify whether to enforce password history, minimum and maximum password age, minimum password length, and whether the passwords must be "strong" (and meets complexity requirements as defined by Microsoft). The account lockout policy, shown in Figure 6.2, defines the account lockout duration and threshold as well as specifies how long to wait before the account lockout counter resets.

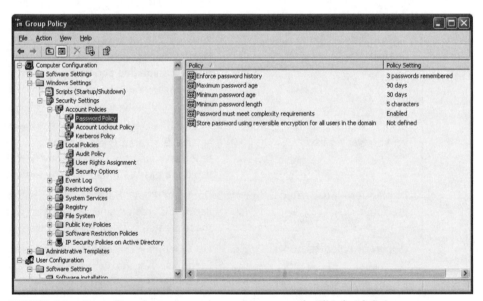

FIGURE 6.1   Centrally enforce your company's password policy by defining a group policy password policy.

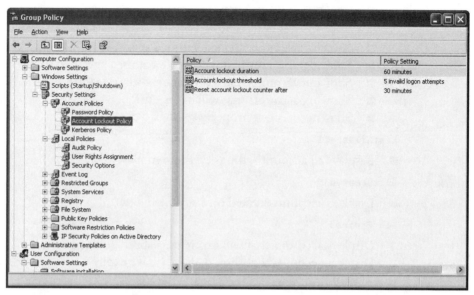

FIGURE 6.2 The account lockout policy protects accounts from repeated unauthorized access attempts.

To ensure that all users receive a consistent login experience, account and password GPO properties are set at the domain level. This effectively means that your ADSI script works predominantly with the domain distinguished name obtained from the RootDSE. After implementing a password policy, you may find your users locking out their accounts (maybe because they forgot to log out of a terminal server session before changing their password) or system administrators setting up exempt accounts with passwords that do not expire. Combining this with the prospect of multiple domains within your environment, you decide the most efficient method of auditing your domains' account status is by scheduling a script to regularly run and send you the results. The script presented in this chapter identifies these distinctive accounts within your domains.

Main Program

- ■ Reads and parses command-line arguments.
- ■ Builds the queries to select the user accounts and domain controllers.
- ■ Loops through all ObjectClass=user and ObjectCategory=person objects to collect account data, including date created, whether the account is disabled or locked out, and whether the password is set to never expire.
- ■ Determines the maximum password age for the selected domain, and calculates the password expiration date.

- Calls the functions to display the script results in a variety of formats, such as:
  - Showing accounts created within a specified date.
  - Showing accounts that will expire within a specified date.
  - Showing accounts that should already have expired.
  - Showing accounts that will never expire.
  - Showing account property settings.

ListAllDetail

- Displays all account settings (verbose mode).

ListCreated

- Displays accounts created on a specified date.

ListExpired

- Displays accounts that will expire on a specified date.
- Displays accounts that should already have expired.

ListPwdExpires

- Displays accounts with passwords that will never expire.

ShowLockouts

- Displays locked-out accounts.

DumpUAC

- Calculates user account settings based on the `UserAccountControl` attribute. This attribute contains more than 20 account settings defined by flags set in the `ADS_USER_FLAG_ENUM` enumeration.
- Displays these account settings in both a matrix and tabular format.

GetMaxPwdAge

- Obtains the ADSI LDAP property `maxPwdAge` and converts it from its stored large integer value in 100-nanosecond intervals to days.

ConvertNanoDays

- Converts 100-nanosecond intervals to days.

CurrentDomain

- Determines the domain of the account running the script.

QueryDSA

- Builds the `Connection`, `Command`, and `RecordSet` objects to execute an ADSI query. Returns the data set to the calling program in an object populated array.

Heading

- Displays a common heading format.

Usage

■ Displays the command-line arguments supported by the script.

ParamError

■ Displays an error message if the command-line arguments are not passed correctly to the script.

Tabulate

■ Vertically aligns the columnar output of script data.

Distribution and Installation

Like the previous scripts *ADSIQuery* and *UserQuery*, this chapter's script *DateQuery.js* is written in JScript. It must be run with the *cscript.exe* engine. Recall, this requirement is because of the use of standard streams (`WScript.StdOut`) that are not supported by the *wscript.exe* engine. Also, the `StdOut.Column` property used by the `Tabulate` function is only supported by Windows Script version 5.6. The command-line parameters for *dateQuery* are shown in Table 6.2.

TABLE 6.2 Command-line arguments of *dateQuery*

Argument	Description
-c	Lists accounts that have been created within [window] days.
-e	Lists accounts that will expire within [window] days.
-l	Lists currently locked-out accounts.
-n	Lists accounts whose passwords are set to never expire.
-u[m]	Enumerates UserAccountControl to display account properties. When used alone, this argument iterates through all properties and displays accounts that satisfy each. When used with the optional m parameter, the data is displayed in a matrix format for every account.
-v	Enables verbose mode, which displays more data than default.
-?	Displays help.
-w [window]	Specifies the number of days that the script should use as boundaries when determining soon-to-expire accounts or recently created accounts.
-d	Specifies an optional delimiter to separate the output data.
-D	User may specify an alternate domain to query.

To display all of the accounts in your current domain that will expire within 14 days, run the script from a command line and specify the type of report you wish to run and the date window you wish to report. Optionally, specify an alternate domain. The following code:

```
DateQuery -e -w 14
```

displays all accounts that will expire within a 14-day window from today, as shown in Figure 6.3.

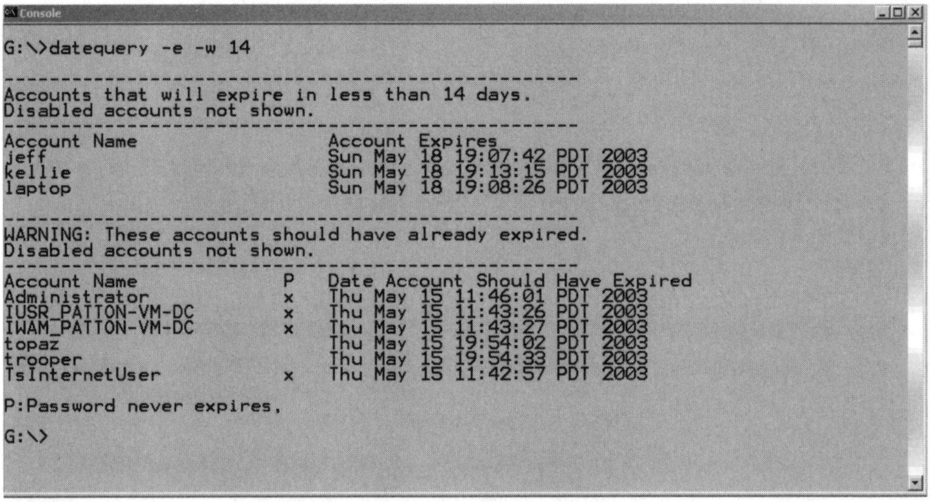

FIGURE 6.3   *DateQuery* shows the names of accounts with passwords expiring within a specified window.

```
DateQuery -c -w 28 -D BLVU
```

returns all accounts created in the last 28-day window from the BLVU domain, as seen in Figure 6.4.

```
DateQuery -u
```

lists all the account properties of the domain in which the script is run, as shown in Figure 6.5.

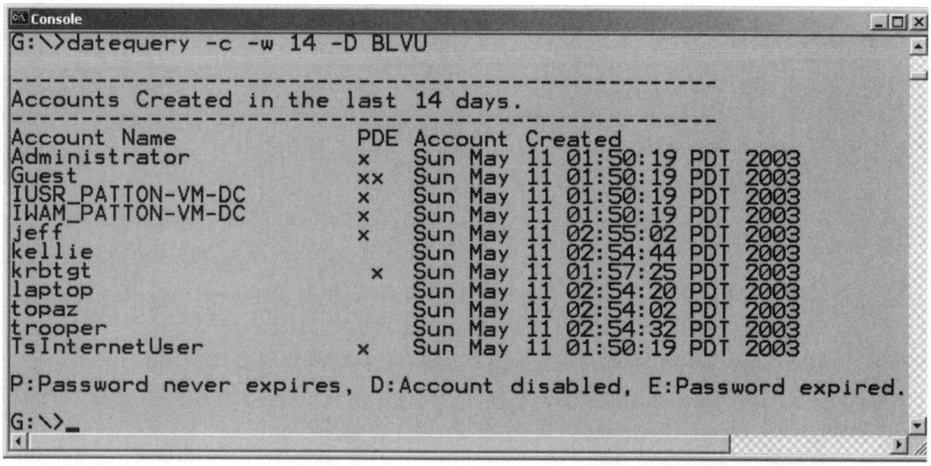

```
Console                                                          _ □ x
G:\>datequery -c -w 14 -D BLVU                                   ▲

----------------------------------------------------------------
Accounts Created in the last 14 days.
----------------------------------------------------------------
Account Name            PDE Account Created
Administrator           x   Sun May 11 01:50:19 PDT 2003
Guest                   xx  Sun May 11 01:50:19 PDT 2003
IUSR_PATTON-VM-DC       x   Sun May 11 01:50:19 PDT 2003
IWAM_PATTON-VM-DC       x   Sun May 11 01:50:19 PDT 2003
jeff                    x   Sun May 11 02:55:02 PDT 2003
kellie                      Sun May 11 02:54:44 PDT 2003
krbtgt                  x   Sun May 11 01:57:25 PDT 2003
laptop                      Sun May 11 02:54:20 PDT 2003
topaz                       Sun May 11 02:54:02 PDT 2003
trooper                     Sun May 11 02:54:32 PDT 2003
TsInternetUser          x   Sun May 11 01:50:19 PDT 2003

P:Password never expires, D:Account disabled, E:Password expired.

G:\>_                                                            ▼
```

FIGURE 6.4 *DateQuery* also displays accounts created within a specified window.

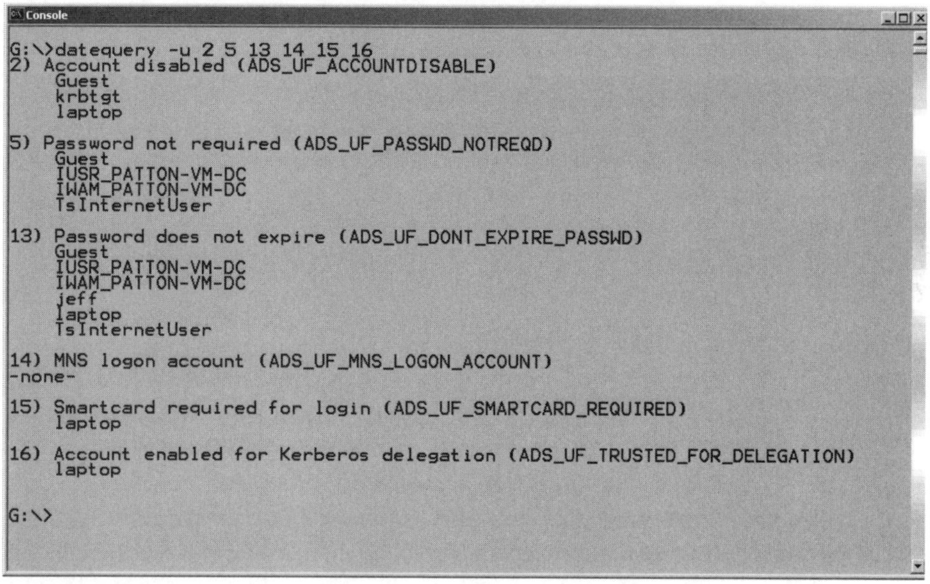

```
Console                                                          _ □ x
G:\>datequery -u 2 5 13 14 15 16                                 ▲
2) Account disabled (ADS_UF_ACCOUNTDISABLE)
    Guest
    krbtgt
    laptop

5) Password not required (ADS_UF_PASSWD_NOTREQD)
    Guest
    IUSR_PATTON-VM-DC
    IWAM_PATTON-VM-DC
    TsInternetUser

13) Password does not expire (ADS_UF_DONT_EXPIRE_PASSWD)
    Guest
    IUSR_PATTON-VM-DC
    IWAM_PATTON-VM-DC
    jeff
    laptop
    TsInternetUser

14) MNS logon account (ADS_UF_MNS_LOGON_ACCOUNT)
-none-

15) Smartcard required for login (ADS_UF_SMARTCARD_REQUIRED)
    laptop

16) Account enabled for Kerberos delegation (ADS_UF_TRUSTED_FOR_DELEGATION)
    laptop

G:\>                                                             ▼
```

FIGURE 6.5 *DateQuery* demonstrates enumeration of the `UserAccountControl`
attribute, which contains settings for 21 account properties, such as whether the account
is locked out.

```
DateQuery -um
```

presents the account properties in a matrix format, as shown in Figure 6.6. To display only selected information, run `DateQuery -um [n1] [n2]` where n1 and n2 are numbers of the account property you wish to view. You probably will not use all of these account properties in the day-to-day administration of your domain, but all have been included in this script to show all the information stored in the LDAP property UserAccountControl, which corresponds to the ADSI enumeration ADS_USER_FLAG_ENUM. An ADSI enumeration is a method of storing many settings, or flags, in a single value. For example, the flags stored in the enumeration ADS_USER_FLAG_ENUM used by this script are shown in Table 6.3.

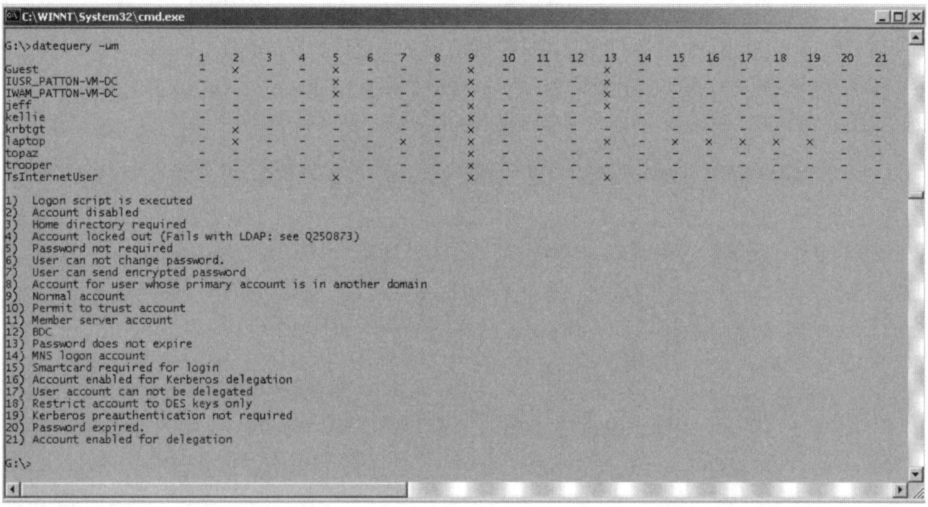

FIGURE 6.6 The user account flags can also be displayed in a matrix format, which provides a quick scan of the properties.

For example, to display all disabled accounts, run `DateQuery -u 2`. You can specify multiple numerical arguments. To see all disabled accounts (2) or those accounts that do not require a password (5), run `DateQuery -u 2 5`. Note the reference to the Microsoft support article (Q250873) noting that the account lockout does not display properly when using the LDAP ADSI provider. Unfortunately, you may find in your scripting endeavors issues or bugs that you must simply acknowledge and script around. In this example, the script bypasses this bug by using an entirely separate function (`ShowLockouts`) to display locked-out accounts. The `-l` argument calls a separate function that uses the ADSI WinNT provider to get the data from the primary directory server. This separate function is required to overcome a known Microsoft issue with trying to get the account lockout status using the LDAP provider.

Although future software versions or updates may fix this problem, remember that when you are coding a larger script you may encounter pitfalls or shortcomings of the tools that you have to work with. Fortunately, more than one way is usually available to solve a problem, so don't be afraid to approach it from different angles.

TABLE 6.3   The `ADS_USER_FLAG_ENUM` ADSI enumeration contains many flags that define user account settings

#	*Flag name*	*Flag description*
1	ADS_UF_SCRIPT	Logon script is executed.
2	ADS_UF_ACCOUNTDISABLE	Account disabled.
3	ADS_UF_HOMEDIR_REQUIRED	Home directory required.
4	ADS_UF_LOCKOUT	Account locked out (fails with LDAP: see Q250873).
5	ADS_UF_PASSWD_NOTREQD	Password not required.
6	ADS_UF_PASSWD_CANT_CHANGE	User cannot change password.
7	ADS_UF_ENCRYPTED_TEXT_PASSWORD_ALLOWED	User can send encrypted password.
8	ADS_UF_TEMP_DUPLICATE_ACCOUNT	Account for user whose primary account is in another domain.
9	ADS_UF_NORMAL_ACCOUNT	Normal account.
10	ADS_UF_INTERDOMAIN_TRUST_ACCOUNT	Permit to trust account.
11	ADS_UF_WORKSTATION_TRUST_ACCOUNT	Member server account.
12	ADS_UF_SERVER_TRUST_ACCOUNT	BDC.
13	ADS_UF_DONT_EXPIRE_PASSWD	Password does not expire.
14	ADS_UF_MNS_LOGON_ACCOUNT	MNS logon account.
15	ADS_UF_SMARTCARD_REQUIRED	Smartcard required for login.
16	ADS_UF_TRUSTED_FOR_DELEGATION	Account enabled for Kerberos delegation.
17	ADS_UF_NOT_DELEGATED	User account cannot be delegated.
18	ADS_UF_USE_DES_KEY_ONLY	Restrict account to DES keys only.
19	ADS_UF_DONT_REQUIRE_PREAUTH	Kerberos preauthentication not required.
20	ADS_UF_PASSWORD_EXPIRED	Password expired.
21	ADS_UF_TRUSTED_TO_AUTHENTICATE_FOR_DELEGATION	Account enabled for delegation.

Output

The script displays matching accounts that fit within the specified input date parameters, as shown in Figures 6.3 and 6.4. In addition to providing account creation and expiration information, the script also enumerates account settings to show properties, such as whether that account is locked out or if it is disabled, as shown in Figures 6.5 and 6.6.

Summary of Solution

The help desk manager schedules the script to run weekly with the arguments to display all accounts that will expire within a window of 14 days. The batch file that starts the script redirects the output to a text file that the help desk manager reviews on a weekly basis. The security manager runs the script ad hoc to display to the console various account property settings and account status. The script uses the ADSI LDAP provider to query and display account information that satisfies date-specific criteria or specified account settings. The script binds to the domain and user account objects to collect domain- and account-specific information. Then, depending on the specified arguments, the script iterates through each user account and displays the appropriate report to the console.

OBJECTIVE

Create a command-line tool to list AD accounts that have been created or will expire within a specified range. Additionally, display account status such as whether the account is locked out or disabled, if the password is set to never expire, and other account properties.

ADODB.Connection

- Provider
- Open

ADODB.Command

- ActiveConnection
- Properties
- CommandText
- Execute

ADSI

- GetObject
- LDAP Provider
- WinNT Provider

Arrays

- ◼ `length`
- ◼ Looping through arrays

Date

- ◼ `getTime`
- ◼ `setTime`

Error Trapping

- ◼ `Try…Catch`

Enumeration Data Types

- ◼ Extracting data from enumeration data types

Large Integers

- ◼ Manipulating large integers

JScript Objects

- ◼ `object={prop: "some prop", value: "some value"}`

RecordSet

- ◼ `EOF`
- ◼ `Fields.Item(x).Value`

RegExp

- ◼ `test`

RootDSE

- ◼ `GetObject("LDAP://RootDSE")`
- ◼ `Get("defaultNamingContext")`

SQL Dialect Query

- ◼ Base Distinguished Name
- ◼ Filter
- ◼ Attributes
- ◼ Scope

Ternary Operator

- ◼ `(statement?true:false)`

WScript.Arguments

- ◼ `colArgs.length`
- ◼ `colArgs.Item`

WScript.StdOut

- Write
- Column

THE SCRIPT

The script begins with the declaration of variables used in the main program. *Date-Query* uses arrays, which contain objects with multiple properties, to store data. For example, the primary array used in this script, aAccounts, contains as its members objects with properties of the account such as its name, ADSPath, expiration date, creation date, and other account information. An array is well suited for this task because the script uses loops to scan and selectively display various properties of all objects contained within the array. Although an array is an object itself, its membership can include other objects.

Populating an array with objects can be tricky to understand at first, but with practice you will find it a useful method to store and retrieve data. Arrays can contain any data type, such as integers or strings. Arrays can also contain objects, as shown in the following example:

```
var objMyObject = new Object();
var aSomeArray = new Array();
aSomeArray[0] = objMyObject;
```

Objects may be used as containers for name value pairs:

```
obj.someName = "someValue";
```

Or another way:

```
obj["someName"] = "someValue";
```

You can also initialize and define an object in one step:

```
var obj = { someName : "someValue" };
```

You can even assign an object to an array index in one step:

```
a[0] = { someName : "someValue" };
```

You can define the array using object literals, like so:

```
aSomeArray[0]= { name: "some name" , value: someValue};
aSomeArray[1]= { name: "another name" , value: anotherValue};
```

Assigning values to an array using object literals lets you recall the object properties. For example, `WScript.Echo(aSomeArray[1].name)` outputs the text `some name`. Similarly, `aSomeArray[2].value` equals `anotherValue`. The array `aSomeArray` contains two items, which themselves are objects. Each object has two properties: `name` and `value`. Stating this in another way, you have just created and defined your own custom object with two properties, and you can use this object and pass it to other functions just as you would any other object.

Consider the following example, which illustrates how one member of an array of objects can be passed to a function and manipulated as the object that it is with its properties intact:

```
//The Main Program
var aSomeArray =new Array;
aSomeArray[0]= { name: "some name" , value: 3};
aSomeArray[1]= { name: "another name" , value: 5};
MyFunction(aSomeArray[1]);
//A Function
function MyFunction(objSomeObject)
{
WScript.Echo(objSomeObject.name); // Displays 'some name'
WScript.Echo(objSomeObject.value); // Displays 3
}
```

To retrieve the data, specify the array element by index number and the desired property. For example:

```
WScript.Echo(aSomeArray[0].name);
```

will output the text `some name`.

Alternatively, loop through the entire array, and display both the name and value properties. (Note that naming the properties is entirely up to you; you don't have to stick with the property names like "name" or "value".)

```
for(i=0;i<aSomeArray.length;i++){
    WScript.Echo(aSomeArray[i].name + "   " +aSomeArray[i].value);
}
```

The functions `QueryDSA` and `DumpUAC` in this script provide additional examples and instruction of how to set and read objects from arrays.

ON THE CD

The script is presented on the CD-ROM as /chapter 6/scripts/DateQuery.js. Before continuing in this next section, open the actual script from the CD-ROM and run it a few times in your test environment to get a feeling of how it works. Explore the entire code on your own and then continue with the walkthrough, which hopefully answers any questions you may have. Plus, seeing the code in its entirety in your script editor gives you a good sense of its scope and also allows you to tweak it as we walk through its function.

The Main Body

Typically, you will find that the variable prefix dt represents date-time variables. In this script, dtToday represents today's date (the date the script is currently being run). The variables dtWindowAhead and dtWindowAgo represent the dates used in determining if an account will expire (in the future) or was created (in the past) within a specified window. The variable iWindow may be set as a parameter using the argument -w [window days], but if not specified, the window defaults to 14 days. The script uses the variables dtPwdExpires and dtPwdLastChg in the date calculations that determine when the password will expire and when it was last changed, respectively. The objState object, which you have seen in the previous scripts, reemerges to track different run options that the user selects through the command-line arguments. The user of the script may specify what account properties the script should display with the -u or -um arguments, and the array aUACTests contains the numbers of these properties to display. The array aAccounts contains all account information in the current domain, and objDomain contains the LDAP and WinNT provider reference to the primary directory services server.

Like the array used to hold the user accounts (aAccounts), the script uses the array aUAC to contain objects that define properties of the ADS_USER_FLAG_ENUM enumeration. The symbolic reference, flag value, and short description are defined for each of the 21 flags and stored in aUAC as properties. When using constants, refer to them using their symbolic name instead of the hexadecimal values. Doing so increases the readability of your code and is generally regarded as good programming practice. Recall that by declaring and initializing this array at the beginning of the script before any other functions are called, you ensure that this array is available to the other functions as a global variable. In this case, aUAC is used both by the main program and the function DumpUAC.

```
//------------------------------------------------------------------
// DateQuery.js
//------------------------------------------------------------------
var i=0,j=0;
var iWindow = 14;
```

```
var iWindowMs;
var iPwdExp, iPwdExpMs;
var dtToday= new Date();
var dtWindowAhead = new Date();
var dtWindowAgo= new Date();
var dtPwdExpires = new Date();
var dtPwdLastChg = new Date();
var strDelimit=" ";
var objState = new Object;
var aUACTests= new Array();
var aAccounts = new Array();
var objDomain = new Object;

var aUAC = new Array();
aUAC[1] = { name: "ADS_UF_SCRIPT",  value : 0X0001,
 desc: "Logon script is executed" };
aUAC[2] = { name: "ADS_UF_ACCOUNTDISABLE",  value : 0X0002,
 desc: "Account disabled" };
aUAC[3] = { name: "ADS_UF_HOMEDIR_REQUIRED",  value : 0X0008,
 desc: "Home directory required" };
aUAC[4] = { name: "ADS_UF_LOCKOUT",  value : 0X0010,
 desc: "Account locked out (Fails with LDAP: see Q250873)" };
aUAC[5] = { name: "ADS_UF_PASSWD_NOTREQD",  value : 0X0020,
 desc: "Password not required" };
aUAC[6] = { name: "ADS_UF_PASSWD_CANT_CHANGE",  value : 0X0040,
 desc: "User can not change password." };
aUAC[7] = { name:
 "ADS_UF_ENCRYPTED_TEXT_PASSWORD_ALLOWED",value : 0X0080,
 desc: "User can send encrypted password" };
aUAC[8] = { name: "ADS_UF_TEMP_DUPLICATE_ACCOUNT",  value : 0X0100,
 desc: "Account for user whose primary account is in another domain" };
aUAC[9] = { name: "ADS_UF_NORMAL_ACCOUNT",  value : 0X0200,
 desc: "Normal account" };
aUAC[10] = { name: "ADS_UF_INTERDOMAIN_TRUST_ACCOUNT",
 value : 0X0800, desc: "Permit to trust account" };
aUAC[11] = { name: "ADS_UF_WORKSTATION_TRUST_ACCOUNT",
 value : 0X1000, desc: "Member server account" };
aUAC[12] = { name: "ADS_UF_SERVER_TRUST_ACCOUNT",  value : 0X2000,
 desc: "BDC" };
aUAC[13] = { name: "ADS_UF_DONT_EXPIRE_PASSWD",  value : 0X10000,
 desc: "Password does not expire" };
aUAC[14] = { name: "ADS_UF_MNS_LOGON_ACCOUNT",  value : 0X20000,
 desc: "MNS logon account" };
```

```
aUAC[15] = { name: "ADS_UF_SMARTCARD_REQUIRED", value : OX40000,
 desc: "Smartcard required for login" };
aUAC[16] = { name: "ADS_UF_TRUSTED_FOR_DELEGATION", value : OX80000,
 desc: "Account enabled for Kerberos delegation" };
aUAC[17] = { name: "ADS_UF_NOT_DELEGATED", value : OX100000,
 desc: "User account can not be delegated" };
aUAC[18] = { name: "ADS_UF_USE_DES_KEY_ONLY", value : 0x200000,
 desc: "Restrict account to DES keys only" };
aUAC[19] = { name: "ADS_UF_DONT_REQUIRE_PREAUTH", value : 0x400000,
 desc: "Kerberos preauthentication not required" };
aUAC[20] = { name: "ADS_UF_PASSWORD_EXPIRED", value : 0x800000,
 desc: "Password expired." };
aUAC[21] = { name: "ADS_UF_TRUSTED_TO_AUTHENTICATE_FOR_DELEGATION",
 value : 0x1000000, desc: "Account enabled for delegation" };

objState.create=0;
objState.expire=0;
objState.verbose=0;
objState.help=0;
objState.domain=0;
objState.neverExpire=0;
objState.userAccountControl=0;
objState.uACMatrix=0;
objState.lockout=0;

if(WScript.FullName.indexOf("WScript")>=0) {
    WScript.Echo("Due to some of the functionality of this script, " +
      "Please run this program using the cscript.exe engine.\n" +
      "for example: cscript.exe UserQuery.js");
    WScript.Quit()
    }
```

The first function call of this script is to the `CurrentDomain` procedure, which determines the domain of the account running the script. This function has been slightly modified from the version in previous chapters. Instead of returning only the name of the domain in LDAP vernacular (for example, the `defaultNaming-Context` is returned to the variable `strDomain`), the function returns an array that contains both the `defaultNamingContext` and the `dnsHostName`. The `dnsHostName` is the fully qualified DNS name of the primary directory server (e.g., *dc.domain.com*). The `dnsHostName` is used by the WinNT provider in the `ShowLockouts` function to bind to the user accounts. These two values are stored in the array `objDomain` as properties named `winnt` and `ldap` and can be retrieved in the script as `objDomain.ldap` and `objDomain.winnt`.

The `GetMaxPwdAge` function returns the number of days until the specified domain's password will expire and stores the result in the variable `iPwdExp`. The function creates the `ADsPath` to the directory server by concatenating the provider string prefix `LDAP://` with the `defaultNamingContext` stored in the array element `objDomain.ldap`. The variable `iPwdExpMs` stores the maximum password age (`iPwdExp`) converted to milliseconds, the preferred time units used by the JScript date functions.

```
objDomain=CurrentDomain();
iPwdExp=GetMaxPwdAge("LDAP://" + objDomain.ldap);
iPwdExpMs=1000*60*60*24*iPwdExp;
var colArgs = WScript.Arguments;
```

DateQuery uses an argument-handling routine similar to those found in both of the command-line scripts presented in Chapters 4 and 5. The `objState` properties track the selections made by the user running the script.

As with previous scripts, if the user does not specify the subsequent argument following the arguments d, w, or D, then the function `ParamError` is called and the script exits.

Any arguments that are not prefaced with a hyphen (-) or are not a part of a subsequent argument are assumed to be one or several `aUACTests`. The script confirms that the argument is a number between 1 and 21 and adds it to the `aUACTest` array.

```
for(i=0;i<colArgs.length;i++) {
    strArg=colArgs.Item(i)
    if(strArg.indexOf("-")==0)            {
        if(strArg.indexOf("c")>=0)  objState.create=1;
        if(strArg.indexOf("e")>=0)  objState.expire=1;
        if(strArg.indexOf("l")>=0)  objState.lockout=1;
        if(strArg.indexOf("n")>=0)  objState.neverExpire=1;
        if(strArg.indexOf("u")>=0)  objState.userAccountControl=1;
        if(strArg.indexOf("m")>=0)  objState.uACMatrix=1;
        if(strArg.indexOf("v")>=0)  objState.verbose=1;
        if(strArg.indexOf("?")>=0)  objState.help=1;
        if(strArg=="-w") {
            if(i+1 > colArgs.length-1) ParamError();
            iWindow=colArgs.Item(i+1);
            i++;
        }
        if(strArg=="-d") {
            if(i+1 > colArgs.length-1) ParamError();
            strDelimit=colArgs.Item(i+1);
            i++;
        }
```

```
            if(strArg=="-D") {
                  if(i+1 > colArgs.length-1) ParamError();
                  objDomain.ldap=colArgs.Item(i+1);
                  objState.domain=1;
                  i++;
            }
            else ParamError();
      }
      else {
            if(strArg>0 && strArg < aUAC.length) {
                  aUACTests.push(strArg);
            }
      }
}
if(objState.help) Usage();
if(objState.verbose){
      Heading("Using Domain");
      WScript.Echo(objDomain.ldap);
}
```

User objects in AD have a `UserAccountControl` property. This property is a bit-field from which you can extract over 20 property settings for a user account. The values contained in this bitfield are described by the ADS enumeration `ADS_USER_FLAG_ENUM`. Because many of these flags may not always be of interest to the script runner, the script supports specifying one or several of the flags to display. For example, `DateQuery -u 5 6` adds to the array `aUACTests` two elements: 5 and 6. These elements correspond to the `ADS_USER_FLAG_ENUM` flags that the script will process and display. Properties 5 and 6, for example, correspond to whether an account's password is not required or if a user cannot change his password. The numbering is not special or a part of ADSI programming—it just so happens that these are the fifth and sixth properties in a list of `ADS_USER_FLAG_ENUM` flags. A description of how to extract the data contained within `ADS_USER_FLAG_ENUM` follows in the description of the main loop and in the description of the function `DumpUAC`. If a user does not specify any specific test, the script builds the `aUACTests` array containing all of the tests, and it fills the array with integers from 1 to the length of the `aUAC` array, which represents each of the account properties.

```
//Fill in the aUACTests if needed
if(aUACTests.length==0) {
      for(i=1;i<aUAC.length+1;i++) aUACTests.push(i);
}
```

The script uses a modified version of the directory service query function presented in Chapter 5. Instead of returning a simple array containing the ADsPath, the function QueryDSA returns an array of objects, with each object containing two properties: the ADsPath and sAMAccountName of all user accounts in the target domain. As previously used, the query defining all user accounts is defined as all objectClass=user and objectCategory=person objects. A more detailed breakdown of the syntax of this query is covered in Chapter 3 under ADSI queries and in Chapter 4 under discussion of the *QueryDS* script.

```
//Find the Users
strCommandText="SELECT ADsPath, sAMAccountName " +
 "FROM 'LDAP://" + objDomain.ldap + "' WHERE objectClass = 'user' and "
+
 "objectCategory = 'person'";
aAccounts=QueryDSA(strCommandText);
```

The two variables dtWindowAgo and dtWindowAhead represent date boundaries from which to determine whether to display recently created or soon-to-expire accounts. The values dtWindowAgo and dtWindowAhead are calculated by subtracting or adding the boundary window (iWindow) to today's date, respectively. Date math can be a bit tricky because you must account for different numbers of days per month, leap years, and possibly different time zones. Fortunately, many programming languages include methods to abstract you from these gritty details. This script uses the JScript method getTime to calculate the number of milliseconds between today's date and time and midnight January 1, 1970 GMT. Next, the script converts iWindow from days to milliseconds (1,000 milliseconds per second, 60 seconds per minute, 60 minutes per hour, and 24 hours per day) and adds this to the previously calculated time value of milliseconds since January 1, 1970. This new value represents the number of milliseconds of the date window from January 1, 1970. Finally, the script uses the setTime method to convert this value back to a true date and time. The script calculates both dtWindowAhead and dtWindowAgo in this manner. This entire process is diagrammed in Figure 6.7.

```
//set the date windows
iWindowMs=1000*60*60*24 * iWindow;
dtWindowAgo.setTime((dtToday.getTime() - iWindowMs));
dtWindowAhead.setTime(dtToday.getTime() + iWindowMs);
```

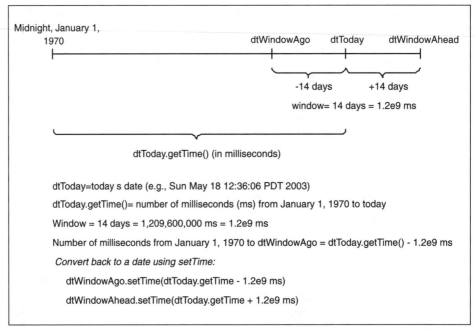

FIGURE 6.7   Using the `getTime` and `setTime` JScript functions allows you to easily perform accurate and reliable date math.

The Primary Loop

Next, let's examine the major loop of the *DateQuery* script that iterates through all user accounts and pulls the creation and expiration date parameters and other account properties. This loop further populates the array `aAccounts` with new object properties containing these new account properties. The array `aAccounts` was created by the function `QueryDSA`, initially containing objects with two properties, the name (LDAP: `sAMAccountName`) and `ADsPath` (LDAP: `ADsPath`). The `name` property provides a short, easy to identify reference to the object; the `ADsPath` property provides a unique string address of the user account, which can be used in the future to reference the object and get data. The script immediately uses the `ADsPath` of the current user object to access the `UserAccountControl` attribute.

As mentioned previously, `UserAccountControl` represents the LDAP property that accesses the `ADS_USER_FLAG_ENUM` enumeration flags. To extract the value of one of these flags you must perform a bitwise comparison of the `UserAccountControl` property and the desired flag to check if that attribute is true. This is because all of 21 flags are stored in a single value represented by the `UserAccountControl` property, and a bitwise comparison can discretely calculate if any of the 21 flags are set. More details on how the `ADS_USER_FLAG_ENUM` enumeration flags are checked are covered under the function `DumpUAC`.

```
//main loop
for(i=0;i<aAccounts.length;i++) {
    objADs=GetObject(aAccounts[i].ADsPath);
//Poll the UAC
    iUAC=objADs.Get("UserAccountControl");
    var ADS_UF_DONT_EXPIRE_PASSWD = 0x10000;
    var ADS_UF_ACCOUNTDISABLE = 0X0002;
```

The script uses a ternary operator to check the value of the `ADS_USER_FLAG_ENUM` flags and then sets new object properties of the `aAccounts` array. Once an object is defined (for example, any of the account objects contained within `aAccounts`), objects with new properties may be added by simply referencing them as `aAccount[x].someNewProperty`. In this manner, the script sets the properties `bPwdExpires` and `disabled`.

```
aAccounts[i].bPwdExpires=
 (iUAC & ADS_UF_DONT_EXPIRE_PASSWD?true:false);
aAccounts[i].bDisabled=(iUAC & ADS_UF_ACCOUNTDISABLE?true:false);
```

The `&` character in the previous example is the bitwise `AND`, and it is used to determine whether a property is set. The bitwise `AND` is described in more detail in the section titled *Displaying Account Properties as a Matrix*.

Active Directory Object Dates

AD contains a creation date for every user object that can be directly accessed by the `whenCreated` attribute (an attribute of the `Top` class).

The `PasswordLastChanged` attribute, however, exists only for objects that have actually had their passwords changed. This means that objects that have never had their passwords changed return an error when the script tries to access this property. To trap for this occurrence, the script encloses this procedure in a JScript `try…catch` statement. If the attempt to get the `PasswordLastChanged` attribute fails, the value is set to a blank string (""").

If the script succeeds at getting the `PasswordLastChanged` attribute, it proceeds to calculate the password expiration date in much the same way that the script calculates `dtWindowAhead` and `dtWindowAgo`. The `maxPwdAge` (stored in `iPwdExpMs` in milliseconds) is added to the millisecond-from-midnight-January-1-1970 converted `PasswordLastChanged` date. This new value is converted back to a date and stored in dtPwdExpires. The following procedure also closes out the main loop iterating through each of the aAccounts array members:

```
//Get account created date
    aAccounts[i].created=new Date(objADs.whenCreated);
//Get password expires date
    try{
        dtPwdLastChg=new Date(objADs.PasswordLastChanged);
        dtPwdExpires=new Date(objADs.PasswordLastChanged);
          dtPwdExpires.setTime((dtPwdLastChg.getTime() + iPwdExpMs));
        aAccounts[i].pwdLastChanged=dtPwdLastChg;
        aAccounts[i].expires=dtPwdExpires;
    }
    catch(e) {
        aAccounts[i].pwdLastChanged="";
        aAccounts[i].expires="";
    }
}
```

Displaying the Results

After the script completes iterating through each of the user account objects and determines the properties of the accounts, it starts the final section of its processing. This section displays the various reports as selected by the user running the script. To improve readability of the code, each report is isolated as its own function, which is called from the main program. The main program calls the functions with all of the arguments required by the function to avoid the use of global variables. Descriptions of the individual reports follow under the description of their code.

After the completion of the reporting functions, the script ends.

```
if(objState.verbose) ListAllDetail(aAccounts, strDelimit, iPwdExp);
if(objState.create) ListCreated(aAccounts, strDelimit, dtWindowAgo);
if(objState.expire) ListExpired(aAccounts, strDelimit, dtWindowAhead,
    dtToday);
if(objState.neverExpire) ListPwdExpires(aAccounts, strDelimit);
if(objState.userAccountControl) {
    DumpUAC(aAccounts, aUACTests, objState);
}
if(objState.lockout) {
ShowLockouts(aAccounts, objDomain);
}
WScript.Quit(0);
```

ListAllDetail

The first report provides a verbose listing of all of the dates calculated by the script and is specified by the verbose (-v) flag. The function displays a heading and the calculated maximum password age (iPwdExp) calculated in the GetMaxPwdAge function. Next, the report iterates through all of the user accounts contained in the array aAccounts and displays the account creation date, when the password was last changed, and when it expires. Additionally, xs are printed if the account's password is set to never expire or if the account is disabled.

```
//--------------------------------------------------------------------
// Display Detailed Information
//--------------------------------------------------------------------
function ListAllDetail(aAccounts, strDelimit, iPwdExp)
{
Heading("Account Date Properties");
WScript.Echo("Domain max password age: " + iPwdExp + " days");
Tabulate("PD",26,strDelimit);
Tabulate("Account Created",30,strDelimit);
Tabulate("Password Last Changed",60,strDelimit);
Tabulate("Password Expires\n",90,strDelimit);

for(i=0;i<aAccounts.length;i++){
    Tabulate(aAccounts[i].name,0,strDelimit);
    if (aAccounts[i].bPwdExpires) Tabulate("x",26,strDelimit);
    if (aAccounts[i].bDisabled) Tabulate("x",27,strDelimit);
    Tabulate(aAccounts[i].created,30,strDelimit);
    Tabulate(aAccounts[i].pwdLastChanged,60,strDelimit);
    Tabulate(aAccounts[i].expires+"\n",90,strDelimit);
}
WScript.StdOut.Write("P:Password never expires, ");
WScript.StdOut.Write("D:Account disabled, ");
WScript.StdOut.Write("\n");
}
```

Like the verbose report, the ListCreated and ListExpired functions iterate through the aAccounts array to display all user accounts that match the specified date criteria.

ListCreated

The ListCreated function lists all accounts created within the specified window.

```
//------------------------------------------------------------------------
// Display Dates Accounts were Created
//------------------------------------------------------------------------
function ListCreated(aAccounts, strDelimit, dtWindowAgo)
{
Heading("Accounts Created in the last " + iWindow  + " days.");
Tabulate("Account Name",0,strDelimit);
Tabulate("PD",26,strDelimit);
Tabulate("Account Created\n",30,strDelimit);
for(i=0;i<aAccounts.length;i++){
    if(dtWindowAgo < aAccounts[i].created) {
        Tabulate(aAccounts[i].name,0,strDelimit);
        if (aAccounts[i].bPwdExpires)
         Tabulate ("x",26,strDelimit);
        if (aAccounts[i].bDisabled)
         Tabulate ("x",27,strDelimit);
        Tabulate(aAccounts[i].created + "\n",30,strDelimit);
    }
}
WScript.StdOut.Write("\nP:Password never expires, ");
WScript.StdOut.Write("D:Account disabled, ");
WScript.StdOut.Write("\n");
}
```

ListExpired

The ListExpired function displays all accounts that will expire within the specified window. The function loops a second time through aAccounts to check whether the password expiration date has already passed. This report is useful for finding accounts that have been inactive for a period of time. Neither of these reports includes accounts that have been disabled, as defined by the aAccounts[i].bDisabled property set previously in the main program.

```
//------------------------------------------------------------------------
// Display Expired or Soon to Expire Accounts
//------------------------------------------------------------------------
function ListExpired(aAccounts, strDelimit, dtWindowAhead, dtToday)
{
Heading("Accounts that will expire in less than " + iWindow  +
  " days. \nDisabled accounts not shown.");
Tabulate("Account Name",0,strDelimit);
```

```
    Tabulate("Account Expires\n",30,strDelimit);
    for(i=0;i<aAccounts.length;i++){
        if( (aAccounts[i].expires < dtWindowAhead) &&
            aAccounts[i].expires > dtToday &&
            !aAccounts[i].bDisabled) {
            Tabulate(aAccounts[i].name,0,strDelimit);
            Tabulate(aAccounts[i].expires + "\n",30,strDelimit);
        }
    }
    Heading("WARNING: These accounts should have already expired." +
     "\nDisabled accounts not shown.");
    Tabulate("Account Name",0,strDelimit);
    Tabulate("P",26,strDelimit);
    Tabulate("Date Account Should Have Expired\n",30,strDelimit);
    for(i=0;i<aAccounts.length;i++){
        if(aAccounts[i].expires < dtToday && !aAccounts[i].bDisabled) {
            Tabulate(aAccounts[i].name,0,strDelimit);
            if (aAccounts[i].bPwdExpires) Tabulate ("x",26,strDelimit);
            Tabulate(aAccounts[i].expires + "\n",30,strDelimit);
        }
    }
    WScript.StdOut.Write("\nP:Password never expires, ");
    WScript.StdOut.Write("\n");
    }
```

ListPwdExpires

The ListPwdExpires report uses the same techniques previously mentioned to iterate through the accounts and displays those accounts with password set to never expire.

```
//--------------------------------------------------------------------
// Display Accounts that Never Expire
//--------------------------------------------------------------------
function ListPwdExpires(aAccounts, strDelimit)
{
Heading("Accounts set to never expire. \nDisabled accounts not shown.")
    Tabulate("Account Name",0,strDelimit);
    Tabulate("Account ADsPath\n",30,strDelimit);
    for(i=0;i<aAccounts.length;i++){
        if(aAccounts[i].bPwdExpires && !aAccounts[i].bDisabled) {
            Tabulate(aAccounts[i].name,0,strDelimit);
            Tabulate(aAccounts[i].ADsPath + "\n",30,strDelimit);
        }
    }
}
```

ShowLockouts

Quickly checking currently locked-out accounts can be a useful troubleshooting tool for a systems administrator. The ADS_USER_FLAG_ENUM enumeration includes a flag for an account lockout; however, there is a bug in its implementation and it does not always update successfully in AD. To reliably get the account lockout status, this script calls the ShowLockouts function that bypasses the LDAP User-AccountControl and finds whether an account is locked out by getting the ADSI WinNT provider IsAccountLocked attribute. Connecting to the primary directory service server using the WinNT provider is fairly straightforward. We already have the DNS host name of the script user's primary directory services server that was retrieved by the function CurrentDomain and stored in objDomain.winnt. We can use this DNS host name to bind to the user object on that directory server, as shown in the following example:

```
WinNT://directoryserver.domain.com/useraccountname
```

The function then uses the IADsUser interface property method to access the IsAccountLocked attribute of the object. It then prints the name of the object if IsAccountLocked returns true.

```
//------------------------------------------------------------------
// WinNT Provider workaround to enumerate account lockouts
//------------------------------------------------------------------
function ShowLockouts(aAccounts, objDomain)
{
Heading("Locked out Accounts");
for(i=1;i<aAccounts.length;i++) {
    objAccount = GetObject("WinNT://" + objDomain.winnt + "/" +
     aAccounts[i].name);
    if(objAccount.IsAccountLocked) {
        WScript.Echo(aAccounts[i].name);
    }
}
}
```

DumpUAC

So far, we have referenced the LDAP UserAccountControl property of a user account object a number of times. Now let's dive into this property a bit deeper. The ADSI LDAP provider UserAccountControl property provides access to the ADS_USER_FLAG_ENUM enumeration, which stores the state of 21 account properties. You can also access these account properties by way of the WinNT provider userFlags property. These account properties can be read and set by a script, and most can be set as account options using the Active Directory User and Computer MMC tool, as shown in Figure 6.8.

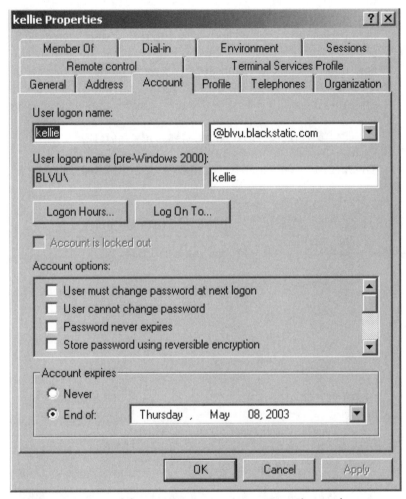

FIGURE 6.8 Many of the `UserAccountControl` properties can be set interactively using the administrative tool Active Directory Users and Computers.

The script displays the `UserAccountControl` properties of all user account objects (such as those contained in `aAccounts`) either as a matrix or in a simple list. Also, the user can specify what properties to display as identified in the array `aUACTests`. Looping through all user objects, and for each user looping through all `UserAccountControl` properties, requires two loops. The sequence of the loops depends on the type of report desired.

Displaying Account Properties as a Matrix

The script generates a matrix report by first looping through every user object and then looping through every ADS_USER_FLAG_ENUM flag. Before the first loop, the function constructs and displays the matrix heading. The script loops through each of the members of the array aUACTests and displays the numbers of the flags that will be displayed. aUACTests contains the flag numbers that the user specified. (It contains all flag numbers if the user did not specify any particular flag.) For example, if a user specified to display only the flags numbered 2, 5, and 12, the aUACTests would contain three members with indices 0, 1, and 2. In this example, the array would look like this:

```
aUACTests[0]=2, aUACTests[1]=5, aUACTests[2]=12
```

The first row of the matrix would be 2, 5, and 12, separated by five characters each.

```
//-------------------------------------------------------------------
// Dump UserAccountControl Information
//-------------------------------------------------------------------
function DumpUAC(aAccounts, aUACTests, objState)
{
var bAccountExists;
if(objState.uACMatrix) {
    for(i=0;i<aUACTests.length;i++){
        Tabulate(aUACTests[i],30+i*5,strDelimit);
        }
    WScript.StdOut.Write("\n");
```

After displaying the matrix heading, the first loop iterating through aAccounts begins. For every member of this array, the script binds to the account using the aAccounts[i].ADsPath property, where i is the index of the current account specified by the loop. The script displays the name of the account and gets the UserAccountControl property.

```
    for(i=0;i<aAccounts.length;i++) {
        objADs=GetObject(aAccounts[i].ADsPath);
        Tabulate(aAccounts[i].name,0,strDelimit);
        iUAC=objADs.Get("UserAccountControl");
```

The method of storing data in the ADS_USER_FLAG_ENUM enumeration is interesting—especially to new programmers. Recall that the LDAP UserAccountControl property contains a single value referencing all of the set ADS_USER_FLAG_ENUM flags. This value is calculated by cumulatively adding all true properties. For example, a normal account (ADS_UF_NORMAL_ACCOUNT = 0x0200) with a password that never expires (ADS_UF_DONT_EXPIRE_PASSWD = 0x10000) would have a UserAccountControl value of 0x10200 (0x0200 + 0x10000). All 21 ADS_USER_FLAG_ENUM flags have a different defined value such that when added together, any combination of true flags sums to a unique number. This number can then be compared to any individual flag using a bitwise comparison (the JScript & operator) to check if that flag is true.

These values and calculations are in hex, and if you print the UserAccountControl using WScript.Echo (or using the ADSIEdit MMC), the value is displayed in decimal.

Supplied with the UserAccountControl property value for the current user account, the second loop iterates through each of the members of the array aUACTests. The UserAccountControl value is compared bit-by-bit against the current ADS_USER_FLAG_ENUM flag. The bitwise AND (in JScript & and in VBScript AND) is used for this comparison. From our previous example, the UserAccountControl (0x10200) & ADS_UF_DONT_EXPIRE_PASSWD (0x10000) would equal 0x10000, which is true. However, if the ADS_UF_PASSWD_NOTREQD flag was not set, then the UserAccountControl would equal only 0x200 (true for ADS_UF_NORMAL_ACCOUNT) and (0x10000 & 0x200) results in zero bit matches, which is false. Using this process, the script compares every member in UACTests against the UserAccountControl value and displays an x to the console for every match.

```
for(j=0;j<aUACTests.length;j++){
    if (iUAC & aUAC[aUACTests[j]].value) {
        Tabulate ("x",30+j*5,strDelimit);
    }
    else { Tabulate ("-",30+j*5,strDelimit);}
}
    WScript.StdOut.Write("\n");
}
WScript.StdOut.Write("\n");
```

Lastly, the script displays a legend for each of the tests that were performed by looping through each of the members of aUACTests and displaying its number and the description.

```
for(i=0;i<aUACTests.length;i++){
    Tabulate(aUACTests[i] + ") ",0,strDelimit);
    Tabulate(aUAC[aUACTests[i]].desc+"\n",5,strDelimit);
    }
}
```

Displaying Account Properties as a List

The second type of UserAccessControl report the script can generate is a list of accounts by ADS_USER_FLAG_ENUM flag. Think of this report as a reverse of the matrix report. The outer loop iterates through all of the specified flags contained in the aUACTests array, and the inner loop cycles through all of the accounts contained in aAccounts. If an account tests true for a specific property, its name is displayed.

```
else {
    for(i=0;i<aUACTests.length;i++){
        bAccountExists=false;
        WScript.Echo(aUACTests[i] + ") " + aUAC[aUACTests[i]].desc
         + " (" +aUAC[aUACTests[i]].name + ")");
        for(j=1;j<aAccounts.length;j++) {
            objADs=GetObject(aAccounts[j].ADsPath);
            iUAC=objADs.Get("UserAccountControl");
            if (iUAC & aUAC[aUACTests[i]].value) {
                Tabulate (aAccounts[j].name + "\n",5,strDelimit);
                bAccountExists=true;
            }
        }
        if(!bAccountExists) WScript.Echo("-none-");
        WScript.StdOut.Write("\n");
    }
}
}
```

GetMaxPwdAge

To calculate the actual password expiration date, the script must determine the maximum password age setting for the domain. This value is domain specific and can be found in the maxPwdAge property of the domain object. The function GetMaxPwdAge returns the maximum password age in days for a specified domain. The function takes as input the distinguished name of a domain (for example, LDAP://dc=domain, dc=com). Using this value, the function binds to the domain and gets the maxPwdAge property. The maxPwdAge value is a negatively valued eight-byte-large integer that represents the maximum password age in 100-nanosecond units.

Before we use this value, we must convert the large integer to a useable number and then convert the value from 100-nanosecond units to days. The script calls the function ConvertNanoDays to make this conversion, takes the absolute value of the number (which makes a positive or negative number a positive), and then rounds the number to the nearest integer. The resulting value is the maximum password age in days.

```
//-------------------------------------------------------------------
// Get the maximum password age
//-------------------------------------------------------------------
function GetMaxPwdAge(strADsPath)
{
var objADs, objMaxPwdAge, iMaxPwdAge;
objADs=GetObject(strADsPath);
objMaxPwdAge=objADs.Get("maxPwdAge");
iMaxPwdAge=Math.round(Math.abs(ConvertNanoDays(objMaxPwdAge)),0);
return(iMaxPwdAge);
}
```

ConvertNanoDays

AD stores some large integer properties as eight-byte (64-bit) objects, which consist of a *high part* and a *low part* of 32 bits each. AD returns an object with a high part and a low part property. Before you can use the values, the parts must be extracted and added together. The high part consists of the upper 32 bits, and the low part consists of the lower 32 bits. To assemble a large integer, use the following math: *assembled value = high part * 2^32 + low part* (2^{32} in JScript is expressed using the pow method, Math.pow(2,32).)

For example, if the high part= -18,105 and the low part = 382,894,080, then:

$$\text{assembled value} = (-18{,}105 * 2^{32}) + 382{,}894{,}080 = -77{,}760{,}000{,}000{,}000$$

The units of this value are in 100-nanosecond intervals, so the script converts the value to days using relationships of 100 100-nanosecond intervals per nanosecond, 1×10^{9} nanoseconds per second, and 86,400 seconds per day. Using the previous example, the original value converts to 90 days:

high part= -18,105

low part = 382,894,080,

*assembled value= -18,105 * 2^32 + 382,894,080 = -77,760,000,000,000*

iOneHundredNanoSeconds: -77,760,000,000,000

iNanoSeconds: -7,776,000,000,000,000

iSeconds: -7,776,000

iDays: -90

```
//-------------------------------------------------------------------
// Convert 8ByteInteger 100 Nanosecond Interval to Days
//-------------------------------------------------------------------
```

```
function ConvertNanoDays(objLargeInteger)
{
var iOneHundredNanoSeconds, iNanoSeconds, iSeconds, iDays;
iOneHundredNanoSeconds=objLargeInteger.HighPart * Math.pow(2,32) +
 objLargeInteger.LowPart;
iNanoSeconds=iOneHundredNanoSeconds * 100;
iSeconds=iNanoSeconds * Math.pow(10,-9);
iDays=iSeconds / 86400;
return(iDays);
}
```

CurrentDomain

The function `CurrentDomain` determines the domain of the user account running the script. It is explained in detail in Chapter 4.

```
//--------------------------------------------------------------------
// Discover the domain of the user context running the script
//--------------------------------------------------------------------
function CurrentDomain()
{
var objDomain=new Array()
var objRootDSE = GetObject("LDAP://RootDSE");
objDomain.winnt=objRootDSE.Get("dnsHostName");
objDomain.ldap=objRootDSE.Get("defaultNamingContext");
return objDomain;
}
```

QueryDSA

The `QueryDSA` function works much like the `QueryDS` function in Chapter 4, but instead of populating the results in a simple array, the `QueryDSA` function defines a new array and populates it with objects that each contain two properties: `ADsPath` and `name`. The array member is an object, and the loop through the `RecordSet` object explicitly defines the object properties. Recall an earlier example of how to define an object as a literal:

```
someObj= { name: "some name", value: someValue};
```

Similarly, the object can also be defined as an array element:

```
asomeArray[i]= { name: "some name", value: someValue};
```

The properties can be variables as well. QueryDSA assigns the properties ADsPath and name from the values of the RecordSet object data returned from the ADSI query. Once the object has been defined, additional properties can be added directly, such as the properties expired, pwdLastChanged, and other properties assigned in the main program.

```
//-------------------------------------------------------------------
// Query Active Directory given an LDAP query parameter - return Array
//-------------------------------------------------------------------
function QueryDSA(strCommandText)
{
var aAccounts = new Array();
objConnection = new ActiveXObject ("ADODB.Connection");
objConnection.Provider = "ADsDSOObject";
objConnection.Open;
objCommand = new ActiveXObject("ADODB.Command");
objCommand.ActiveConnection = objConnection;
objCommand.Properties("Page Size") = 999;
objCommand.Properties("Sort On") = "Name";
objCommand.CommandText=strCommandText;
objRecordset=objCommand.Execute();
j=1;
while (!objRecordset.EOF) {
    aAccounts[aAccounts.length] = {
     ADsPath : objRecordset.Fields.Item("ADsPath").Value,
     name : objRecordset.Fields.Item("sAMAccountName").Value};
    j++;
    objRecordset.MoveNext;
}
return aAccounts;
}
```

Heading, Usage, ParamError, Tabulate

The Heading, Usage, ParamError, and Tabulate functions are all carried over from Chapters 4 and 5, and details of their construction can be found in these previous discussions.

```
//-------------------------------------------------------------------
// Format headers in the output
//-------------------------------------------------------------------
function Heading(strMessage)
```

```
{
  WScript.Echo("\n---------------------------------------------------");
  WScript.Echo(strMessage);
  WScript.Echo("---------------------------------------------------");
}


//----------------------------------------------------------------------
// Display the help screen
//----------------------------------------------------------------------
function Usage()
{
  WScript.Echo("Tool to list user account date information");
  WScript.Echo("Usage:");
  WScript.Echo("DateQuery -cenu[m]v -D [Domain] " +
    "-d [delimiter] [criteria] -w [time window]:");
  WScript.Echo("-c list dates accounts were created");
  WScript.Echo("-e list dates accounts will expire");
  WScript.Echo("-n list accounts with passwords set to never expire");
  WScript.Echo("-l list locked out accounts");
  WScript.Echo("-u list user account settings");
  WScript.Echo("-um list user accounts settings in a matrix format");
  WScript.Echo("-v verbose mode");
  WScript.Echo("-D [domain] specify domain other than default");
  WScript.Echo("-d [delimiter] specify delimiter to separate fields.");
  WScript.Echo("-? Help");
  WScript.Echo("");
  WScript.Echo("Example:");
  WScript.Echo("DateQuery -c -w 14 : Displays accounts created in last");
  WScript.Echo(" 14 days.");
}


//----------------------------------------------------------------------
// Display An error if a dependant script parameter is not included
//----------------------------------------------------------------------
function ParamError()
{
  Heading("Parameter Error Received. Please check syntax.\n");
  Usage();
  WScript.Quit();
}


//----------------------------------------------------------------------
// Display output in left-aligned columns
//----------------------------------------------------------------------
```

```
function Tabulate(strData, iColumn, strDelimit)
{
while(WScript.StdOut.Column<iColumn)
    {WScript.StdOut.Write(strDelimit);}
WScript.StdOut.Write(strData);
}
```

SUMMARY

The *DateQuery* script described in this chapter completes the three ADSI-focused console scripts, and, as you can see, although each of the scripts focuses on a different task and even uses different techniques, fundamentally their designs follow a similar pattern.

DateQuery examines how to query, manipulate, and display both date-specific account information and user account properties. ADSI contains a wealth of information, and this script examines a few of the more complicated ADSI properties to extract, manipulate, and display. The techniques used for working with dates and large integers can be extended to other ADSI properties not used in this script. For example, the ADSI lastlogon property is also stored in a large integer format, and ADSI uses over 30 enumerations to hold other directory data. You can access them using similar techniques used to get data from ADS_USER_FLAG_ENUM.

KEY POINTS

- Most data in ADSI can be retrieved and displayed using straightforward properties and methods. However some AD data is stored in large integers or bitfields, which require extraction and manipulation to interpret the results.
- Decode the ADS_USER_FLAG_ENUM ADSI enumeration to access 21 properties of ADSI, such as whether the password will expire or if the account is disabled.
- Date-time math can be tricky; look for a method that converts a date into an integer (such as milliseconds since January 1, 1970), and then you can safely add or subtract appropriate units to or from that number and convert it back to a date.
- Many time values are stored in AD as large integers (eight bytes long). Add the converted high part to the low part, and then convert to your preferred units.

7

Move Away from Batch Files—A New Logon Script

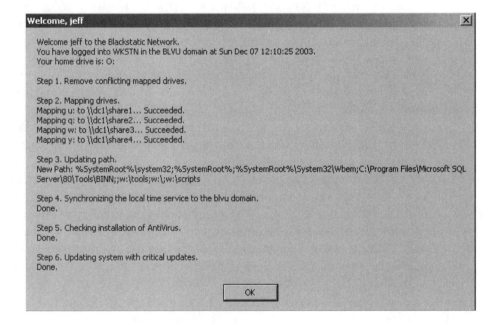

Welcome, jeff ×

Welcome jeff to the Blackstatic Network.
You have logged into WKSTN in the BLVU domain at Sun Dec 07 12:10:25 2003.
Your home drive is: O:

Step 1. Remove conflicting mapped drives.

Step 2. Mapping drives.
Mapping u: to \\dc1\share1... Succeeded.
Mapping q: to \\dc1\share2... Succeeded.
Mapping w: to \\dc1\share3... Succeeded.
Mapping y: to \\dc1\share4... Succeeded.

Step 3. Updating path.
New Path: %SystemRoot%\system32;%SystemRoot%;%SystemRoot%\System32\Wbem;C:\Program Files\Microsoft SQL Server\80\Tools\BINN;;w:\tools;w:\;w:\scripts

Step 4. Synchronizing the local time service to the blvu domain.
Done.

Step 5. Checking installation of AntiVirus.
Done.

Step 6. Updating system with critical updates.
Done.

[OK]

OVERVIEW

The network logon script reliably provides a centrally managed method for distributing common local and network settings to all clients logging in to your network. Utilitarian scripts map network shares to local drives, update environment variables, run program updates or patches, load critical software such as antivirus programs, set the local computer time, and a host of other functions. Scripts leaning towards the frivolous side greet the users and welcome them to the network and perhaps offer a bit of advice akin to a Chinese fortune cookie vis-à-vis the message of the day.

Even with Windows 2000 Group Policy and other migrations towards enterprise centralized-management operations, logon scripts continue to play a key role as a primary means of distributing arbitrary code or settings to many clients.

Logon scripts provide logic and structure around processing a user logon, and batch files that may have begun as a simple NET USE statement to map a drive have evolved into scripts that can tweak or report on much of the system's operation. Logon scripts provide myriad functionalities. For example, instead of repeatedly trying to patch a system, every time the user logs in, a smarter logon script first checks a registry setting or other source to see whether the update has already been applied. Consider a more elaborate logon script that not only maps drives but also performs a basic diagnostic on the computer, checks for running services, and warns the user (or system administrator) if disk space is low. The possibilities of how far you can extend a logon script are only limited by your imagination (and time, of course).

Popular enterprise programs, including Microsoft Systems Management Server® and Symantec Norton Anti-virus Corporate Edition®, leverage logon scripts as one of several installation mechanisms for their products. As another benefit, logon scripts run on any client logging in to the network—from Windows 3.1 to Windows Server 2003. Remember to code your logon script with recognition of what platforms might log in to your network or provide down-level scripts compatible with older clients. But if you run a Windows 2000 network or later, you not only will be able to take advantage of the procedures within this chapter but also will be well positioned to extend and utilize many of the latest technologies—such as WMI and ADSI—within your logon scripts.

TIP

The logon script runs before Explorer loads and control is handed over to the user. This means that an improperly configured logon script could hang your computer at logon. Be sure to thoroughly test your logon scripts and include error handling to reduce the risk of a hung script. Also, test your scripts in a test lab that permits you to edit a problematic script from a separate computer. If you configure a logon script via a local computer policy and your computer hangs (possibly meaning that you cannot log in to it at all), remember that the network connections for the computer are already active. Use a second computer, and remotely access the file system of the hung computer to remove or edit the offending script, then reboot (for example, for a Windows 2000 server, access the administrative share c$, and edit the script located in \\machinename\c$\winnt\system32\grouppolicy, and then edit either the directory User for logon/logoff scripts or Machine for startup/shutdown scripts).

SCENARIO

Having migrated from a predominantly Windows NT 4.0 domain, your company has recently begun to leverage AD and Windows 2000. As the systems administrator of a medium-sized engineering company, you look forward to more discrete control over your computing environment. You plan to implement group policy, disk quotas, and other Windows 2000 features. However, today you have a more immediate desire. You need to deploy a set of patches across various computers in your environment. Also, you have the feeling that updates will continue to roll out from the vendor, and you need a convenient method for updating the deployment mechanism to include the latest patches. Eventually, you will deploy Software Update Services for Windows, but for the time being you are looking for a quick method for deploying cross-vendor updates (not just for Microsoft products). You also want to check the registry of the computer before attempting to install a software update to see if it has been previously installed.

You have relied on a batch file logon script in the past, and now you wonder whether you can update your logon script to deploy the patches and provide new logon script functionality for your new Windows 2000 domain. Your current batch file logon script maps network drives, modifies the search path, sets the current time, and installs an antivirus software client. You want to recode this into JScript and increase its functionality. In addition to providing basic patch deployment services, you want track the actual login time and date of your network users and what machine they logged on and record this in a text file on a centralized share.

ANALYSIS

Converting your batch file logon script to a script authored in a Windows scripting language provides more control over the program's logic while also providing access to many of the Windows object models. At first you may think that a batch file logon script is simpler and shorter. Although this is perhaps true for basic scripts, both the code manageability and flexibility to accommodate future changes becomes much easier using JScript or VBScript.

Plus, you can usually convert batch file calls to external commands to a WSH script in fairly short order using the WSH exec or run method. These methods enable your script to fire off external applications—a common function of a logon script. Use these methods to call third-party or vendor-provided batch files or command-line utilities not supported by WSH. Your needs map into the following technical requirements listed in Table 7.1.

TABLE 7.1 Translating logon script objectives into technical requirements

Business objectives	Technical requirements
Deploy cross-vendor patches to all systems in your domain that do not already have the patch.	Check the registry for a specific update key signaling whether the update has already been installed. If it has not, install the update. Store all of the updates and their registry installation checks in an Extensible Markup Language (XML) file.
Map network drives.	Use WSH Network object to remove conflicting drive mappings, and add new drive mappings stored in an XML configuration file.
Modify the search path.	Read the desired path additions from an XML configuration file, and add them to the user path if they do not already exist.
Set current time.	Run the external command NET TIME to synchronize Windows NT 4.0 computers to a domain Network Time Protocol (NTP) server.
Install antivirus software.	Run an external command to check whether the company's antivirus software is installed.
Log login information and errors.	Create or append to a text file on a network share accessible by everyone, and update with the login time and any deployed updates.

These technical requirements map into the following functions:

Stub Program: Batch file launcher of the main JScript logon script.

- Centrally records the logon of the user.
- Ensures execution of main logon script using desired WSH engine.

Main Program: Contains the main code of the logon script.

- Builds the welcome message.
- Notifies the user of his home directory.
- Maps home and network drives.
- Extends the user's directory search path to include other common directories.
- Synchronizes the time to a Network Time Protocol (NTP) server.
- Calls the installation program for an antivirus software package.

`LogFile`

- Provides file-based logging service to network share or local folder.

`TabulateFile`

- Similar to previous tabulate functions, `TabulateFile` formats output to log text files.

`PatchSystem`

- Provides registry-checking service for deploying program updates and patches.

`EvalCriteria`

- Helper function for `PatchSystem`.

`checkregistry`

- Provides registry lookup functions for deploying updates and patches.

Distribution and Installation

Logon scripts can be assigned to domain or local computer users. Domain logon scripts can be specified in either AD Users and Computers or in a GPO. Specifying the logon script using AD Users and Computers, as seen in Figure 7.1, makes your logon script accessible to Windows NT 4.0 computers, and the scripts are stored in the `NETLOGON` share of the Windows 2000 domain controller, which physically maps to the domain controller `%Windir%\SYSVOL\sysvol\[domainname]\scripts` directory. This directory is replicated to all other domain controllers.

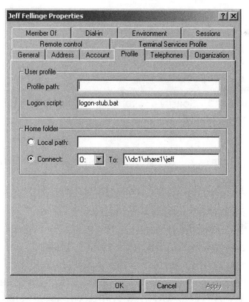

FIGURE 7.1 Individually specify the logon script for each user via the user's Profile tab in the AD Users and Computers MMC console.

Assign *local* user logon scripts by opening up the Computer Management applet on the target computer and navigating to Local Users and Groups and then Users. Double-click the user, click the Profile tab, and assign the name of the preferred logon script. The local logon script is typically stored in `%Windir%\System32\Repl\Import\Scripts`.

Using Group Policy to Deploy Logon Scripts

Logon (and logoff, startup, and shutdown) scripts can also be assigned via GPO, as seen in Figure 7.2, which allows you to set a particular logon script to a domain, site, OU, or anywhere else group policy is defined.

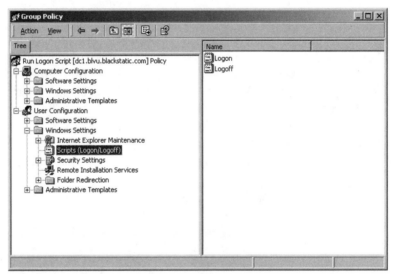

FIGURE 7.2   Specify a logon script to apply to a domain, site, or OU using AD group policy.

Keep in mind that the domain controller that authenticates a user to the domain also provides the user with the logon script. After you have copied a new logon script to your domain controllers, allow time for the script to replicate to the other domain controllers in your network.

Recall that Windows defaults to using the *wscript.exe* WSH engine, and some users may change this default to the console-based *cscript.exe* engine. As we have seen in several prior scripts, some scripting objects and methods are not available to both engines (for example, *wscript.exe* does not support standard streams used to input and output data). To ensure that your scripts run reliably on all systems in your environment, we specify the engine used to execute your scripts. To ensure that the script is run by the preferred engine every time, we use a logon-stub. The logon-stub is a batch file that calls the main logon script using the preferred engine and then exits.

This batch file need not be complex—possibly only a single line (although the actual *logon-stub.bat* file used in the following example provides a bit more functionality):

```
call cscript %0\..\logon-body.js
```

This command simply calls the JScript file *logon-body.js* using the *cscript.exe* engine. The prefix `%0\..\` provides that the directory path of *logon-body.js* is the same as that of the calling batch file (that is, the two files are contained in the same directory). The variable `%0` in this batch file contains the name and location of the batch file. So, if called from `\\dc1\NETLOGON`, then `%0` would contain: `\\dc1\NET-LOGON\logon-stub.bat`. The entire argument `%0\..\logon.js` expands to `\\dc1\netlogon\logon.bat\..\logon.js`. The parameter `..` refers to the object's parent, and so the parameter `\..\` effectively backs off the filename and returns only the path (from our previous example, `\\dc1\netlogon\`). Note that this is only the effective result, meaning that if you expand `%0\..\logon.js`, you see `\\dc1\netlogon\logon.bat\..\logon.js`. By appending the new batch or script filename, we can always ensure the exact location of this second file, and we can rest assured that any subsequent files called from this batch file will work if they exist in the same folder as the calling script. (Alternatively, if you don't care to keep the scripts on the domain controller that the user logs on to, you could alternatively hardcode the location of the second script to a network share accessible to all users.)

Operating System Compatibilities

Every device that logs on to your domain runs your logon script, which requires that you consider operating system and application version compatibility. For example, if someone logs on to your network with a computer running Windows NT 4.0, will it process your logon script? And if it doesn't, will it hang, crash, or exit gracefully? When coding your own scripts that must be deployed to a wide variety of systems, consider whether they recognize these technologies:

- Active Directory Group Policy—not recognized by Windows NT 4.0.
- ADSI—ADSI extensions must be installed on pre-Windows 2000 computers providing some (not all) compatibility.
- WMI—WMI Core Components can be installed on earlier operating systems, but even later Windows operating systems (such as Windows XP and Windows Server 2003) offer new classes not available on Windows 2000.
- WSH—Windows 2000 installs with WSH 5.1; current WSH version at 5.6.
- XML—Windows 2000 installs with MSXML 2.5. MSXML supports concurrent installations and is currently at version 4.0.

The logon script presented in this chapter has been designed for use in an AD-based domain running Windows 2000 computers and later; however, the scripts include error handling and software version checking to exit gracefully if a feature is not supported (although functionality beyond the error will be aborted). For example, if you choose to use AD group policy (instead of the Profiles tab in User Account Management), computers running Windows NT 4.0 will not process those scripts, but they should not lock up or otherwise adversely affect the system. The script requires both WSH 5.6 and MSXML 3.0 but does not use WMI or ADSI technologies.

Output

The logon script displays a dialog box welcoming the user to the network and displaying user-specific information. The user can close the box, or the box will close automatically after 30 seconds.

Summary of Solution

You will create two files, *logon-stub.bat* and *logon-body.js*, to process logon script functions for your domain users. The *logon-stub.bat* file is a MS-DOS batch file that simply calls the *logon-body.js* WSH script using the *cscript.exe* engine. The *logon-body.js* script file performs the real work of the logon script. Both of these files will be copied to the domain controllers NETLOGON share or specific group policy logon script folder, which will then be replicated to all other domain controllers in your domain.

Logon information and status and error logs will be recorded to a centralized location. This information keeps a record of who logged in, when, and on which computer.

OBJECTIVE

This script demonstrates using Windows scripting technologies for logon scripts to provide more functionality and flexibility than found in traditional MS-DOS logon scripts. The script uses the following objects:

Arrays
- `length`
- Looping through arrays

Error Trapping
- `try...catch`

MSXML
- `async`
- `getAttribute`

- getElementsByTagName
- item
- length
- load
- parseError
- resolveExternals
- selectSingleNode
- setProperty

RegExp

- match

Scripting.FileSystemObject

- OpenTextFile
- Column
- CreateTextFile
- WriteLine

WScript.Arguments

- Length

WScript.Network

- ComputerName
- ExpandEnvironmentalStrings
- UserDomain
- UserName

WScript.Shell

- Environment("Process")
- Environment("System")
- Environment("User")
- Run
- RegRead
- Popup

THE SCRIPT

The logon script contained in this chapter actually consists of two scripts and a configuration file: *logon-stub.bat*, *logon-body.js*, and *logon-config.xml*. The main purpose of *logon-stub.bat* is to ensure that *logon-body.js* is called using the appropriate WSH engine (*cscript.exe*), because the script *logon-body.js* uses methods and properties only available to this engine. This script introduces XML through the use of this configuration file and begins to describe some of the benefits that using XML brings to your scripts. XML continues to grow in popularity, and many new IT tools now support exporting data in XML format. Learning the basics of XML, even through a basic script such as this, will help you in the future with manipulating XML data from other programs using your own scripts. Before we look at the logon script, let's briefly introduce XML by examining the XML configuration file used by this script. Although this configuration file contains specific references to the logon script we have not yet covered, reviewing the function and form of the XML file sets the stage for later discussions in the actual logon script.

Why XML?

The logon script uses an XML data file for storing logon-script configuration information. XML is an ideal method for storing this data because it is self-describing (that is, it does not need a separate, external structure or definition like a database does, although you can choose to define an external structure, which is called a *schema*), and the objects and methods for reading and accessing the data lend themselves to handling various quantities of data. Plus, when using XML, you can define XML any way you wish, as long as it's *well formed*. Well-formed XML follows the XML standard—basically a set of rules. Some of these rules are described in the section *XML Configuration File*, which explains the layout of the configuration XML data file. XML lets you define your own data hierarchy. For example, in addition to configuration information, this script's XML data file defines what updates should be installed. Using XML, you can easily add new updates. Each update could have one or several associated criteria to evaluate whether an update is needed. In XML, adding new updates or criteria is as easy as adding new lines defining the new data.

Because XML is hierarchical in nature, you don't need to worry about coding elaborate parsing routines to handle new or special data as you might if you used a text file to store this data. The MSXML Document Object Model (DOM) provides a built-in set of methods and properties that generally take care of the parsing of the data for you. During your review of the following code, reference the actual XML configuration file and notice how the script hierarchically accesses the data. Understanding XML might be a bit confusing at first, but think about how you might approach this problem using a text file and `FileSystemObject` coding. You will agree that using XML can save you a lot of time and effort because it handles most of the data organization and parsing for you. Many utilities are beginning to support XML data output in addition to more traditional CSV, TSV, or other formatted text. Understanding how to manipulate XML data proves to be useful in manipulating and displaying the output from these applications. This script example covers loading data from a specific configuration file, but the concepts can be extended to parsing any type of XML data.

XML Configuration File

The XML data file provides a location to store configuration data used by the logon script that might change from domain to domain. XML data looks similar to HTML in that it uses tags (`<tag>`) to separate objects. The XML presented in this script is basic and does not include an XML schema. An XML schema describes the structure and rules by which the data must abide (for example, if data is mixed alphanumeric, its length, whether it must be unique, and so on). An explicitly defined XML schema is a good idea if you plan to work with larger or more complicated data sets or wish to provide clarity to others who may use your XML data.

The structure of the XML data is fairly straightforward and follows a basic set of rules. Elements are described with tags, such as `<element>`, and they have a beginning tag and ending tag. Ending tags are denoted with a forward slash, such as `</element>`. An element text is described between the tags—`<element>some data text</element>`—and does not require quotations. Attributes are additional name-value pairs assigned to an element. For example, in the XML element `<domain ntp="blvu" name="Blackstatic">`, the element name is domain and its two attributes are `ntp` and `name`, which have values of `blvu` and `Blackstatic`, respectively.

A single top node must be defined to encompass all other elements; in our logon script example, the top node is `<config>`. Beneath `<config>` are two elements, `<domain>` and `<updates>`. The `<domain>` element contains the configuration information for the logon script, including the location of the various logs, definitions for the home drive and mapped drives, and the path statements that should be added. You can have multiple, differently named elements in parallel, and you are generally free to assign attributes to any element. Although an explicit schema is not defined, if it were, it would permit only three logs: one of type `error`, one of type `logon`, and the last of type `patch`. It would permit only one home drive but any number of mapped drives and path elements. Now you may begin to see why using an XML configuration file really improves the flexibility of the script. To add another mapped drive, just copy and paste a `<mapdrive>` element and add the `drive` attribute and the share text, and the script automatically includes it the next time it runs.

The second part of the XML data file defines the updates that we want the logon script to deploy. The `<updates>` XML structure permits many `<patch>` elements that can have one or several `<criteria>` elements. Consider this flexibility when deploying one of the Microsoft product updates. For a given update bulletin, you might have multiple versions of the update—one for Windows NT and one for Windows XP/2000/2003. Next, you might need to check a particular product version. For example, Internet Explorer 5.01, 5.5, 6.0, and 6.0sp1 might each have separate updates. By adding more `<criteria>` elements you can add as many clauses as needed. In the sample update in the following XML code, one patch is defined with a `qfeid="sample"`. If the criteria is satisfied, then the logon script runs the file *applypatch.js* located on the *wkstn**patch* share. In this sample, only one criteria is defined. The script looks in the registry for a key HKLM\SOFTWARE\sample\Sample Patch for a value of 1. If the key exists and if it does have a value of 1, then, according to the installIfTrue attribute, the script will not install the patch.

```
<config>
<domain ntp="blvu" name="Blackstatic">
    <log type="error">\\wkstn\logon\logon-errors.txt</log>
    <log type="logon">\\wkstn\logon\logon-log.txt</log>
    <log type="patch">\\wkstn\logon\patch-log.txt</log>

    <homedrive drive="p:">\\dc1</homedrive>

    <mapdrive drive="u:">\\dc1\share1</mapdrive>
    <mapdrive drive="q:">\\dc1\share2</mapdrive>
    <mapdrive drive="w:">\\dc1\share3</mapdrive>
    <mapdrive drive="y:">\\dc1\share4</mapdrive>

    <path>w:\tools</path>
    <path>w:\</path>
    <path>w:\scripts</path>
</domain>
<updates>
    <patch qfeid="sample" patchfile="\\wkstn\patch\applypatch.js">
        <criteria regpath="HKEY_LOCAL_MACHINE\SOFTWARE\Sample\Sample
Patch" value="1" installIfTrue="no"/>
        <criteria regpath="HKEY_LOCAL_MACHINE\SOFTWARE\Sample\Version
Number" value="10.0" installIfTrue="yes"/>

    </patch>
</updates>
</config>
```

Launching the Stub

The simple batch file code used to launch the main WSH logon script follows.

The script consists of four actual lines of code. In the following text, the first two lines and last two lines are actually a single line but are shown on two lines because of page-width constraints of this text. They have been offset for readability but should be a single line in your code.

The first line writes the date, time, username, and computer name to a text file named *logon.txt* located in a central location on the network share *\\wkstn\logon* using the MS-DOS batch echo command and the redirect and append operator (>).

(You will want to change this share to reflect your own environment.) The next line displays the actual location running the script and is generally for information only. Knowing this location is useful when troubleshooting why an external command or batch file called from your script is not behaving properly. Sometimes a problem can be traced to a missing file or incorrect path. The third line calls the main logon script using the *cscript.exe* engine, as previously described. The argument to the *logon-body.js* script is an XML logon-script configuration file. The last line reports any errors from calling the *cscript.exe* logon script to the same *logon.txt* file.

```
echo %date%-%time% :: %USERNAME% logged in to %computername% >
\\wkstn\logon\logon.txt
echo - Running logon script from: %0 > \\wkstn\logon\logon.txt
call cscript %0\..\logon-body.js logon-config.xml
If Errorlevel 1 echo Error: Check WSH installation and script location
> \\wkstn\logon\logon.txt
```

ON THE CD

This script is presented on the CD-ROM as /chapter 7/scripts/logon-stub.bat. It is designed to work in conjunction with the second script of this chapter, logon-body.js, *which is also located in this folder on the CD-ROM: /chapter 7/scripts.logon-body.js. As previously noted, remember to test this script in a test environment before deploying to any production systems.*

Activating the Logon Script

After you have copied the scripts to the domain controller NETLOGON share or AD group policy logon script directory of your test domain, activate the logon script either through AD Users and Computers or AD Group Policy.

Activating the Logon Script at the User Level

Using AD Users and Computers to configure your logon script for each user is the most straightforward method—but also the most time consuming to set up as you must do this for every user in your domain. (Of course you should script it!) Navigate to the user object for which you wish to set the login script and double-click it to bring up its properties. Click the Profile tab, and type the name of the logon script, in this case *logon-stub.bat*, into the logon script field, as shown in Figure 7.1. Also, optionally enter information about the *home folder* for the user. We will cover the home folder and home drive in the section *Mapping Drives versus Home Drives*, which covers the logon script drive-mapping functionality. Copy any files called by the logon script into one of the domain controller's NETLOGON shares, and they will be replicated to the other domain controllers and are available to the logon script. That's it. Log out and back in to a member server (or domain controller), and the logon script should run automatically.

Activating and Using Active Directory Group Policy

Using group policy to configure and deploy logon scripts is a bit more complicated. You should understand and feel comfortable with group policy because changes here could affect some or all of the users within your domain. As with all of these scripts, deploy into a test domain first to ensure that all wrinkles have been ironed out of your scripts, including calls to external scripts, utilities, or programs.

NOTE

Applying a logon script to discrete users can be approached in a number of ways, and how you might implement something similar in your own environment will depend on your own AD design and policy. For example, we could create several OUs and apply custom GPOs to each. Or we could create a domain-wide GPO that defines a logon script policy for the entire domain and then block policy inheritance for an OU containing the excluded accounts. Another approach might be to create a domain-wide GPO called run logon *script and then set security permissions that deny access to a global group containing excluded user accounts. The possibilities for implementing the GPO defining the logon script vary widely and recommendations differ across AD implementations.*

In the following examples, we configure a logon script that applies only to user accounts of actual people. Other user accounts, such as those built in or service accounts, are excluded from this group policy. We do this by moving all people user accounts into a separate OU and then configuring a GPO specifically for that OU that runs the logon script.

Create a new GPO for the OU, and navigate to the User Configuration, Windows Settings, Scripts (Logon/Logoff) node, as seen previously in Figure 7.2. In the right pane, double-click the Logon node to bring up the Logon Properties dialog box. Click the Show Files button to open an Explorer window showing the location for the logon scripts for this specific GPO. The path looks something like `\\domain\sysVol\domain\Policies\GUID\User\Scripts\Logon`, where `domain` is the fully qualified domain name (FQDN) of your domain and the GUID is a Globally Unique Identifier for this GPO. (A GUID is a number guaranteed to be unique— for example {6AC1786C-016F-11D2-945F-00C04fB984F9}. Every GUID is unique, and yours will be different.) Copy the *logon-stub.bat, logon-body.js,* and *logon-config.xml* files into this folder, and make a note of its location, as shown in Figure 7.3. When you make future updates to your logon scripts, you must update the files in this location for them to take effect.

Return to the Logon Properties dialog box, and click the Add button. Click the Browse... button, and add the script *logon-stub.bat*, as seen in Figure 7.4. In this dialog box, you can add multiple logon scripts, but do not add the *logon-body.js* script because the *logon-stub.bat* file will call it directly. Click OK twice to exit the dialog boxes and close the Group Policy MMC.

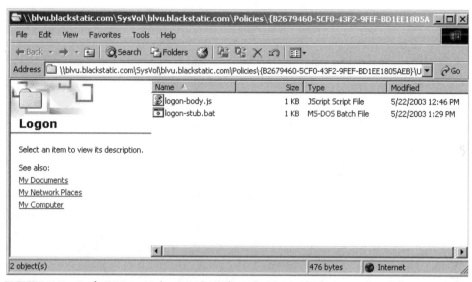

FIGURE 7.3   Each GPO contains a unique location into which to copy your logon script.

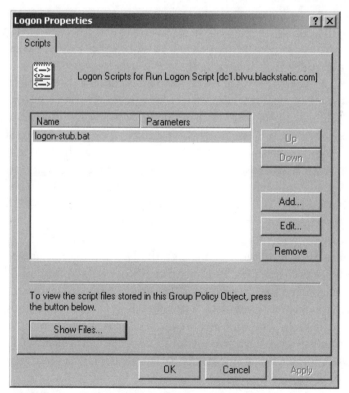

FIGURE 7.4   Specify the name of your logon script in the Logon Properties dialog box of the target GPO.

Next, refresh the machine policy on a member computer (either reboot or run *gpupdate* on Windows XP/Windows Server 2003 or *secedit /refreshpolicy machine_policy* on Windows 2000) to ensure the group policy has been applied. Log on as a network user, and verify that the share that you configured to write your logon information successfully recorded your logon and processed your script.

The Main Body–*logon-body.js*

The main logon script consists of several sections. The first section loads configuration data from an XML configuration file. The next section processes the steps of the logon script. The last section consists of the auxiliary functions used to support the script.

The logon script uses arrays of objects to keep track of information loaded from an XML configuration file. (The XML configuration file is described toward the end of this chapter after the body of the script.) The variable strStatusText provides a running tally of the progress of the updates and ultimately is displayed in a popup window. The arrays aDrives, aPath, and object objLog contain user-defined script data for mapping drives, adding path statements, and specifying the location of log files used during script processing. strNTPDomain contains the domain name used to synchronize the client to a NTP time service. iResult is used to assess whether an externally called application runs successfully via the Shell object Run command. The script uses iStartIndex to specify the aDrives index used to begin mapping drives, which corresponds to whether or not to include the home directory. strDomain, strUser, and strComputer refer to the Network object's properties that specify the domain, user, and computer running the logon script.

A logon script that errors out could hang the logon session and prevent the user from even logging in. For this reason, this script includes a large try...catch statement encompassing most of the script. Some of the functions also use this error handling. If an error occurs in the script, the catch statement tries to record the error to a central log file, which can be used in identifying possible problems. Another useful feature of JScript error handling is the ability for nested try...catch statements. This means that you can include a larger try...catch statement in your main program and also create smaller try...catch statements in your various functions called from the main program. The try...catch statements in the functions might handle the errors in a manner unique to the function, but you can optionally re-throw the error (using the throw statement), which will trigger the next higher try...catch statement. The following example shows how this nesting works.

The main program consists of a try..catch statement. The function MyFunction also uses its own try...catch statement. In MyFunction, an error is purposely thrown (using the throw statement). The MyFunction try...catch statement responds to the error using its own local error handling. However, it then re-throws the error. This triggers the main program's try...catch statement, which responds using its try...catch error handling routine.

```
try {
    MyFunction()
    }
catch(error) {
    WScript.Echo("Error triggered in main program: " + error)
    }

function MyFunction() {

try {
    throw "Error in my function"
    }
catch(errorInMyFunction) {
    WScript.Echo("Error triggered in My Function: " +
errorInMyFunction);
    throw errorInMyFunction
    }
}
```

The output of this example code, when saved as *trycatch.js* is:

```
C:\>trycatch
Error triggered in My Function: Error in my function
Error triggered in main program: Error in my function

C:\>
```

Next, the script creates two WScript objects, `Network` and `Shell`. The script uses the `Network` object to get the name of the domain, user, and computer running the logon script. Although not as robust as other technologies such as WMI or ADSI, using this object is more compatible with older systems, which may not have the latest technologies installed.

Technology selection becomes more and more important as the distribution of any script widens. This logon script runs on Windows NT 4.0 with the latest service packs, but it does not use ADSI or WMI technologies, which require additional software. If your later-generation environment consists of Windows XP, 2000 or Server 2003 computers, or if you have loaded WMI core components or ADSI Client Extensions, you could use these technologies in your logon script to provide even more functionality.

The script uses an XML file to store logon-script configuration information. Storing this data externally from the script makes initial or future updates to the configuration data easy. For example, to add additional mapped drives, add new `mapdrives` elements to the configuration file. The script requires the name of this configuration file as a command-line runtime argument. It parses the arguments using the

`WScript.Arguments(0)` and stores the name of this configuration script in the variable `strUpdateFile`. The script returns an error and exits if no configuration file is provided.

```
// logon script
var strStatusText = "";
var i;
var aDrives = new Array();
var aPath= new Array();
var objDrives=new Object;
var objPath= new Object;
var objLog= new Object;
var strNTPDomain;
var iResult;
var iStartIndex;
var strDomain, strUser, strComputer;
var strUpdateFile;

try {
objNet=new ActiveXObject("WScript.Network");
objShell= new ActiveXObject("WScript.Shell");

var colArgs=WScript.Arguments;
if(colArgs.length == 0) {
    iResult=objShell.Popup("Please specify the config XML file.",30,
     "Logon Script Initialization Error");
    WScript.Quit(0);
    }
strUpdateFile=WScript.Arguments(0);
```

This script uses the XML Document Object Model (DOM) to access and manipulate XML data. Another popular method for using XML data is through Extensible Stylesheet Language transformations, or XSLT. XSLT provides a mapping of XML data into most any other format. For example, an XSLT document might transform XML data into an HTML document. XSLT is a terrifically efficient means to convert data to a variety of formats.

Tap your script into XML using automation by instantiating a new MSXML object. Over the past few years, Microsoft has released many new versions of the Microsoft XML Core Services (MSXML), and compatibility varies between versions. To see what versions your system supports, you can look in %windir%\ system32 for *msxml*.dll* files. For example, if you have *msxml.dll* and *msxml3.dll*, then your system supports XML versions 1.x–2.x and 3.0 (The Microsoft Knowledge Base article 269238, *http://support.microsoft.com/default.aspx?scid=kb;en-us;269238*, maps MSXML version number to name and file version number of the

MSXML DLLs.) You can choose to run XML in two modes: side-by-side mode or replace mode. Running in side-by-side mode offers the most compatibility because different XML versions coexist on a single computer. Different versions support different modes, for example, MSXML 3.0 and earlier install in side-by-side mode, but MSXML 3.0 SP2 installs only in replace mode. MSXML 4.0 installs in side-by-side mode. But why does this matter? If you use XML in your scripts and you deploy these scripts to run on multiple machines (such as in the case of a logon script), you need to understand what versions of MSXML are installed on your target systems and tailor your script to use an appropriate version for compatibility.

To ensure your script's compatibility and stability when running on other systems, specify the version of XML to use when you instantiate the MSXML object. In the following code, the script creates an MSXML object using version 3.0. If you want to use new features of MSXML 4.0, simply change the code to `Msxml2.DOMDocument.4.0`. However, if you try to run this code on a system that does not have MSXML 4.0 installed, you receive a server automation error. Specifying the MSXML version like this is known as specifying a version-*dependent* progID. (The progID is the name of the object, such as e.g. `Msxml2.DOMDocument.4.0`.) Since MSXML 4.0, version-*independent* progIDs are no longer supported. For more information, search the Microsoft MSDN site; one good article on MSXML parser modes is Microsoft Knowledge Base article 303207 (*http://support.microsoft.com/default.aspx?scid=kb;en-us;303207*) and MSDN GUID and ProgID Information article (*http://msdn.microsoft.com/library/en-us/xmlsdk/htm/sdk_guidprogid_367i.asp.*)

The XML DOM describes the objects, methods, and properties used to manipulate XML data. After an XML file is loaded, it is parsed by an MSXML parser into a *DOM tree*. The DOM tree represents the data in a hierarchal format, accessible from your script. The data is contained in logical groupings, or nodes. The nodes contain the content and structure of the XML data.

NOTE

An alternative method to DOM for manipulating XML data is using the Simple API for XML (SAX or SAX2). This method provides direct control over the XML data. Whereas DOM parses the entire data set into a hierarchical DOM tree upon loading, SAX lets you control the loading and parsing of the object; for example, you could stop parsing after finding a specific value. Using SAX, your program manages the XML parsing. SAX is more memory efficient than DOM because it does not need to load the entire XML data into memory, but it does require more coding on your part. DOM provides a structure, objects, and methods for accessing data, performing complex searches, saving data, and transforming data using XSLT. SAX is not covered in this text.

The script defines two XML first-level properties before it loads the XML configuration file. The `async` property instructs the code to not permit downloading the XML asynchronously. This means that when the XML is loaded, control does not return to the script until the entire XML document has been loaded. Designed to work across the Internet and be related to other Web-friendly protocols such as HTML, XML supports external links for additional content or structure definition. This script does not rely on external links for information, so the second property, resolveExternals, is set to false.

The script loads and parses the XML file into the DOM using the load method and assigns the XML data to the XML document object xmlDoc. You can specify an XML file location by URL or file path. Check for any errors that may have occurred during the loading and parsing process by inspecting the parseError object. Using the DOM, the XML document is automatically parsed at load time for errors and accesses information about the errors via the parseError property. The parseError object of an XML DOM returns an object that, in turn, exposes other properties that describe the actual error. For example, parseError.reason describes the error, and parseError.line returns the line number of the error. parseError also includes errorCode, filepos, linepos, srcText, and url, which further describe the specific error.

A handy method of quickly validating your XML data file is to simply open it up in an XML-supported Web browser such as Internet Explorer. Any errors are displayed in the browser. If everything looks good, the browser shows you your XML file in a collapsible, hierarchical view. Internet Explorer 5.0 and later make terrific XML viewers.

```
var xmlDoc=new ActiveXObject("Msxml2.DOMDocument.3.0");
xmlDoc.async = false;
xmlDoc.resolveExternals = false;
xmlDoc.load(strUpdateFile);
if (xmlDoc.parseError.errorCode != 0) {
    iResult=objShell.Popup("XML Parse Error: " +
     xmlDoc.parseError.reason,60, "Logon Script Config Error");
    WScript.Quit(0);
    }
```

Now loaded, the XML data exists in a DOM tree and is accessible using DOM objects, methods, and properties. The tree comprises nodes, which represent different XML entities, such as elements, attributes, and text (among others). Data can be accessed from this tree using a variety of methods, such as iterating through the nodes or querying nodes for specific elements, attributes, or text. This script demonstrates several of these methods. Querying XML can be done using the *XML Path Language* (XPath) or the depreciated *XSLPattern* selection language. This script uses XPath, which provides a query syntax to access XML data by hierarchically referencing data and specifying pattern-matching criteria. Specify this language by setting the SelectionLanguage DOM property to XPath using the MSXML method `setProperty`. First-level properties can be assigned directly (such as the previous `xmlDoc.async=false` and `xmlDoc.resolveExternals = false`). However, second-level properties must be set using the `setProperty` method.

With the properties defined and the XML data loaded and parsed, we can begin extracting data from the configuration and populating variables used throughout the script. First, the script uses the `selectSingleNode` method to set `xmlConfig` to the node `/config/domain`, which the script references for subsequent data requests. This method returns the first node that matches the XPath query. As previously mentioned, the XPath syntax for accessing data is hierarchical in nature. The syntax `/config/domain` references any domain nodes located under the `config` node. The method `selectSingleNode("/config/domain")` returns the first node (a node object) matching the element named `domain`. Next, the script pulls the name of the domain that the client's time synchronizes to using both the `selectSingleNode` and `getAttribute` method. The method `.getAttribute("ntp")` returns the value of the attribute named `ntp`.

TIP

You can directly manipulate and display data using the method `selectSingleNode` *as it returns a node object (for example,* `selectSingleNode(XPath).xml` *or* `select-SingleNode(XPath).text`*). However, when working with node lists (containing multiple nodes), you must first iterate through the list and select single nodes before you can display the data. The methods* `selectNodes` *or* `getElementsByTagName` *return a node list. For example: WScript.Echo (xmlDoc.selectSingleNode("/config/domain").xml); displays the XML for a single node and its children. To display the XML of a node list, you can iterate through the list and display the individual nodes as you would other collections:*

```
objNodeList=xmlDoc.selectNodes("/config/domain");
for (i=0;i<objNodeList.length;i++)
  { WScript.echo(objNodeList.item(i).xml);}
```

Filtering a selection by a particular attribute still uses the `selectNodes` *method and still returns a node list that must be iterated, even if it has only one member:*

```
objNodeList=xmlDoc.selectNodes("/config/domain[@ntp='blvu']");
for (i=0;i<objNodeList.length;i++) {
WScript.echo(objNodeList.item(i).xml);}
```

The domain name attribute, `name`, is pulled using the same technique used to get the `ntp` domain member. The object `objLog` contains the path and filename for the three different logs used by the script. The script records errors to the error log, denoted by `objLog.error`. `objLog.logon` records logon information, and `objLog.patch` records patch update information. The locations for these logs are stored as XML data in the `xmlDoc` object. As described in the previous tip, this code uses the method `selectSingleNode` and an XPath attribute filter to return the text of a single node. Even though multiple log elements are available, there should be only one log element that also has an attribute type of `error`, `logon`, or `patch`. This code sets the `objLog` object to the text of the specified log element.

```
xmlDoc.setProperty("SelectionLanguage", "XPath");
var xmlConfig = xmlDoc.selectSingleNode("/config/domain");
strNTPDomain= xmlConfig.getAttribute("ntp");
strNetwork= xmlConfig.getAttribute("name");
objLog.error =
xmlConfig.selectSingleNode("log[@type='error']").text;
objLog.logon =
xmlConfig.selectSingleNode("log[@type='logon']").text;
objLog.patch =
xmlConfig.selectSingleNode("log[@type='patch']").text;
```

Learning to navigate the data hierarchy dramatically aids your XML programming. When possible, create new subsets of data to simplify programming. For example, consider the following XML:

```
<objects>
    <car>
        <name>Boxster</name>
        <color>Black</color>
    </car>

    <airplane>
        <name>Spitfire</name>
        <era>WW2</era>
    </airplane>
</objects>
```

The procedure:

```
objNodeList=xmlDoc.getElementsByTagName("name");
for (i=0;i<objNodeList.length;i++)
  { WScript.echo(objNodeList.item(i).xml);}
```

displays the following data:

```
<name>Boxster</name>
<name>Spitfire</name>
```

Notice that it returned any (and all) matching elements—even across sibling nodes. To return just the elements from the car node, use the following code. (To accommodate multiple cars, this example uses the `selectNodes` method.)

```
objNodeList=xmlDoc.selectNodes("/objects/car");
    for (i=0;i<objNodeList.length;i++) {
WScript.echo(objNodeList.item(i).xml);}
```

If you want to return just the names of the cars, this slightly inefficient code does the trick. First, we get a subset of nodes from /objects/car. *Then we get a second node list containing all tags named* name *from which we finally print the XML of the resulting nodes.*

```
objNodeList=xmlDoc.selectNodes("/objects/car");
for (i=0;i<objNodeList.length;i++) {
    objNodeList2=objNodeList.item(i).getElementsByTagName("name");
    for (j=0;j<objNodeList2.length;j++) {
    WScript.Echo(objNodeList2.item(j).xml)
    }
}
```

A cleaner approach is to simply dive straight into /objects/car/name, *as the following code illustrates:*

```
objNodeList=xmlDoc.selectNodes("/objects/car/name");
for (i=0;i<objNodeList.length;i++)
  { WScript.Echo(objNodeList.item(i).xml) }
```

These examples demonstrate how hierarchy and basic node-searching methods can be combined, and results can vary depending where in the DOM tree you begin.

Mapping Drives versus Home Drives

This logon script differentiates the home drive from a generic mapped drive. The script associates the home drive to the Windows *Home folder*, as defined on a user object's Profile tab, as seen in Figure 7.1. The AD home folder offers characteristics distinct from a typical mapped drive. For example, when you launch a command shell, it defaults to your home folder. Also, when you define a home folder for a user using AD Users and Computers, this tool automatically creates the folder and sets the appropriate permissions to restrict access to that particular user. The home folder is represented in three parts: the homeDrive (p:), homeShare (\\dc1\users), and homePath (\username). This separation of share and path permits a Windows NT 4.0 computer to benefit from a home drive, even though Windows NT 4.0 does not support mapping a drive to a share and path.

Because of these distinctions, the logon script differentiates the home drive from any mapped drives. Because only one home drive exists, the script uses the select-SingleNode method to populate the two properties of the objDrives object stored as the first element of the aDrives array. The script sets the properties objDrives.drive and objDrives.share and then sets this to aDrive[0]. These properties are set as literals using the following nomenclature: objDrives = { drive: "drive letter" share: "\\server\share" }. The *drive letter* is an attribute of the <homedrive> element, and the *server**share* is the text property of the <homedrive> element.

TIP

This script explicitly sets the properties of an object, then assigns the object as a member of an array. Optionally, you can take a shortcut method of implicitly making this assignment. For example, instead of setting obj.property1 and obj.property2 and then setting aArray[x]=obj, you could set aArray[x]. property1 and aArray[x].property2.

```
//homedirectory
objDrives={
 drive: xmlConfig.selectSingleNode("homedrive")
  .getAttribute("drive"),
 share:xmlConfig.selectSingleNode("homedrive").text};
aDrives[0]=objDrives;
```

The script reads the mapped drives from the XML data file and populates the aDrives array in a similar fashion; however, because any number of mapped drives can exist (and not just one), the script uses the method getElementsByTagName to return a NodeList object containing all mapdrive nodes. You may notice that the getElementsByTagName method is used with the xmlDoc object, and it returns any and all elements named mapdrive, no matter where in the hierarchy they might reside. Because we control the XML data we absolutely know that there can be no other <mapdrive> elements in the XML.

The script sets the objDrives.drive property to the <mapdrive> attribute drive, and the objDrives.share property to the <mapdrive> element text. It then sets the array member aDrives[i+1] to objDrives. The getElementsByTagName begins at index 0, whereas the aDrives begins at 1 to allow for the addition of the home drive, if needed.

```
//other mapped drives
objMapDrive = xmlDoc.getElementsByTagName("mapdrive");
for(i=0;i<objMapDrive.length;i++){
    objDrives={drive: objMapDrive.item(i).getAttribute("drive"),
     share: objMapDrive.item(i).text};
    aDrives[i+1]=objDrives;
}
```

Setting the Path

The logon script also extends the path for the user, ensuring that no matter what computer they log on to, their path statement will include the paths specified in the XML configuration file. The additional paths are extracted from the xmlDoc object using the getElementsByTagName method, parsed, and then stored in the aPath array.

```
objPath = xmlDoc.getElementsByTagName("path");
for(i=0;i<objPath.length;i++){
    aPath[i]=objPath.item(i).text;
}
```

The Shell object provides access to the WshEnvironment object environment variables. Windows supports three types of environment variable: System, User, and Process. This logon script uses the System and User variables to retrieve the current path for the user, and it uses the Process type to get information about the status of the home folder.

The strStatusText variable tracks the progress of the logon script and is ultimately displayed to the user. The first text of this message welcomes the user to the network and informs them of what domain and computer they logged on to.

```
strDomain=objNet.UserDomain;
strUser=objNet.UserName;
strComputer=objNet.ComputerName;
objEnvSys=objShell.Environment("System");
objEnvUsr=objShell.Environment("User");
objEnvPro=objShell.Environment("Process");
strStatusText+="Welcome " + strUser + " to the " +
  strNetwork + " Network.\n";
strStatusText+="You have logged into " + strComputer +
                " in the " + strDomain +
                " domain at " + Date() + ".\n";
```

Previously, we mentioned some the unique characteristics of a user's home folder. Using ADSI, this script could set the homePath, homeShare, and homeDrive properties for the user, which effectively configures the home folder. However, simply setting these properties does not create the folder or set its appropriate permissions as the AD Users and Computers tool does. The script could do that, too, but it would begin to encumber the script and take it off track from its primary function of initializing the environment for the user.

However, the script does check to see whether or not a home folder is already set by looking for the presence of the HOMESHARE environment variable. This variable only exists if the home folder has already been specified. If a home folder has been specified, then the script displays the letter of that drive and skips processing the XML-specified `<homedrive>`. It does this by setting the `iStartIndex` variable to `1`. `iStartIndex` is used as the starting point when iterating through the mapped drive `aDrives` array. (Remember that `aDrives[0]` holds data about the `homedrive` and setting `iStartIndex` to `1` skips this element of the array.)

TIP

Another caveat for Windows NT 4.0 users: Windows NT 4.0 does not support mapping a drive letter to a share and path (for example, \\server\share\path). This means that you must specify your `homedrive` as a server (for example, \\server). The script then appends the username to this (for example, \\server\username). You need to share out every user's home directory as a share itself. Windows 2000 permits mapping to a \\server\share\path, and you can specify a `homedrive` as \\server\share. The script appends the username (for example, \\server\share\username), but this will be accessible so long as \\server\share, and under that share the path \\username, exists.

If the `HOMESHARE` environment variable does not exist, then the script concatenates the username to the specified `homedrive` value and sets it to the `aDrive[0].share` property. For computers running Windows 2000 and later, this method provides a mapped drive to a user-customized share and path.

TIP

The home folder supports three properties: drive, share, and path, which are compatible with Windows NT 4.0 as well. This characteristic represents another benefit of using AD Users and Computers to specify a user home folder because it maps a drive to \\server\share but will also set the environment variable `homePath` to \\ username. Thus, when you launch a command shell, it opens to drive:\\username.

```
//Determine the users home directory
if(!objEnvPro("HOMESHARE")) {
        objDrives.share+= "\\" + strUser;
        aDrives[0]=objDrives;
        strStatusText+="No AD profile home directory set. " +
         "Your locally mapped home directory is: " + objDrives.share;
        iStartIndex=0;
        }
else {
    strStatusText+="Your home drive is: " + objEnvPro("HOMEDRIVE");
    iStartIndex=1;
    }
```

Step 1. Remove Conflicting Mapped Drives

Before mapping drives to those specified in the XML configuration, the script un-maps any existing drives that may conflict. The script loops through each of the currently mapped drives and compares them to the drives that the script plans to map. The script uses the `Network` object `EnumNetworkDrives` method. `EnumNetworkDrives` returns an array whose members contain alternating drive and share mappings. For example, `aMappedDrives[0]` equals the drive letter of the first mapped drive, and `aMappedDrives[1]` equals the mapped *server**share* of that same drive. For this reason, the iteration through the existing mapped drives steps through by 2.

The script uses the regular expression `match` method to compare the existing drive mapping to a proposed drive mapping. The script creates a new `RegExp` object that contains both the regular expression (in this case, the proposed drive letter) and the parameter `i`, which specifies case insensitivity. If the XML configuration data contained either `h:` or `H:` for a proposed drive map, the script would match it against an existing drive map of `H:`. Performing this comparison and removing only conflicting drives ensures that other drive mappings that the user may have previously configured remain.

The script removes any matching drives using the `Network` object `RemoveNetworkDrive` method. This method accepts two Boolean parameters. The first parameter, `bForce`, instructs the method to force the removal of the mapped drive. The second parameter determines if the method should remove the drive mapping from the user profile. This script sets both to `true`; the logon script removes a previous mapping and also removes the mapping from the user's profile.

```
//Unmap conflicting drives
aMappedDrives=objNet.EnumNetworkDrives();
strStatusText+="\n\nStep 1. Remove conflicting mapped drives.\n";
for (e=0;e<aMappedDrives.length-1;e=e+2) {
    for(i=iStartIndex;i<aDrives.length;i++) {
        objDrives=aDrives[i];
        var reDrive= new RegExp(objDrives.drive ,"i");
        if(aMappedDrives.item(e).match(reDrive)) {
            strStatusText+=objDrives.drive + " is mapped to " +
             aMappedDrives.item(e+1) +". Removing drive.\n";
            objNet.RemoveNetworkDrive(objDrives.drive,true,true);
        }
    }
}
```

Step 2. Mapping Drives

The mapping of the actual drives uses the `WScript.Network` `MapNetworkDrive` method, which requires the drive letter and share to be mapped as input parameters. Because any conflicting drives were previously removed, the script does not need to check for possible conflicts. However, the script does check first to see if the share exists using the `FileSystemObject` `FolderExists` method. This problem might occur if the XML data file is misconfigured or if the server hosting the share is inaccessible. Even though the earlier `try…catch` statement would have handled any errors resulting from trying to map a drive to a nonexistent share, using the `FolderExists` method provides a more elegant method to report the possible misconfiguration or problem and keeps the script running.

```
//Map Drives
strStatusText+="\nStep 2. Mapping drives.\n";
for(i=iStartIndex;i<aDrives.length;i++) {
    objDrives=aDrives[i];
    strStatusText+="Mapping " + objDrives.drive + " to " +
     objDrives.share;
    var objFSO = new ActiveXObject("Scripting.FileSystemObject");
    if (objFSO.FolderExists(objDrives.share)) {
            objNet.MapNetworkDrive(objDrives.drive,objDrives.share);
            strStatusText+="... Succeeded.\n";
        }
    else {
        strStatusText+="... Failed.\n"
        LogFile(objLog.error, strComputer, strUser,
            "Mapped Drive not Accessible: " + objDrives.share);
        }
    }
```

Step 3. Updating the Path

The logon script also adds any number of additional paths to the user's path statement. The path statement consists of two parts—the *system* and the *user* path. Both of these parts are stored as environment variables. (You can manually set the path by editing the environment variables in the System applet on the Advanced tab, as seen in Figure 7.5.) You can also set the path manually via the command prompt by simply typing `path=%path%;c:\some\new\path`. However, this persists only in that command shell session.

FIGURE 7.5 Manually edit the path user or system variable from the System applet. (Windows 2000 is shown; Windows XP and 2003 differ slightly.)

As seen in the previous example, adding to the path can be cumulative, meaning that if path=something and you want to add something, the new path redundantly equals path=something; something. (For example, if path= c:\windows\system32; c:\windows, and you add c:\windows to it, the new path will equal path=c:\perl\bin\; c:\windows\system32;c:\windows;c:\windows.) To work around these redundancies, the script uses the IndexOf method to compare proposed path additions to the existing path. Although path statements are not case-sensitive, IndexOf method does distinguish case, which means that c:\WINDOWS would not be found in the string c:\windows. However, we desire these to match so the script converts both strings to lowercase using the toLowerCase() method.

Be careful when working with backslashes in JScript. In Windows, paths are typically represented with backslashes (\\), such as c:\\windows. However, the backslash is used as an escape character in JScript, and so to represent an actual backslash, you must use two backslashes (\\\\). Depending on how you manipulate data in JScript that rely on backslashes, such as a path or URL, watch for how JScript interprets your \\; you may need to use two.

Don't forget about the backslash when troubleshooting some problems in JScript. Consider the following code:

```
strBackSlash="c:\test";
reOne=new RegExp(strBackSlash);
WScript.Echo(reOne);
```

JScript interprets the \\t in c:\\test as a tab character and displays:

> */c: est/*

The code that displays this correctly includes double backslashes, as follows:

```
strBackSlash="c:\\test";
reOne=new RegExp(strBackSlash);
WScript.Echo(reOne);
```

which displays:

> */c:\test/*

The script loops all of the proposed path additions contained in the array aPath and compares each to the existing path. If there is no match, the script adds the new path directly to the user environment variable, objEnvSys("Path").

```
//Modify the Path
strStatusText+="\nStep 3. Updating path.\n";
strCurrentPath=objEnvSys("Path") + objEnvUsr("Path");
for(i=0;i<aPath.length;i++) {
   objPath=aPath[i];
    if(strCurrentPath.toLowerCase()
     .indexOf(objPath.toLowerCase()) < 0) {
    strStatusText+= "Adding to path: " + objPath + "\n";
        objEnvUsr("Path")=objEnvUsr("Path") + ";" + objPath;
    }
}
strStatusText+="New Path: " + objEnvSys("Path") + ";"
 + objEnvUsr("Path") + "\n";
```

Step 4. Synchronizing the Local Time Service

The logon script attempts to synchronize the client computer's time with a domain Simple Network Time Protocol (SNTP) server. Whereas Windows 2000 and later computers support the time synchronization service (W32Time), no direct method is available for synchronizing the time programmatically with script for Windows NT 4.0 computers, so the logon script simply runs a command shell and executes the net utility. The parameter to set time to the domain is net time /domain:some-domain /set /yes. This sets the time to the domain somedomain and does not inter-actively prompt you if you want to change the time. Extend or customize this example using net time to any other utility that you wish to run at logon.

The Shell object Run method calls the cmd command to open a command shell from which the net time command is issued. When launching the command shell, the /C parameter instructs the command shell to execute the command specified in the string and then terminate. WSH runs the cmd command, which executes the net time command. When it completes, the cmd window closes. Note that the termination of the command shell differs from the termination of the Shell object Run method.

Two parameters govern the execution of the Run command. The first parameter defines the window style in which you can hide the window (0), activate and display the window (1), or select from up to 10 styles. The second Boolean parameter, bWaitOnReturn, instructs the script to wait until the command is terminated or con-tinue processing. If true, the script waits until, in this case, the cmd window is termi-nated. If false, the logon script immediately continues processing regardless of the outcome of the spawned command shell. In our code, the Run method activates and displays the window and does not wait for it to exit before resuming processing.

The most conservative approach to working with external applications in your logon script is to display the window but allow the script to continue processing. Doing so ensures that the logon script does not hang because of some hidden prompt, script, or batch file that paused for a command or acknowledgment. For these reasons, the logon script consistently displays a window and never waits for the completion of a command for it to resume processing. There may be times that you need the script to wait for the completion of an external command, such as if the script depends on the results of that command, but in these cases, be careful and test your scripts before deploying to many users—especially in a logon script.

```
//Set the client to the domain time server
strStatusText+="\nStep 4. Synchronizing the local time service " +
  "to the " + strNTPDomain + " domain.\n";
iResult=objShell.Run("cmd /C net time /domain:" +
 strNTPDomain + " /set /yes",1,false);
if(iResult) {
    strStatusText+= "Error synchronizing time service.\n";
```

```
        LogFile(objLog.error, strComputer, strUser, "NTP Error");
        }
else
    { strStatusText+= "Done.\n"; }
```

Step 5. Checking Installation of an Antivirus Program

Similar to the running of the net time command, the logon script demonstrates shelling to a command prompt to run some other batch file; in this example, a fictitious antivirus installation batch file. The antivirus installation batch file is designed to run every time the user logs in and checks to install or upgrade the user's antivirus software program. Like the previous use of the Run method, this code displays the command shell that executes the batch file in a window but does not wait for it to complete before continuing with additional processing.

```
//Make sure Antivirus software is installed
strStatusText+="\nStep 5. Checking installation of AntiVirus.\n";
iResult=objShell.Run("cmd /C \\\\dc1\\share1\\avinstall.bat",1,false);
if(iResult) {
    strStatusText+= "Error configuring Antivirus software\n";
    LogFile(objLog.error, strComputer, strUser, "AV Install Error");
    }
else
    { strStatusText+="Done.\n" }
```

Step 6. Updating System with Critical Updates

The last main functionality of the logon script offers rudimentary patch-installing services. Although basic, the code presented here may give you an idea of how to use XML along with Windows scripting techniques to assess a machine and take action based on observed results. The functionality of this code is limited to checking the registry for specified values to determine whether a patch has been installed or should be installed; imagine extending this processing to report on the service pack level, installed applications, or anything else for which you might want a regular report.

The script calls the function patchsystem to initiate the patch management features. The function requires the location of the patch log, name of the computer and user, and the XML object containing the patch management data.

```
//Check for and Install Patches
strStatusText+="\nStep 6. Updating system with critical updates.\n";
PatchSystem(objLog.patch, strComputer, strUser, xmlDoc);
strStatusText+= "Done.\n";
```

Displaying the Results

At its conclusion, the logon script calls the Shell object PopUp method to display the summary of the actions taken by the script, which are contained in the string strStatusText. The title of the message box welcomes the user by name, and the main box contains the status text. The dialog box remains for 30 seconds before automatically closing, which gives the users time to look at it but does not force them to manually close it every time they log in to the domain.

At the end of the main body, the script also completes the initial try...catch error handling. If an error occurs while running the main body of the script, the code execution continues at the catch statement, and the error is stored in the error variable, which represents an error object. The script tries to write the original error to the error log and also displays the error to the user logging in to the system via the PopUp method before it quits. The error object supports four properties: description, method, name, and number. The properties description and message display the error in the most readable terms (for example, one error message might read, *"The object does not support this method or property"*). The name property contains the name of the error (such as the name *"TypeError"*).

NOTE

Typically you will look up the error number in a development SDK or on the Internet via its decimal or hexadecimal value. The error object number property is a 32-bit number that contains both the facility (upper 16 bits) and error code (lower 16 bits). Use a bitwise AND (in JScript, this is the & character) to extract the components. For example:

```
error.number>16 & 0x1FFF  = facility
error.number & 0xFFFF  = error code
```

The facility represents a number category for the error, and the error code is the actual error code number. Using these properties you can extract useful information from errors caught by try... catch.

```
iResult=objShell.Popup(strStatusText,30,"Welcome, " + objNet.UserName);


} //end of try Statement
//bail gracefully
catch(error) {
    LogFile(objLog.error, strComputer, strUser,
      "General Script Error: " + error + " : " + error.description);
    iResult=objShell.Popup("General Script Error: " + error +
      "\nName: " + error.name +
      "\nDescription: " + error.description, 30, "Logon Script Error");
    WScript.Quit(0);
    }
```

LogFile

The function LogFile provides file-level logging for the logon script. This generic logging function accepts the location and filename of the log, as well as the computer and username and the message that should be displayed in the log. A try…catch statement handles errors in case the user-specified log file cannot be created (such as if the log file location path is bad). Specify the location of the log files via the XML data file either on a local server or a remote network share. Using a network share makes centralized logging to a single file both useful and convenient. The function creates a new FileSystemObject object and creates a new or opens an existing text file using the OpenTextFile method for appending. The last Boolean parameter in the OpenTextFile method set to true instructs the method to create a new file if one does not already exist.

Similar to the Tabulate function used in the scripts in Chapters 4, 5, and 6, this script uses a modified version to output columnar-formatted text to a text file instead of to the console window. The methodology is the same; however, instead of using the console WScript.StdOut method, it writes the data to the file object (objFile).

```
//--------------------------------------------------------------------
// Write to the Event Log
//--------------------------------------------------------------------
function LogFile(strLog, strComputer, strUser, strMessage)
{
try {
var FOR_APPENDING = 8;
objFSO=new ActiveXObject("Scripting.FileSystemObject");
objFile=objFSO.OpenTextFile(strLog, FOR_APPENDING,true);
TabulateFile(Date(), 0, " ", objFile);
TabulateFile(strComputer, 30, " ", objFile);
TabulateFile(strUser, 45, " ", objFile);
TabulateFile(strMessage, 60, " ", objFile);
objFile.WriteLine();
objFile.close();
} //end of try statement
catch(error) {
    iResult=objShell.Popup("Error Writing Logfile: " + error +
        "\nName: " + error.name +
        "\nDescription: " + error.description,30, "Logon Script
Error");
    }
}
```

TabulateFile

```
//------------------------------------------------------------------
// Write text file output in left-aligned columns
//------------------------------------------------------------------
function TabulateFile(strData, iColumn, strDelimit, objFile)
{
while(objFile.Column<iColumn)
    {objFile.Write(strDelimit);}
objFile.Write(strData);
}
```

Checking the Registry and Patching Systems

The patching procedures consist of three functions: PatchSystem, EvalCriteria, and CheckRegistry, which work together to loop through each of the patches and evaluate whether the specified criteria matches in the system registry. The functions PatchSystem and EvalCriteria set up and iterate through the different patches and criteria defined in the XML configuration file. PatchSystem loops through each of the <patch> elements and creates a new NodeList comprising all criteria for that specific patch. The script sends this NodeList to EvalCriteria, which iterates through each of the criteria for that particular patch. The criteria consists of a registry key and value as well as instructions of what to do if the key matches the value or not. The actual registry lookup and comparison is handled by the CheckRegistry function, which returns a true or false, depending if the key exists and matches the specified value. (Consider extensions to this script, which might include an additional function that checks for the existence of a file—named, say, *checkFile*. New criteria could be added to include a filename and location for the file in question. The XML data would be easy to extend: just add a new element <criteria-file>, and look for it, or add a new attribute to <criteria> that defines a criteria type of either registry or file.)

PatchSystem

The function PatchSystem requires as its parameters the log file name and location, the computer and username, and the XML object (xmlDoc) that contains the patch data. Like previous functions that handle possible user-configured data, all of the PatchSystem code uses a try...catch system for handling errors.

The function first constructs a NodeList of all of the <patch> elements using the method getElementsByTagName and stores it in objPatches. This process essentially trims the XML object of the configuration information that we used previously and provides a subset of patch-related data. The script then loops through and processes each of these nodes one at a time. For each individual <patch>, the script makes a second getElementsByTagName call to pull the <criteria> and stores this in the NodeList objCriteria. The script then calls the function EvalCriteria via a conditional. This means that the script sends objCriteria to the EvalCriteria for additional processing. The function EvalCriteria returns a Boolean value (true or false), and the PatchSystem function installs the patch or not.

If EvalCriteria returns true, then the patch is applied. If a system needs the patch applied, the script calls the getAttribute method to retrieve the path and filename of the patch stored in the attribute named patchfile from the current objPatches item. The function uses the Shell object Run method, as was previously used to synchronize the network time and launch the antivirus installation software, to apply the patch. Lastly, the log file is updated with a line of code designating that the particular patch was applied.

If the function EvalCriteria returns false, then the patch does not need to be applied, and the log file is updated simply stating this.

```
//------------------------------------------------------------------
// Patch the system
//------------------------------------------------------------------
function PatchSystem(strLog,strComputer,strUser,xmlDoc)
{
try {
i=0;
objPatches = xmlDoc.getElementsByTagName("patch");
for(i=0;i<objPatches.length;i++){
    objCriteria = objPatches.item(i).getElementsByTagName("criteria");
    if(EvalCriteria(objCriteria)) {
        strPatchFile=objPatches.item(i).getAttribute("patchfile");
        iResult=objShell.Run("cmd /C " + strPatchFile,1,false);
        LogFile(strLog, strComputer, strUser, "-Applied: " +
         objPatches.item(i).getAttribute("qfeid"));
        }
    else {
        LogFile(strLog, strComputer, strUser, " Need not apply: " +
         objPatches.item(i).getAttribute("qfeid"));
        }
    }
} // end of try statement
catch(error) {
    LogFile(objLog.error, strComputer, strUser,
     "Patch System Error: " + error + " : " + error.description);
        iResult=objShell.Popup("Patch System Error: " + error +
    "\nName: " + error.name +
    "\nDescription: " + error.description,30, "Logon Script Error");
```

```
        WScript.Quit(0);
    }
}
```

EvalCriteria

The `EvalCriteria` function uses a `NodeList` consisting of `<criteria>` element tags as preprocessed by `PatchSystem`. The criteria tags contain three attributes: `regpath`, `value`, and `installIfTrue`. For each `<criteria>` element, the function checks if the specified `regpath` equals the specified `value` for the system running the logon script.

The variable `bEval` represents what to do if the criteria matches or not and translates from the `installIfTrue` attribute. For example, suppose a `<criteria>` element checks whether the computer is running Windows 2000 by examining a product version registry key for a value of version 5.0 (which represents Windows 2000). If it finds this key and value, then the criteria is true, and the patch would apply. Therefore, `installIfTrue` should be `yes`, which translates using the ternary operator to a `bEval` value of `true`.

Contrast this example with a `<criteria>` check of whether a particular update has been applied. Often (not always), a Microsoft update records a successful installation in the registry. In this case, if the registry check is true, then the patch has already been installed. We would not want to reinstall it, so we would set `installIfTrue` to `no`, which translates to a `bEval` of `false`. Notice that the registry entry exists for both of these examples; however, one indicates that the patch is needed or applicable, and the other indicates it is not required.

To actually test the registry values, the function calls the `CheckRegistry` function and sends the registry key and value to confirm. The results of this test are also Boolean, and they are compared to `bEval`. If the result equals `bEval`, then the function returns `true`, indicating to the calling function `PatchSystem` that the patch should be installed.

```
//-------------------------------------------------------------------
// Check the criteria for any matches
//-------------------------------------------------------------------
function EvalCriteria(objCriteria)
{
var j;
var objShell=new ActiveXObject("WScript.Network");
for(j=0;j<objCriteria.length;j++) {
    var strRegPath=objCriteria.item(j).getAttribute("regpath");
    var strValue=objCriteria.item(j).getAttribute("value");
    var bEval=(objCriteria.item(j)
      .getAttribute("installIfTrue")=="yes"?true:false);
```

```
        var bResult=CheckRegistry(strRegPath,strValue);
        if(bResult == bEval) return(true);
        }
    return(false);
    }
```

CheckRegistry

The CheckRegistry function reads a value from a registry location specified by strRegPath. This value is compared to strValue, and, if they match, the function returns true. The function returns false if the values do not match. If the registry key does not exist, WSH returns an error. However, this possible error is expected and handled by the try...catch statement, which returns false, indicating to the calling function that the value does not exist.

```
//--------------------------------------------------------------------
// Check Registry
//--------------------------------------------------------------------
function CheckRegistry(strRegPath,strValue)
{
var reValue=new RegExp(strValue,"i");
var objShell = WScript.CreateObject("WScript.Shell");
try{
    if(objShell.RegRead(strRegPath).match(reValue)) return(true);
    return(false);
    }
catch(error) { return(false); }
}
```

SUMMARY

Beyond the functions of the logon script, the design around this potentially broadly deployed script also provides an interesting look at compatibility and error handling that may not factor as important with individually run utility scripts such as those presented earlier in this text. Microsoft continues to improve enterprise services such as AD group policy to centralize the management of many users. But as we have seen, by turbocharging the simple and ubiquitous logon script, you can add your own customized features and provide a richer experience for the user. Using centralized logging, you have a definite record of who accesses your network and when and can also respond to users' issues when installing updates or logging in to the domain. Granted, features like centralized logging may not be a requirement and do add a bit to the complexity, but these features do help in troubleshooting and tracking who accesses your network and when.

When updating this script for your own test deployment, remember to configure the following files:

logon-stub.bat—Configure batch file to output to a network share that is write-accessible for all users.

logon-body.js—Although the XML file includes most of the custom configuration, customize or remove the antivirus logon installation procedure. It was included to demonstrate how to interweave your own legacy applications into a Windows scripting-based logon script.

logon-config.xml—This XML file contains a majority of the configuration information. Be sure to customize this file to suit your own environment and add any updates that you wish to scan for and deploy.

KEY POINTS

- Be careful when developing a group policy-based logon script so that any programming errors do not hang your client computers.
- Use the `WScript.Shell` object `Run` or `Exec` command to run third-party tools or batch files required at logon.
- You have probably heard about XML; consider using it for storing script input or output data. Although only a bit more complicated to use than text files, XML provides hierarchical structure and superior parsing tools.
- The XML Document Object Model (DOM) provides XML reading and parsing capabilities and returns an XML object that is fairly straightforward to access and manipulate.
- Logging script output to a central location may provide additional troubleshooting tools to help figure out odd behavior.
- The logon script runs locally under the context of the to-be-logged-on user, which means that you have more scripting flexibility than when working with a remote computer (such as local registry queries using `WScript.Shell` registry methods).

Next, we step away from command-line JScript utilities and examine how to leverage Office applications and Visual Basic for Applications (VBA) as useful, robust, and easy-to-program platforms from which to launch myriad scripts.

8

Foundation: A Directory Computer-Dump Tool

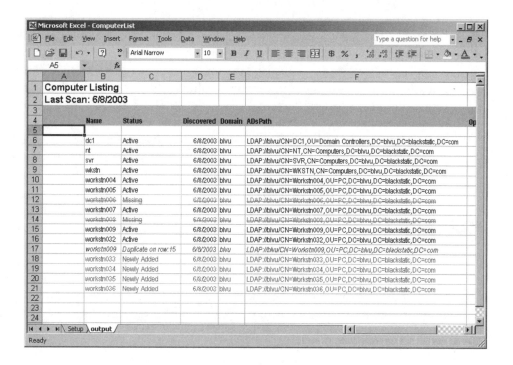

OVERVIEW

As you probably see by now, scripting offers an excellent opportunity to automate widespread or repeatable changes across many systems. An essential foundation of these changes often begins with a complete, reliable list of all computers (or users or other objects) that you wish to individually process. We have seen in the earlier scripts *ADSIQuery* and *DateQuery* that using ADSI ADO quickly generates a list of preferred objects from which we can base subsequent actions. But sometimes a dynamically generated list used one time is not ideal. There may be times that you wish to maintain a persistent list of objects and their specific attributes over time. For example, you could keep a running list of the service state of antivirus software on all members of your domain or of all of the local administrator usernames and passwords for a set of systems. Or maybe you would like to scan a set of computers for all local shares and then annotate what those shares contain. But when you run the scan again in the future, you might like to append the data to an earlier run without wiping out your previous annotations. One tool that works particularly well for managing both script output and execution is Microsoft Excel. Much like pages in a book journal, Excel worksheets provide a terrific canvas to record script configuration information as well as output from previous and current runs. For script writing and execution, Excel supports Visual Basic for Applications (VBA), which offers a good development environment and supports robust debugging capabilities. Through automation and inclusion of external references, Excel (VBA) supports most of the previous Windows scripting technologies that we have used thus far. In addition, when you harness your scripts to Excel, you can also take advantage of the Excel object model, which includes access to Excel functions, formatting, and other Excel- and Office-specific objects and methods not available to Windows Script Host.

SCENARIO

You have been learning more about scripting and already have a list of several new projects that you want to begin working on, particularly in the area of remote administration. Many of these projects require testing, probing, or searching attributes from remote computers in your domain on a reoccurring basis. After the list of computers is generated, you want to perform the tests and annotate the list of computers with the results. A week later, you will perform another batch of tests. You would like to not only keep a record of the results from previous runs but also add new computers (or mark computers no longer found) to the list and then record the results in incremental columns. You plan to e-mail the results to others and want to keep the computer list and results document portable and formatted.

ANALYSIS

Microsoft Excel provides an easy-to-use platform for scripts such as this. Because the data and script is stored in an Excel workbook format (.xls), it is easily portable, and the workbook or individual worksheets can be formatted once and then copied and e-mailed to any other Excel user. This VBA script references a setup worksheet for its runtime arguments, queries Active Directory (AD) for a list of computers using ADSI, and records the results on a user-specified output worksheet. Once this output worksheet is created, future runs of this script will add new computers to the bottom of the existing list and not alter existing data already entered on the worksheet. The script also highlights the names of computers that it can no longer find in the directory. As a foundation for future scripts, this script is designed to allow easy plug-in of new functions providing new features.

You realize that this script will be mainly a stepping-stone to future scripts that can more or less plug into the lists created by this script. As such, your needs map into the technical requirements listed in Table 8.1.

TABLE 8.1   Translating script objectives into technical requirements

Script objectives	Technical requirements
Generate list of computers in your domain.	Construct an ADSI query to return all computer objects from the directory services server of the specified domain. The user may specify the domain.
Append and update results from previous runs.	Scan through the output worksheet and read the list of all previously recorded computers. Store the computer name and its location (by row number) for future reference by this script or a follow-on script.
Add new computers.	Compare the results of a current query of AD against previous results. Add any new computers not found in a previous run to the bottom of the output worksheet list.
Mark missing computers.	Highlight computers on the output worksheet from a previous run that are not found in the current scan.
E-mail results to others, and keep results in a portable document.	The script writes output to an Excel worksheet, which can be e-mailed to others or copied into a separate workbook.
Be able to flexibly format the results.	The script provides basic formatting, and the user of the script can format the output worksheet using any of the Excel formatting tools. Because the output worksheet is reused, subsequent runs of the tool retain the formatting of a previous run.

The functions and subroutines used by this script follow:

DumpComputers: Starting point of main program.

- Reads script input arguments from worksheet *Setup*.
- Creates new output worksheet if one does not already exist.
- Builds ADSI query to return computers from AD.
- Cycles though ADSI query results looking for duplicates, missing computers, and newly added computers.
- Updates the output worksheet with newly found and updated data.

LoadComputerList: Builds aComputers array from a previous output worksheet.

- Cycles through specified output worksheet looking for previous computers.
- Builds the aComputers array with the newly found data.
- Marks any duplicate computers it finds while iterating through the output worksheet.

QueryDS: Performs ADSI query.

- Executes an ADSI query from a provided CommandText.

FindPrevious: Looks for previous matches.

- Looks for any previous computer and domain matches in the aComputer array, given a computer name and domain.

MarkMissingComputers: Marks computers not found in AD.

- When a computer on the output worksheet is found in AD (such as from a previous run) or newly added to the output worksheet, the aComputer(n).Exists property (a custom data type) is updated. This script loops through the aComputer array and marks all computers that aComputer(n).Exists is False (indicating that the ADSI query did not find the computer).

PreFormatTable: Formats a newly created output worksheet.

- Formats the worksheet font sizes, type, and size when the program creates a new output worksheet.
- Adds the table headings to the output worksheet.

PostFormatTable: Formats output worksheet at conclusion of run.

- Adjusts the column sizes to show data.

FormatComputerMissing: Formats the row of a missing computer.

- Turns the row of a missing computer red and strikethrough.
- Updates the status to *Missing*.

FormatComputerDuplicate: Formats the row of a duplicate computer.

- Turns the row of a duplicate computer blue and italic.
- Updates the status to *Duplicate on Row: n.*

FormatComputerAdded: Formats the row of a newly added computer.

- Turns the row of a missing computer teal.
- Updates the status to *Newly Added.*

FormatClear: Clears the formatting of a row.

- Turns off previous bold, italic, or strikethrough formatting.
- Sets color to black.
- Updates the status to *Active.*

Distribution and Installation

To distribute scripts created in a Microsoft Office application, share or e-mail the document containing the script. Remember, though, if you use technologies outside those provided by the Office application—such as ADSI, WMI, or WSH—you must remember to update the recipient computers with these technologies, if needed. This practice is no different from distributing a script that uses *cscript.exe* or *wscript.exe.*

Installation of this script consists of two parts: configuring Excel to recognize the *References* used in the script and then opening and configuring the startup parameters of the script listed on the worksheet named *Setup.*

Specifying References

Development environments and script interpreters may access object libraries differently. For example, WSH provides access to any registered object using the new `ActiveXObject` or `CreateObject` syntax. When using Excel or another Microsoft Office application you can instantiate a new object as in WSH. You can also make an object available by adding its library to a list of references. When a library is added as a reference to an Office document, it provides access to all the objects, methods, and properties contained within that library, and it allows Office to use the objects directly (that is, you can declare variables as that type of object). For example, this script uses `ADODB` objects to perform the ADSI query. From previous scripts, we instantiated a new `ADODB` connection object using code such as the following:

```
Set objConnection = CreateObject("ADODB.Connection").
```

By adding the ADODB library as a reference, we can declare a new connection object within the code as:

```
Dim objConnection As new ADODB.Connection.
```

When a variable is declared in this way, Excel supports syntax completion and you are required to set a reference to the object library. This means that if you begin to use a previously defined variable—as in the previous example, if we typed `objConnection.`—then after the period, Excel would provide a list of all the different methods and properties available to a `Connection` object. This feature makes a handy and useful method of remembering possible methods and properties.

However, when you run this code on another computer, you must make sure that the appropriate references exist on the target computer as well. Without the references, you get an error message: *Compile error: Can't find project or library.* To check the missing and available references, first load the Excel document containing the script, and then press Alt-F11 to enter the Visual Basic Editor. Next, from the menu Tools, select References. You see a list of all selected references as well as any that might be missing, as seen in Figure 8.1. Notice in this figure that the highlighted reference to *Microsoft ActiveX Data Objects 2.7 Library* is missing because the script was created in Microsoft Excel 2003 and a reference to ADO 2.7 was added. The figure was taken on a computer running Microsoft Excel 2000, and ADO 2.7 is not available on this older system. Fixing this problem is easy; simply uncheck the missing reference, and then select a replacement. In this case, we first unselect *Microsoft ActiveX Data Objects 2.7 library.* Next, we scroll the list of available references to the Ms until we find *Microsoft ActiveX Data Objects 2.1 Library,* and then select it to make it available. Be careful when selecting an older library because any objects, methods, or properties used in your script must be compatible with the older version—or else you will have to add that DLL containing the newer library to your system (or upgrade Office). This script uses basic ADO commands available in earlier libraries, so we are OK with adding downgrading to 2.1.

To avoid backwards compatibility issues from the beginning, if we knew we were going to deploy this script on computers running Office 2000, we could have selected the ADO 2.1 Library in Excel 2003.

TIP

Configure the Setup Worksheet

On the setup worksheet, specify the domain name that you wish to enumerate as well as the name of the output worksheet.

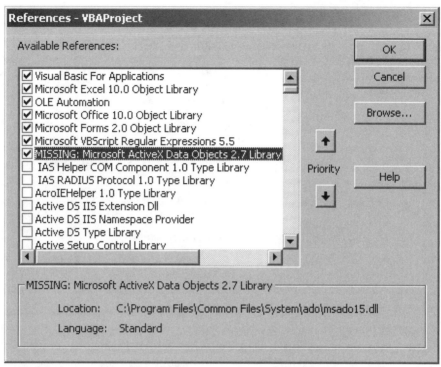

FIGURE 8.1   The Visual Basic Editor References dialog box allows you to specify object libraries to expose and use them in your Excel VBA scripts.

Output

The script lists all computers within the chosen domain on the output worksheet specified in the *setup* worksheet. The script creates a new output worksheet if one does not already exist; otherwise, it appends data to a previous run.

The primary purpose of the script is to search for new computers in a specified domain and record them on the specified output worksheet. Additionally, if computers recorded from a previous run are removed from AD, a subsequent run of the script marks these computers as missing. Table 8.2 lists the data recorded on the output worksheet by column. In VBA you can reference an Excel cell by either its alphanumeric address, such as A1, or its numerical address, such as Cells(1,1). Iterating through loops is made easier by referencing the row and column reference as their numerical values.

TABLE 8.2   Script recorded data on output worksheet

Column	Column, Index	Displayed Data
A	1	Will be used in future scripts—Used as a signal to identify rows for future processing by other scripts.
B	2	Computer Name—ADSI name property.
C	3	Status—Lists a system's status as of the last run, including: newly added, active, duplicate, or missing.
D	4	Date that the system was discovered by the script (when a system was newly added).
E	5	Domain name.
F	6	ADsPath—the ADSI ADsPath property for the computer.

Column A will be used by future scripts to denote specific computers to process. Column B lists the name of the computer, and Column C lists the status of the computer as of the last run of the script (for example, newly added, active, missing). Computers added during the last run are marked as *Newly Added*. The script records computers from a previous run as *Active*. Computers listed on the spreadsheet that cannot be found in AD are marked as *Missing*, and computers that are listed multiple times on the output worksheet are marked with a status of *Duplicates*, including the row number of the duplicate entry. Column D lists the date that the script recorded the computer as newly added. Column E lists the domain, and Column F lists the ADsPath of the computer.

Figure 8.2 shows the output worksheet from a sample script run identifying active, missing, newly added, and duplicate computers. In this example, the script has been run twice. The first time generated the list seen in rows six through 17. Before the second run, the computers in row 12 and 14 (*workstn006* and *workstn008*) were deleted from AD. Also, the computer in row 15 (*workstn009*) was copied to row 17, and the second script run flagged it as a duplicate. The computers in rows 18 through 21 were added to AD after the first run and before the second run. As you can see, running this script on a weekly basis—even without additional functionality or computer scanning—can give you a good history of what is coming and going in your AD.

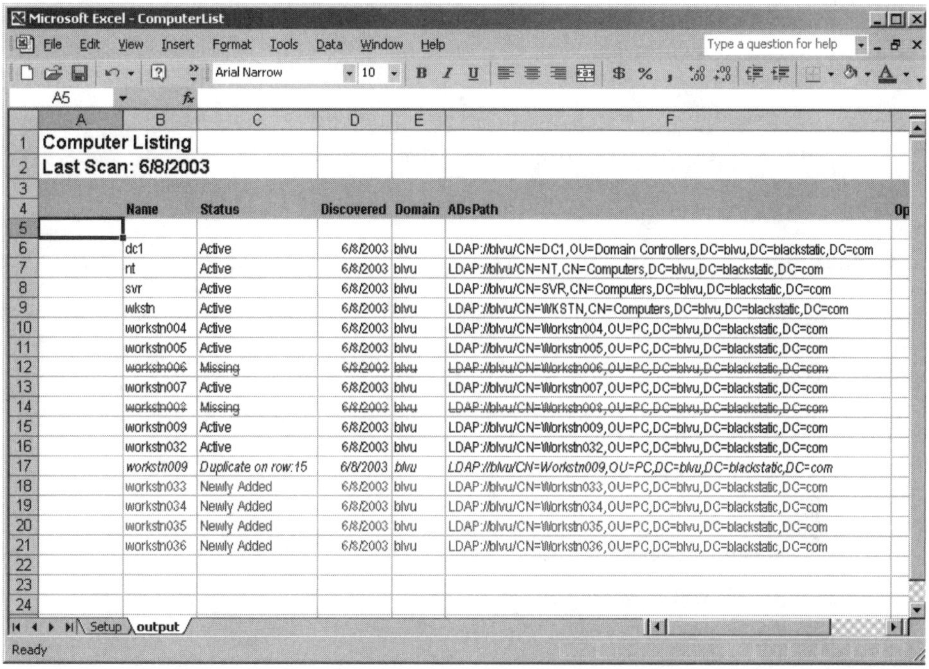

FIGURE 8.2 A sample output worksheet showing active, missing, newly added, and duplicate computers.

Summary of Solution

The script presented in this chapter provides a foundation for future scripts in this text. Essentially, this script scans a domain and lists all computer objects on a specific worksheet. However, the computer scan is nondestructive, and new computers are appended to the list generated from a previous run. In this manner, future scripts can leverage the information and actually use the output worksheet to add new columns containing script-specific data.

The operation of the scripts might look something like this:

Step 1. Run this script to generate a list of all computers in a domain.

Step 2. Run an add-on (future) script that processes each of these computers and adds additional data to the worksheet generated in step 1.

Step 3. Rerun the computer-scanning script days or weeks later to add new systems to the worksheet.

Step 4. Tag the new computers and rerun the add-on script to process only these computers.

Through these steps, you can generate and maintain a list of computers and future data together in a single Excel document.

OBJECTIVE

The purpose of this script is to provide an Excel-based tool that adds and displays the computer members of a specified domain. The script uses the following objects:

`ADODB.Connection`

- Provider
- Open

`ADODB.Command`

- ActiveConnection
- Properties
- CommandText
- Execute

`Arrays` (VBA)

- `ReDim Preserve`
- `Ubound`

Excel Object Model

- Worksheet manipulation
- Cell/range formatting

`Recordset`

- `EOF`
- `Fields.Item(x).Value`

VBA Options

- `Option Explicit`
- `Option Compare Text`

VBA Error Handling

- `On Error Resume Next`
- `On Error Goto 0`

User Data Types

- Using `Type` to declare user-defined data types

THE SCRIPT

The script is contained within a VBA module of an Excel document. Think of the Excel document as a notepad in which you keep a persistent record of input parameters to the script as well as script output. Shortcut keys assigned to macros or Visual Basic clickable buttons on your worksheets initiate your scripts. (Think of a macro as any subroutine that you can code.) The first step to working with VBA is to launch Excel and open the document containing the script.

The script is presented on the CD-ROM as /chapter 8/scripts/dump computer list.xls. This script is contained within an Excel document that requires Excel 2000 or later as well as Windows Script Host 5.6. Before continuing in this next section, open the actual script from the CD-ROM and run it a few times in your test environment to get a feeling of how it works. Explore the entire code on your own and then continue with the walkthrough, which hopefully answers any questions you may have asked yourself. Plus, seeing the code in its entirety in your script editor gives you a good sense of its scope and also allows you to tweak it as we walk through its function.

Setting Macro Security

Microsoft refers to a VBA script in Excel as a macro. When you load a document containing VBA scripts or code, depending on your macro security settings, you may be prompted whether you want to load the document with macros enabled or disabled, as seen in Figure 8.3. (You must enable macros for this script—or any script in Excel—to run.) To review your macro security settings, select Tools, Macro Security. From this screen, you can choose to set high, medium, or low settings, as shown in Figure 8.4. Medium offers a good balance, and on this setting, Excel prompts you anytime you open a document that contains a macro (or VBA code module).

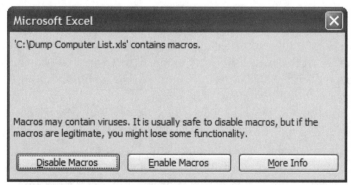

FIGURE 8.3   You must enable macros when opening Microsoft Office documents containing scripts you wish to run.

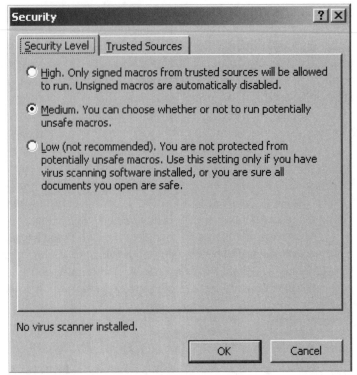

FIGURE 8.4   When set to medium security, Excel prompts every time you open a document that contains a VBA module.

The Script Architecture

The framework of this script consists of the VBA code and two worksheets. The first worksheet, named *setup*, contains script-operating parameters, as seen in Figure 8.5. This simple worksheet contains the parameters used to process the script—much like the command-line arguments we use to control the execution of console scripts. On this worksheet, specify the name of the domain and the name of the worksheet to which the output data will be written. When you are ready to run the script, click the Visual Basic button *1. Pull Computers from Directory*, which executes a procedure that calls the script.

The second worksheet used by the script stores the script output, and you name it on the *setup* worksheet as the Output worksheet. In this example, the Output worksheet is aptly named *output*, and a sample of a run that the script generated is shown in Figure 8.6. An advantage of being able to specify any name for the Output worksheet is that you can create a number of script runs (say, one for every domain in your environment) and store them on different worksheets.

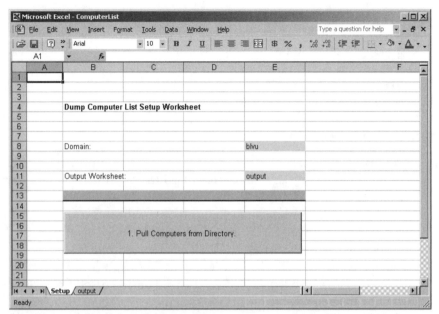

FIGURE 8.5 An advantage to using a VBA-supported Office application as your development platform is its document management capabilities, which let you easily store a record of setup information.

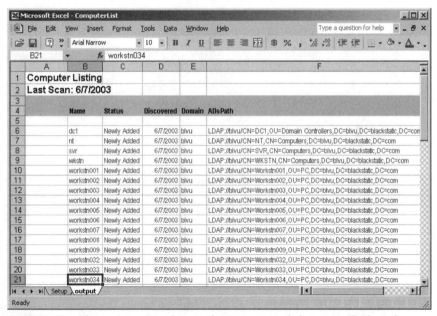

FIGURE 8.6 The script writes data to the Output worksheet specified on the *Setup* worksheet—in this case, data is written to the worksheet named *output*.

The script supports multiple, subsequent runs over time and captures newly added computers and duplicate computer entries and also marks computers that no longer can be found in AD. In this manner, you can manage a running inventory and history of all computers in your AD. The real advantage lies in the ability to extend this script as a foundation for future scripts that might run against some or all of the computers stored in this spreadsheet.

Where to Find the Code in Excel

ON THE CD

Scripts in Excel can reside in a number of places, but they can all be accessed through the Visual Basic editor. First load the document *Dump Computer List.xls* from the included CD-ROM, and then access the Visual Basic Editor by selecting Tools, Macro, and Visual Basic Editor (or press the shortcut key, Alt-F11), and you see a screen similar to that shown in Figure 8.7. Notice the Project Explorer located in the left side of the screen. (If it's not there, activate it using View, Project Explorer.) Notice the VBA Project (*ComputerList.xls*), which has two nodes below it: *Microsoft Excel Objects* and *Modules*. Expand the Microsoft Excel Objects node, and you see the names of two worksheets and the workbook. Double-click the worksheet named *Setup*. In the main code menu you see the callout to a private subroutine called CommandButton1_Click(). Inside this is a single piece of code:

```
Call DumpComputers
```

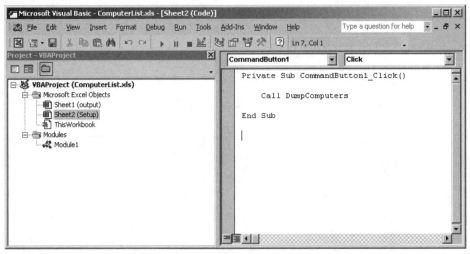

FIGURE 8.7   The Visual Basic editor provides a robust development environment for the Microsoft Office applications.

What you are looking at is a subroutine defining the action associated with the click event of the CommandButton1 object located on the *Setup* worksheet. When this event occurs (that is, when the button CommandButton1 is clicked), the subroutine named CommandButton1_click runs, and it calls the subroutine Dump-Computers, which is the actual name of the main script that does the real work. This button and click-event code provide an entry point to starting this script. (Another means of starting a script in Excel is by using Tools, Macro, but adding a button that the user can click is a more user-friendly way of launching your script.) But how does the function CommandButton1_click correlate to the button? Let's look at this now.

Switch back to the Excel application, and click (activate) the *Setup* worksheet. Besides some basic script configuration text is a large button named *1. Pull computers from Directory*. We saw that the code associated with CommandButton1_click tells Excel what to do when you click the button on the *Setup* worksheet. However, if you right-click the button, no menus appear and you can't access properties of the button. You must first enter the Visual Basic design mode. From the Visual Basic menu (if you don't have this menu, right-click in the menu bar to list all available menus and turn on the menu named Visual Basic), click the Design Mode button to activate the Design mode, as seen in Figure 8.8. Now right-click our large button and select Properties to display all of the properties of the button. From here you can change the caption of the button (what it displays), its name, and other attributes such as size and color. To change the action associated to clicking it, close the Properties dialog box and right-click the button again. This time, click View Code. Notice that we are looking at the same code for CommandButton1_Click that we were looking at just before. But how do we create a new button or Visual Basic control? That's fairly straightforward as well.

FIGURE 8.8   The Design mode in Microsoft Excel lets you edit Visual Basic controls.

Return to the *Setup* worksheet, and this time, turn on the Control Toolbox (as before, right-click the menu bar and enable the menu Control Toolbox), as seen in Figure 8.9. This menu provides you with a palette of tools, which you can drag onto your worksheet. Select Command Button, and drag an outline on the *Setup* worksheet where you want this new button to go. When you release the mouse button, notice the button is created and is named something like CommandButton*n*. You should still be in Design mode from before, so you can right-click this new button, and click properties or View Code to edit its behavior. If you look at the code window of this new button, you see that a new CommandButton*n*_click subroutine has been created, and the code editor has placed your cursor at the position within the block of code. Any code you place here is activated when that button is clicked.

FIGURE 8.9 The Control Toolbox provides access to controls such as buttons and checkboxes that you can add to your Excel documents and control with scripting.

The main script in the *Dump Computer List.xls* document is initiated in this manner. The click event for the main button calls the DumpComputers subroutine using a simple call statement, as follows:

```
Call DumpComputers()
```

Let's look at DumpComputers, the script's main code, next. Independent code not associated with a particular worksheet or Excel object can be stored in a module. From the Visual Basic editor Project Explorer, expand the Modules node below the

Microsoft Excel Objects node. Double-click the DumpComputerList icon. (This module has been renamed—when you start with a new workbook, the first module is named *Module1*.) In the code window you see all of the code that makes up the *DumpComputerList* script. As in previous scripts, this code is presented in a linear fashion with calls from a primary subroutine (we'll call it the main program) DumpComputers to other subroutines and functions. But first, let's look at the code at the Module level—code that is not a part of any other subroutine or function.

Module-level Code

Module-level code provides definitions and instructions available to all code within a particular VBA project. Our script defines options, type definitions, and public constants at the module level.

The Option keyword defines specific operating parameters of the script and supports four statements. Option Base declares the lower bound for an array subscript—either 0 or 1. As in JScript, arrays in VBA by default are zero based. However, with Option Base 1 declared, you can set the base to 1, although you generally benefit from sticking with the recognized standard of zero-based arrays.

Option Compare [method] sets the default comparison method for methods in the script as either Text, Binary, or Database. For example, this script uses the method StrComp to compare two strings. Because this method recognizes the Option Compare setting, instead of setting each comparison in StrComp to Text, the script sets it once at the beginning with the statement Option Compare Text.

When Option Explicit is set, all variables must be declared (for example, with the Dim statement) or the compiler throws an error. This statement troubleshoots possible variable misspellings and also encourages the developer to think more about the variables used.

Lastly, the statement Option Private Module prevents external projects from accessing public data within the project for which it is set.

The VBA Type statement allows for custom, user-defined data types, similar to creating objects in JScript, or structs in C. A user-defined data type contains one or more elements. (You can think of a new data type as similar to a JScript object with one or more properties.) In the next code, we create a new data type named ComputerData, along with five elements: Name, Row, Domain, Exists, and ADsPath. User-defined data types can be defined as arrays, and the elements accessed similarly to properties in a JScript object. (For example, if aComputer() is defined as an array of type ComputerData, then aComputer(5).Name returns the name of the fifth element of the aComputer array.)

```
Option Explicit
Option Compare Text
```

```
Type ComputerData
    Name As String
    Row As Integer
    Domain As String
    Exists As Integer
    ADsPath As String
End Type
```

The script uses constants to define the column values of the output worksheet. Referencing the values as a constant improves readability of the script and means you need to update the definitions only in one place. Declaring the constants `Public` makes them visible to all modules within the VBA project.

```
Public Const COL_NAME = 2
Public Const COL_STATUS = 3
Public Const COL_DATE = 4
Public Const COL_DOMAIN = 5
Public Const COL_ADSPATH = 6
Public Const COL_OPTION1 = 7
Public Const COL_OPTION2 = 8
```

DumpComputers

The subroutine `DumpComputers` queries AD for all computers within a specified domain and then outputs the results to a user-specified output worksheet. As a subroutine, `DumpComputers` does not return any values and cannot be used in an expression (unlike a function).

Primary data storage of the script occurs in an array defined as the user data type `ComputerData`. The array `aComputer` stores all of the computer information pulled from AD and previous runs of the script.

Because we have included an explicit reference to ADO within Excel, we can instantiate a new `ADODB` object as a part of the declaration of the object. The code `Dim objRecordset As ADODB.Recordset` instantiates a new object `objRecordset`, which represents the `ADODB Recordset` object. The script represents the two worksheets used in the script as the objects `objWsSetup` and `objWsOutput`. These objects are instantiated and refer to the worksheet *Setup* and to the output worksheet named by the user in the Setup worksheet.

TIP

You do not have to run Excel or VBA within Excel to work with Excel objects. Even though the worksheet object is native to Excel, it can be called externally, just like any other object. For example, you could create a cscript.exe console script that performs some actions and then creates a new Excel object and dumps the output data there instead.

`strCommandText` and `strDomain` represent strings of text used to represent the ADSI query command and current domain, respectively. The integers `i`, `iRowNum`, and `iCount` represent various counters used by the subroutine.

```
'------------------------------------------------------------------
' Main Function: DumpComputers
'------------------------------------------------------------------

Sub DumpComputers()
Dim aComputer() As ComputerData
Dim objRecordset As ADODB.Recordset
Dim objWsSetup As Worksheet
Dim objWsOutput As Worksheet
Dim strWsOutput, strCommandText, strDomain As String
Dim i, iRowNum, iCount As Integer
```

Whereas previous scripts parsed command-line arguments for operating parameters, this script reads its operating parameters from the *Setup* worksheet using the `Cells(row,column).Value` properties. The `Worksheets` object is native to Excel, and you can use it without instantiating your own object (such as with a `GetObject` method that we have used previously in WSH scripts). Creating or opening an Excel document exposes these Excel-specific objects.

```
Set objWsSetup = Worksheets("Setup")
strWsOutput = objWsSetup.Cells(11, 5).Value
strDomain = objWsSetup.Cells(8, 5).Value
```

The script checks for the existence of the specified output worksheet. If it doesn't exist, then it creates and formats a new output worksheet and names it according to the name specified on the *Setup* worksheet. The script supports multiple output worksheets—simply change the name of the output worksheet on the setup page and rerun the script. For example, if you had three domains, you could run the script three times, one for each domain. And you could specify the name of the output script as *Output-A*, *Output-B*, and *Output-C* for each of the domains. In this manner, you could manage the computers in all of these domains separately on their own worksheets.

An `On Error Resume` statement is used to trap any error generated when trying to set the output worksheet object. The first time the script runs (or any time a new output worksheet is specified), an error occurs and is set to the variable `Err`. `On Error Resume Next` instructs the script to continue processing if an error does occur, and it skips the failed code. In anticipation of this error, the script immediately tests whether `Err` exists. If it does, the script creates the new output worksheet, names it to the name specified on the *Setup* worksheet, and calls the `PreFormatTable` subroutine to format it. You may have remembered the previous usage of the Call statement to run a subroutine or function. This statement is not required; you can call the procedure directly. The On Error GoTo 0 statement disables error handling.

```
On Error Resume Next
Set objWsOutput = Worksheets(strWsOutput)
If Err Then
    Set objWsOutput = Worksheets.Add(After:=Worksheets("Setup"))
    objWsOutput.Name = strWsOutput
    PreFormatTable
End If
On Error GoTo 0
```

You can access data on a specific worksheet directly, such as objWsOutput. Cells(1,1).Value, or you can access it through the active worksheet. The script sets objWsOutput as the active worksheet, and any general methods called execute on this active worksheet. Using the previous example, once objWsOutput is activated, then you can simply refer to objects on it as Cells(1,1).value.

The script calls the function LoadComputerList to load the array aComputer with each of the computers that might already exist from a previous run on the output worksheet. LoadComputerList returns the row number of the last computer object it found. The variable iRowNum represents the new row number of a computer added to the output worksheet. Row number 6 is the first row used by the output worksheet, so if no computers exist, then the script sets iRowNum to 6.

```
objWsOutput.Activate
iRowNum = LoadComputerList(aComputer, objWsOutput)
If iRowNum < 6 Then iRowNum = 6
```

Next, the script creates the CommandText to be executed by the command object in the ADO query called in the QueryDS function. This query returns the Name and ADsPath of all computer objects in the specified domain. The QueryDS function returns a Recordset object (objRecordset) containing the results.

```
strCommandText = "SELECT ADsPath,Name FROM 'LDAP://" & strDomain & _
    "' WHERE objectCategory = 'computer'"
Set objRecordset = QueryDS(strCommandText)
```

Unlike in JScript, dynamic arrays in Visual Basic must be declared to a set value before they are used, and there is no length property to return the size of an array. (For example, if you declare an array as aArray(100), and you use only 65 elements, no property returns the number of elements used.) Visual Basic does support the method UBound, which returns the upper bounds of an array. In the previous example, UBound(aArray) would always return 100, no matter how many elements are actually used. To use this method, this script redimensions the array to the current number of elements plus 1, using the ReDim Preserve statement. In this manner, UBound(aComputer)-1 always returns the exact number of elements in the array.

The script cycles through each of the objRecordset records and calls the Find-Previous function to see if any of the computers returned from the ADO query already exist in the output worksheet. If the computers already exist, the script does nothing with them and continues to the next record.

However, if the computer does not exist, the script must add it to the output worksheet at the row number iRowNum. Try not to confuse iRowNum with iCount. Whereas iRowNum represents the current row number of a new computer, the iCount variable represents the number of computers in the aComputer array. When adding a new computer, the script increments the iCount parameter and redimensions the aComputer array using the Redim Preserve statement. The Preserve keyword ensures that the data in that aComputer array remains intact; otherwise, redimensioning the data clears the array. Next, the script adds query results to the aComputer array. The name of the computer is converted to lowercase and put into the user-defined data type ComputerData element named Name. The current row number is added to the ComputerData Row element. The script sets aCompute.Exists to True, and aComputer.Domain to strDomain.

After updating the aComputer array, the script records the new computer on the output worksheet. The script uses the Cells(iRowNum, specifiedcolumn) property to update the output worksheet to include the newly discovered computer data, including its name, date, domain, and ADsPath. The script calls the FormatComputer-Added subroutine to annotate the status as newly found and highlight the addition. The script increments iRowNum and cycles to the next record.

```
iCount = UBound(aComputer)
While Not (objRecordset.EOF)
    If FindPrevious(objRecordset.Fields.Item("Name").Value, _
      aComputer(), strDomain) = 0 Then
        iCount = iCount + 1
        ReDim Preserve aComputer(iCount)
        aComputer(iCount).Name = _
         LCase(objRecordset.Fields.Item("Name").Value)
        aComputer(iCount).Row = iRowNum
        aComputer(iCount).Exists = True
        aComputer(iCount).Domain = strDomain
        Cells(iRowNum, COL_NAME) = aComputer(iCount).Name
        Cells(iRowNum, COL_DATE) = FormatDateTime(Now(), vbShortDate)
        Cells(iRowNum, COL_DOMAIN) = strDomain
```

```
            Cells(iRowNum, COL_ADSPATH) = _
              objRecordset.Fields.Item("ADsPath").Value
            FormatComputerAdded iRowNum
            iRowNum = iRowNum + 1
        End If
        objRecordset.MoveNext
    Wend
```

After cycling through the ADO query Recordset, the script timestamps the output worksheet and then calls the MarkMissingComputers subroutine to cycle through the output worksheet and highlight all missing computers. Lastly, the script calls the PostFormatTable subroutine to adjust the column widths to accommodate the new computer information and then exits.

```
    Cells(2, 1) = "Last Scan: " & FormatDateTime(Now(), vbShortDate)
    MarkMissingComputers aComputer(), strDomain
    PostFormatTable

End Sub
```

LoadComputerList

The function LoadComputerList cycles through the output worksheet, parses computer information from each of the rows, and stores the information in the aComputer array. The script uses the variable i as a counter to indicate the current aComputer array element. The variables strDomain and strComputer represent domain and computer names on the current row.

The function supports skipping up to 10 blank lines without halting processing. This means that after a run is completed, you could sort the results, add spacing, and categorize or group the computers, and the script still processes the list. iRowNum represents the current row number of the parsing script, and iBlankLines provides the counter variable of up to 10 blank lines. While loading the data, the function checks all new entries against what it has already found and stored in aComputer. Any duplicates are flagged, and their row number is stored in iDupeRow. The script also records the row number of the duplicate to aid the user in tracking the duplicate entry. To begin, the script redimensions the dynamic array aComputer to a single member. (It will be redimensioned again after adding data.)

```
    '------------------------------------------------------------------
    ' Load Existing Computer List
    '------------------------------------------------------------------
```

```
Function LoadComputerList(aComputer() As ComputerData, objWsOutput) _
 As Integer
Dim i As Integer
Dim strDomain, strComputer As String
Dim iRowNum, iBlankLines, iDupeRow As Integer
ReDim aComputer(1)

i = 1
iRowNum = 6
iBlankLines = 0
iDupeRow = 0
```

The parsing function consists of two loops. The first loop runs until it finds 10 sequential blank lines. The second loop looks for a single blank line represented by an empty cell in the output worksheet where a computer name should be. If the cell is empty, then the inner loop exits, and both the `iRowNum` and `iBlankLines` variables increment. If `iBlankLines` is greater than 10, then the outer loop exits. Otherwise, a new inner loop is created to examine whether the new row number is empty.

Assuming that computer data exists, the current row computer name will *not* be empty. The script resets the `iBlankLines` to zero and clears any previous formatting. The script resets `iBlankLines` because we are looking for 10 blank lines in a row, not 10 blank lines in total. The script sets `ScreenUpdating` to `False` during the execution of this loop to decrease running time. The function `FormatClear` sets cell format attributes (such as turning off bold and italic), which causes the worksheet to redraw, slowing the script considerably. At the end of this function, `ScreenUpdating` is enabled again.

To read previous data from the output worksheet, the script sets the variables `strComputer` and `strDomain` to the cell value of the current row number and `COL_NAME` and `COL_DOMAIN`, respectively.

```
Application.ScreenUpdating = False
While iBlankLines < 10
        While IsEmpty(Cells(iRowNum, COL_NAME)) = False
            iBlankLines = 0
            FormatClear iRowNum
              strComputer = Cells(iRowNum, COL_NAME).Value
              strDomain = Cells(iRowNum, COL_DOMAIN).Value
```

Armed with the name and domain, the script calls the `FindPrevious` function to check if it has encountered another computer or domain pair previously in this loop. (Notice that the script reuses the same `FindPrevious` function used to see if an ADO `RecordSet` computer matches a computer in `aComputer`.)

If no duplicate is found, the function loads the current element (i) of the aComputer array with the data about the computer object, including the name, its row number, its domain, and ADsPath. The script increments i and redimensions the aComputer array by one. In this manner, a tight rein is kept on the aComputer boundary. With the processing of this computer complete, the script increments iRowNum.

If, however, the newly found computer is a duplicate of an earlier computer, the script marks it as such (including the row number of the original) and calls the function FormatComputerDuplicate to highlight the entry as a duplicate. This entry is not added to the aComputer array, and the script increments the iRowNum variable to proceed to the next row.

```
            iDupeRow = FindPrevious(strComputer, aComputer(), strDomain)
            If (iDupeRow = 0) Then
                aComputer(i).Name = Cells(iRowNum, COL_NAME).Value
                aComputer(i).Row = iRowNum
                aComputer(i).Domain = Cells(iRowNum, COL_DOMAIN)
                aComputer(i).ADsPath = Cells(iRowNum, COL_ADSPATH)
                i = i + 1
                ReDim Preserve aComputer(i)
                iRowNum = iRowNum + 1
            Else
                FormatComputerDuplicate iRowNum, iDupeRow
                iRowNum = iRowNum + 1
            End If
        Wend
    iRowNum = iRowNum + 1
    iBlankLines = iBlankLines + 1
    Wend
```

Lastly, as a bit of cleanup, the script redimensions aComputer one last time to a dimension of i-1, which represents the true number of computers in the array. (Remember, the script kept dimensioning the array at one more to keep ahead of the script adding the next computer.) Also, because of the 10 permissible blank rows, the script subtracts 10 from the iRowNum and returns that value to the calling function, which represents the row number of the last computer found.

```
    ReDim Preserve aComputer(i - 1)
    LoadComputerList = iRowNum - 10
    Application.ScreenUpdating = True
End Function
```

QueryDS

The QueryDS function essentially represents a VBA port of the JScript *QueryDS* used in the script in Chapter 5. The function takes a string representing the CommandText as input and returns the query results as a RecordSet to the calling procedure. The procedure includes a short bit of error handling to trap the error if the user input the name of a domain that cannot be reached. Because we added a reference to Active X Objects 2.x, we can handily instantiate the returning data type as an ADODB.Recordset.

```
'-------------------------------------------------------------------
' Query Active Directory given an LDAP query parameter
'-------------------------------------------------------------------
Function QueryDS(strCommandText) As ADODB.Recordset
Dim iRowNum, iComputerCount As Integer
Dim objConnection As New ADODB.Connection
Dim objCommand As New ADODB.Command
objConnection.Provider = "ADsDSOObject"
objConnection.Open
objCommand.ActiveConnection = objConnection
objCommand.Properties("Sort On") = "Name"
objCommand.CommandText = strCommandText
On Error Resume Next
Set QueryDS = objCommand.Execute()
If Err <> 0 Then
    MsgBox "Error Connecting with the Domain Controller. " & _
    "Please check the domain name and try again.", _
    vbCritical, "Check Domain Name"
    End
End If
On Error GoTo 0
End Function
```

FindPrevious

The function FindPrevious compares a computer name and domain pair against each computer name and domain pair stored in the aComputer array. The function cycles through all elements of the aComputer array and uses the StrComp method to compare the strings. The syntax for StrComp is StrComp(string1, string2, compare), where string1 and string2 are the two strings to compare and compare is an optional third argument specifying the comparison method (vbBinaryCompare, vbTextCompare, or vbDatabaseCompare). We do not use the compare argument in the next example because we set the module-level statement Option Compare Text, which specifies the comparison property module-wide.

If a duplicate is found, then it is marked as existing, and the row number of the matching computer in aComputer is returned to the calling function.

```
'--------------------------------------------------------------------
' Check if Computer Name is already in aComputer()
'--------------------------------------------------------------------
Function FindPrevious(strComputer, aComputer() _
 As ComputerData, strDomain) As Integer

FindPrevious = 0

For i = 1 To UBound(aComputer)
    If StrComp(aComputer(i).Name, strComputer) = 0 And _
     StrComp(aComputer(i).Domain, strDomain) = 0 Then
        FindPrevious = aComputer(i).Row
        aComputer(i).Exists = True
        Exit Function
    End If
Next i

End Function
```

MarkMissingComputers

Recall from the subroutine DumpComputers the call to the MarkMissingComputers subroutine. This subroutine scans through each of the computers in aComputer and examines the Exists element (which we defined as a part of our user-defined data type ComputerData). Any elements that do not have Exists set to true were not discovered in the AD scan, which signifies that the computers could have been deleted from the AD. This subroutine processes computers only in the domain specified in the *Setup* worksheet and calls the subroutine FormatComputerMissing to highlight any computers that no longer exist (or more correctly, that do not show up in the ADSI query).

```
'--------------------------------------------------------------------
' Mark all computers on the Worksheet no longer in Active Directory
'--------------------------------------------------------------------
Sub MarkMissingComputers(aComputer() As ComputerData, strDomain)
Dim i As Integer
For i = 1 To UBound(aComputer)
    If aComputer(i).Domain = strDomain And Not aComputer(i).Exists _
     Then FormatComputerMissing aComputer(i).Row
Next i

End Sub
```

PreFormatTable

Formatting data in Excel using scripts can be a real time-saver and can assist in expediting reviews of voluminous data. For example, imagine writing a function that scans a range spanning hundreds of rows looking for a specific attribute or value. If it finds it, it highlights the cell in green. The formatting subroutines of this script perform a similar function to highlight categories of data.

The subroutine `PreFormatTable`, as its name suggests, formats the newly created output worksheet. The function uses mostly Excel methods and properties to set the font name, size, color, and style, as well as shading rows and setting values. As we've become accustomed to seeing in our previous Windows scripting examples, objects and their methods and properties are hierarchical in nature. The same is true with Excel formatting objects. To format a cell, first, we specify the `Cells` property (which when called alone references the active worksheet), and we then drill down and set each of the individual object properties such as `Font.Name` and `Font.Size`.

TIP

The script uses the `With` statement to bundle the execution of a series of statements. If we know that we want to perform multiple operations on an object, instead of referencing that object over and over, we can define it using a `With` statement to open a block. The following statements, such as `.Name`, `.Size`, and `.FontStyle`, can be added without having to respecify the object. For example, the following example illustrates two blocks of code that accomplish the same thing, but using the `With` statement improves readability and is a bit more elegant.

With

```
With Range("A1:A2").EntireRow.Font
    .Name = "Arial"
    .Size = 12
    .FontStyle = "Bold"
End With
```

Without

```
Range("A1:A2").EntireRow.Font.Name = "Arial"
Range("A1:A2").EntireRow.Font.Size = 12
Range("A1:A2").EntireRow.Font.FontStyle = "Bold"
```

The script uses the property `Range().Interior.ColorIndex` to set the cell shading of the heading.

```
'-----------------------------------------------------------------
' Format new output worksheet.
'-----------------------------------------------------------------
```

```
Sub PreFormatTable()
Cells.Font.Name = "Arial Narrow"
Cells(1, 1) = "Computer Listing"
With Range("A1:A2").EntireRow.Font
    .Name = "Arial"
    .Size = 12
    .FontStyle = "Bold"
End With

Cells(4, COL_NAME) = "Name"
Cells(4, COL_STATUS) = "Status"
Cells(4, COL_DATE) = "Discovered"
Cells(4, COL_DOMAIN) = "Domain"
Cells(4, COL_ADSPATH) = "ADsPath"
Cells(4, COL_OPTION1) = "Option 1"
Cells(4, COL_OPTION2) = "Option 2"

Range("A3:H4").Interior.ColorIndex = 15
Range("A3:H4").Font.FontStyle = "Bold"
End Sub
```

PostFormatTable

After its completion the `DumpComputers` subroutine calls the `PostFormatTable` subroutine to `Autofit` the columns. Specifying the `Cells` property without any parameters sets the property for all of the cells within the current active worksheet. The code `Cells.EntireColumn.AutoFit` says to take all `Cells` in the worksheet and their associated columns (essentially all columns) and `AutoFit` their contents. `Autofit` adjusts the column widths so that all data is displayed without hiding behind a column too narrow.

```
'-------------------------------------------------------------------
' Format results after DumpComputer completes
'-------------------------------------------------------------------
Sub PostFormatTable()
Cells.EntireColumn.AutoFit
Columns(1).ColumnWidth = 10
End Sub
```

The Formatting Subroutines

The subroutines `FormatComputerMissing`, `FormatComputerDuplicate`, `FormatComputer-Added`, and `FormatClear` conceptually perform the same operation but specify different formatting. They each accept a row number as an argument and set the formatting for that specific row to something associated with the function. For example, `FormatComputerMissing` sets the row font color to red and sets strikethrough to true for any row specified. Separating these procedures from the calling functions shortens the calling functions and makes them a bit easier to read and debug. Plus, making changes to these functions can be easier when they are all together and isolated from their calling procedures.

```
'-------------------------------------------------------------------
' Format Rows: Computers in worksheet but not in Active Directory
'-------------------------------------------------------------------
Sub FormatComputerMissing(iRowNum)
With Cells(iRowNum, 1).EntireRow.Font
    .Strikethrough = True
    .ColorIndex = 3 'Red
End With
Cells(iRowNum, COL_STATUS).Value = "Missing"
End Sub


'-------------------------------------------------------------------
' Format Rows: Duplicate computer/domain in worksheet
'-------------------------------------------------------------------
Sub FormatComputerDuplicate(iRowNum, iDupeRow)
With Cells(iRowNum, 1).EntireRow.Font
    .Italic = True
    .ColorIndex = 5 'Blue
End With
Cells(iRowNum, COL_STATUS).Value = "Duplicate on row:" & iDupeRow
End Sub


'-------------------------------------------------------------------
' Format Rows: Computers in Active Directory but not in the Worksheet
'-------------------------------------------------------------------
Sub FormatComputerAdded(iRowNum)
With Cells(iRowNum, 1).EntireRow.Font
    .ColorIndex = 14 'Teal
End With
Cells(iRowNum, COL_STATUS).Value = "Newly Added"
End Sub


'-------------------------------------------------------------------
' Format Rows: Clear formatting for subsequent runs.
'-------------------------------------------------------------------
Sub FormatClear(iRowNum)
With Cells(iRowNum, 1).EntireRow.Font
    .Italic = False
    .Bold = False
    .Strikethrough = False
    .ColorIndex = 0 'Black
```

```
End With
Cells(iRowNum, COL_STATUS).Value = "Active"
End Sub
```

SUMMARY

In this chapter we moved away from coding console applications in JScript to using VBA in Excel. Whereas the syntax may differ slightly, the techniques and design of these programs remains remarkably similar. Coding your scripts in a Microsoft Office application such as Excel lets you access many of the same Windows scripting functions available to WSH, plus all of the VBA Office scripting support. Additionally, scripting within a tool like Excel lets you format and save your data in a variety of formats. Whereas a console-based script might prove handy when run from a scheduled task or batch file, Excel provides robust formatting options and an enriched development environment.

KEY POINTS

- The Microsoft Office suite of applications provides a fairly robust VBA development environment for script writers.
- Microsoft Excel offers an ideal canvas for prototyping, as well as quick-and-dirty reporting and archiving of voluminous scripting information, such as lists of computers or users.
- Most scripting technologies work directly in Excel VBA—including ADSI, WMI, and even WSH.
- Separating formatting subroutines from the main functions increases the readability of your code and makes changes easier.
- By loading previous data and comparing it to a current run, this script highlights additions and deletions to the list of computers in a domain, making it much easier to identify changes and new computers to process by future scripts.
- The LoadComputerList function is tolerant of multiple blank rows and does not care about row order, which improves its flexibility in that you can reorder or group rows of computers (for example, group domain controllers or servers together).

This script provides a running tally of all computers within a specified domain and serves as a foundation for future scripts, such as the Worm Scanner and Local Account Enumeration tool upcoming in the next chapters.

9 Browser-based Computer Information Diagnostics

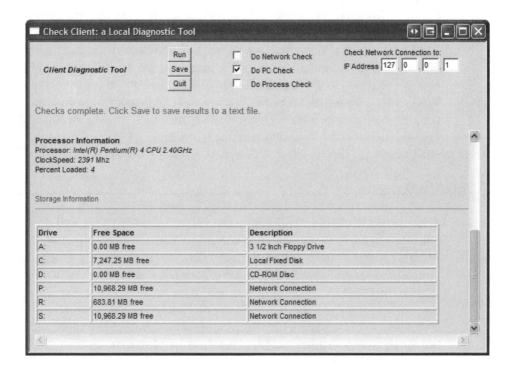

OVERVIEW

Many organizations rely on their help desk or technical support departments to provide user support ranging from general assistance with a computer system to specific application support. Remote-control tools built into the Windows operating system including Terminal Services®, Remote Assistance®, and NetMeeting® have supplemental, long-time, commercial remote-control applications. However, in many cases, these tools may be too invasive or otherwise not practical to use. For example, not many users want a software company's technical support technician remote-controlling their private machine. Plus, use of remote-control software necessitates a higher degree of computer or application and security knowledge. Checking registry entries and network connections and performing other diagnostics requires knowledge of the operating system and familiarity of how that user has configured his system.

The scenario in this chapter outlines the requirements for an easy-to-use tool that collects basic local computer configuration information and does not require an invasive software installation. This tool can perform a number of diagnostics of the machine—such as determining machine type, processes running, and network connectivity—and then present these findings to the user, who then can e-mail the results to the help desk or technical support.

This script demonstrates how to write a basic client diagnostics application based on a *hypertext application (HTA)* shell and use a WMI and WSH engine to collect the data. An HTA provides script functionality in an HTML GUI—sans Web server. This means that you can open an HTA directly from your local file system through your browser. This script assumes that you are familiar with basic HTML, such as creating and formatting tables and text.

SCENARIO

Your company offers an application service provider (ASP) solution that provides recruiting services to subscribing customers. Your customers include human resources (HR) staff from companies of all sizes and backgrounds. Using an Internet connection, your customers upload job descriptions, which you post on their behalf, to their own Web sites. Candidates upload their own resumes for specific positions on your site, and you make those resumes privately available to your customers' HR staffs. Due to some proprietary forms and controls within the application, you require that your customers use the latest version of Internet Explorer.

Business is good, and your company is growing rapidly. Your company has a small team of customer support representatives who answer inbound phone and e-mailed questions about the application. Advanced technical questions are escalated to the IT department and software developers who manage the Web sites and developed the ASP code.

In an attempt to offset the rising costs of your own customer support department, you want to create a diagnostics tool to assist in collecting information about customers' systems. This tool must be easy for customers to run and must not compromise their privacy. Output from the tool must be easily readable by the customer, your customer support department, and the IT department. Additionally, the tool must be easily expandable to allow for new features and future tests.

ANALYSIS

You meet with your customer support department and talk with them about the requirements of this tool. They provide you with their Frequently Asked Questions (FAQ) of the application as well as basic troubleshooting tips to help solve the most frequent problems. Armed with this information, you put together a requirements list. This list guards against *feature creep* and ensures that your application is scoped appropriately—both in terms of features and time to produce. We have all experienced feature creep—the extension of the scope of a project as the project is underway. For example, let's say you set out to paint one room in your house. As you are painting, you decide to paint the hallway and the kitchen. When all is done, you have spent more time (and possibly more money) to deliver, ideally, a superior product. Feature creep is not necessarily bad, but it can wreak havoc on a schedule or resources if not recognized and managed.

From your interview with the customer support department, you boil down the business objectives into a list of technical requirements, presented in Table 9.1.

TABLE 9.1   Mapping business objectives to technical requirements

Problem	Diagnostic tool
General troubleshooting.	Get the OS version, installed service packs, disk free space, make and model, and the network configuration.
The application is slow.	Check the processor speed, memory, and latency of the Internet connection.
Some of the ASP Web pages do not load correctly (or at all).	Check the version of Internet Explorer. Check all running processes for possible conflicts.
User cannot access the application.	Check network connectivity to the application Web server.
The application times out while waiting for the results of a page.	Check the latency and consistency of the Internet connection.

From these requirements, build a list of technical diagnostics checks by function. Notice that the requirements are morphing into a technical specification from which you could pattern the actual script. Next to the technical description is the technology that can perform the diagnostic.

DoPCCheck: Checks the operating system.

- Check the version of the operating system—WMI, get the OS from `Win32_OperatingSystem`.
- Check the make and model—WMI, get from `Win32_ComputerSystem`.
- Check the processor speed—WMI, get the speed from `Win32_Processor`.
- Check the memory—WMI, get the memory from `Win32_ComputerSystem`.

DoNetworkCheck: Checks the network.

- Check network connectivity to the application Web server—`Ping` the ASP gateway.
- Check the consistency of the Internet connection—`Ping` the ASP gateway multiple times.
- Check the latency of the Internet connection.—Perform a `traceroute` to the ASP gateway.
- Check the network configuration—Run the `ipconfig` command.

DoProcessCheck: Checks running processes.

- Check all running processes for possible conflicts—WMI, get processes from `Win32_Process`.

This script draws on tools and technologies from WSH and WMI. No language requirements are specified, so you are free to choose the language in which you are most comfortable.

Distribution and Installation

The script will be run by customers on their computers, and the requirements emphasize ease of use and ease of installation. Because your product to which the customers are subscribing is a Web-based ASP, we can deduce that they have both a functioning Internet browser and an Internet connection. The simplest tool would be one that can be downloaded and run from any computer without an installation process. Hosting this tool directly from a Web page would be more complex because of the necessity to run the tool outside of the secured Web browser. (Web browser security is designed to contain a Web page within a sandbox that prevents it from accessing local system information.) Downloading multiple files would require bundling them into a single self-installer (such as with WinZip Computing® WinZip® or InstallShield® Software Corporation InstallShield®) and making installation more complicated (and invasive).

A technology that satisfies these requirements is a hypertext application, or HTA. An HTA is an HTML-based script that runs locally in a Web browser and is not served via HTTP from a Web server. An HTA also runs under the security context of the currently logged-in user who executes the script, which widens the script operational features beyond those limited to the browser sandbox. An additional benefit is that the tool supports user interaction and is easily formatted using widely known HTML.

Output

The HTA tool displays the output of the tool in an HTML-formatted window. An HTA supports the scripting environment available on the computer on which it is run. This means that the script can write output to a text file using the WSH FileScriptingObject. Also, per the original specifications, to ensure that the output is readable by everyone, the script will incorporate an option to save the results to a text file.

To forward the results to customer support, the customer can either copy and paste the output directly from the results window or they can e-mail the output text file as an attachment.

Summary of Solution

Customers will download an HTA application and then run it on their local systems. The HTA will run under its own context but will not be connected to any other system, and thus the privacy is maintained. Customers will navigate an HTML front-end to select and start the diagnostic tests. The HTA will present summary output of the tests to users and will allow them to save the results to a text file.

OBJECTIVE

The purpose of this script is to demonstrate the process of gathering a variety of information using WSH and WMI scripting from within an easily formatted HTA script. The script uses the following objects:

WScript.Shell
- ExpandEnvironmentalStrings
- Run
- RegRead

Scripting.FileSystemObject
- DeleteFile
- OpenTextFile
- ReadLine
- CreateTextFile
- WriteLine

WbemScripting.SwbemLocator
- ConnectServer
- ExecQuery

Hypertext Application

■ HTML, Styles, Classes, Buttons

The script is contained within a hypertext application (.hta) file that combines both the programming functions and HTML formatting instructions in a single script. An HTA can be authored with a simple text editor and saved with an HTA extension. By default, all HTA documents are associated with a Web browser, so to run it, simply execute the HTA and it will open in your browser.

THE SCRIPT

Figure 9.1 shows the interface of the tool. The script accepts input from the user through HTML form controls, including buttons, checkboxes, and text fields. The user can select which tests to perform, specify an IP address to test network connectivity, and ultimately save the test results to a text file. The user initiates the test by clicking the Run button and can exit the application by clicking the Quit button.

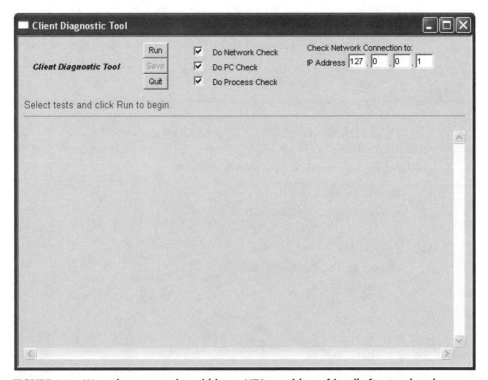

FIGURE 9.1 Wrapping your script within an HTA provides a friendly frontend and user interaction using HTML.

ON THE CD

The script is presented on the CD-ROM as /chapter 9/scripts/ClientCheck.hta. Before continuing in this next section, open the actual script from the CD-ROM and run it a few times in your test environment to get a feeling of how it works. Explore the entire code on your own and then continue with the walkthrough, which hopefully answers any questions you may have. Plus, seeing the code in its entirety in your script editor gives you a good sense of its scope and also allows you to tweak it as we walk through its function.

In the following pages, we walk through the actual script and discuss inline how it works and why certain decisions were made.

A Hypertext Application

A hypertext application (HTA) is essentially a Web page written in HTML that does not require hosting on a Web server to run. Instead, an HTA uses the client machine's Web browser to run, but it differs significantly from opening a simple Web page locally (such as if you opened *somefile.html* into your browser). The normal browser interface is not visible (for example, the navigation buttons and address bar). However, you can define how the HTA window looks, such as whether to use scroll bars, the size and location of the window, and other GUI characteristics. An HTA has read and write access to both the local file system and the registry using the security context of the locally logged-in user (or as otherwise specified in the underlying script). The structure of the file is similar to a standard Web page, and HTML is used to format the output. These characteristics make HTAs a powerful platform to deliver good-looking and interactive scripts of all types—including the previously discussed technologies WMI, ADSI, and WSH.

NOTE

HTAs are powerful tools that allow scripts to run through your web browser under the privilege of the script runner. Because Internet Explorer and web pages are popular targets and transports for viruses, some antivirus programs may warn you when opening an HTA that it could contain a virus. You should be cautious of any HTA that you run on your computer. It is easy to inspect an HTA file before you run it. Open the HTA in Notepad (or your favorite text editor) and review the source code. As a best practice, you should review the source code of any script (from this book or elsewhere) before running it on your computer.

An HTA begins with the standard HTML framework but will likely include additional `<script>` tags to define the functionality of the application, as the following pseudocode presents. Pseudocode is an outline of a program that you can later convert to actual programming statements. Using pseudocode can sometimes be helpful in scoping out a problem or communicating ideas for your program to others.

```
<html>
<HTA:APPLICATION options/>
<head>
    <title>Untitled</title>
    <style> Include Cascading Style Sheet information here </style>
    <script> Include the program functionality here </script>
</head>
<body>
    Include the HTML document defining the HTA format here and
    call functions or subroutines from this section.
</body>
</html>
```

The `<html>` container envelops the entire application. Following this tag, define the operational aspects of your application using the `HTA:APPLICATION` object. Table 9.2 lists the attributes that you can set in the `HTA:APPLICATION`.

TABLE 9.2 The `HTA:APPLICATION` attributes define the operational aspects of your HTA

Name	Value (default)	Description
APPLICATIONNAME		Name of the HTA. (Different than `<title>`.
BORDER	*thick*, dialog, none, thin	Defines the border of the HTA window. The border type also includes whether to include a title bar with a close box. Note that you can only resize an HTA with a thick border.
BORDERSTYLE	*normal*, complex, raised, static, sunken	Specifies the look of the border.
CAPTION	*yes*, no	Turn on the title bar.
CONTEXTMENU	*yes*, no	Turn the context menu (right-mouse click) on and off.
ICON	File name	Specify the 32x32 icon for the HTA.
INNERBORDER	*yes*, no	Turn on the inner 3D border.
MAXIMIZEBUTTON	*yes*, no	Turn on a maximize button in the title bar.
MINIMIZEBUTTON	*yes*, no	Turn on a minimize button in the title bar.

TABLE 9.2 The `HTA:APPLICATION` attributes define the operational aspects of your HTA *(continued)*

Name	*Value (default)*	*Description*
NAVIGABLE	yes, *no*	Open links inside the HTA window (instead of opening in a new browser window).
SCROLL	*yes*, no, auto	Turn on scroll bars. *Auto* displays scroll bars if the content exceeds the client area.
SCROLLFLAT	yes, *no*	The scroll bars are flat and not 3D.
SELECTION	*yes*, no	Allow the HTA content to be selected by mouse or keyboard.
SHOWINTASKBAR	*yes*, no	Show the HTA in the Windows taskbar alongside other running applications.
SINGLEINSTANCE	yes, *no*	Only one version of the HTA can run at a time.
SYSMENU	*yes*, no	Turn on the system menu (the upper left icon or menu in the title bar).
VERSION	a string	Set or read the version number of the HTA.
WINDOWSTATE	*normal*, minimize, maximize	Specify to open the HTA in a minimized, maximized, or normal state.

Notice that in the `HTA:APPLICATION` definition for the following *ClientCheck* script, the scroll bars are turned off for the HTA. This means that the entire window will not have scroll bars, but internal elements, such as the `<div>` element, which contain the output of the script, can still have scroll bars, as shown in Figure 9.2. The use of multiple dedicated areas within the HTA is ideal for separating an application configuration area from the actual results output. The *ClientCheck* HTA frontend GUI consists of an HTML table made of three rows: the top row contains the configuration data (where the buttons and checkboxes reside); the smaller middle row provides the status during the application execution; and the large bottom row contains the output of the application.

```
<html>
<!------------------------------------------------------------
* ClientCheck.hta
------------------------------------------------------------>
```

```
<HTA:APPLICATION
      id="ClientDiagnostics"
      APPLICATIONNAME="Client Diagnostic Tool"
      SCROLL="no"
      SINGLEINSTANCE="yes"
      WINDOWSTATE="normal"
      BORDER="thick"
      BORDERSTYLE="normal"
      MAXIMIZEBUTTON="yes"
      MINIMIZEBUTTON="yes"
      SYSMENU="yes" />
```

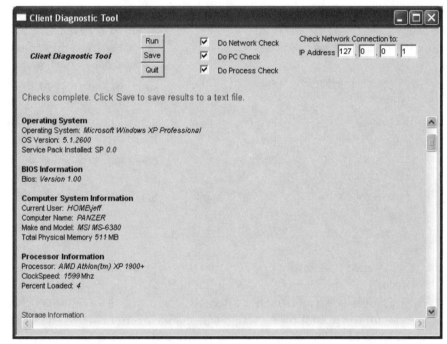

FIGURE 9.2 The output of the *ClientCheck* HTA is contained within its own container managed by that container's scroll bars.

The <head> element is where the title, styles, and script are defined. (A Web page may also define the metatags and search information here as well.) The requirements of this tool stipulated easy execution without an installation process. The entire tool is contained within a single file, including the cascading style sheet (CSS), which defines the look and feel of the application, including font style, size, and color. The CSS styles are defined within the <style> tags.

Alternatively, if you design many HTAs, you can pull an external CSS from a file location accessible to all scripts. This step ensures that all of your HTAs have a similar look and feel.

Defining Styles

Styles control the format of HTML elements and can be applied to most HTML elements. Inline text can also be assigned styles by enclosing them within division <div> or span <span> tags. Styles that begin with a dot (.) are inherited from the previously assigned style. Styles that begin with a hash sign (#)are applied to the element of the same id. Elements can be assigned IDs, which can be referred to programmatically. In the next example, the styles #output and #update apply only to the elements with id= "output " and id= "update ".

```
<head>
<title>Client Diagnostic Tool</title>

<style>
body
{background-color:#e8e8e8; font-family:Arial; font-size: 8pt;
margin-top:5px; margin-left:5px; margin-right:5px; margin-bottom:5px;}
.tbox
{font-family:Arial; font-size:8pt; width:25px;}
.button
{font-family:Arial; font-size:8pt; width:35px;}
.table
{font-family:Arial; font-size:8pt;}
#output
{font-family:Arial; font-size: 8pt;}
#update
{font-family: Arial; font-size: 10pt;color="Red";}
</style>
```

The Programming Core

The <script> element holds the subroutines and functions of the application and declares the scripting language (for example, VBScript). The Option Explicit command forces all variables to be explicitly declared. Declaring variables is required in strongly typed languages like C#, and using Option Explicit to force variable declaration (even in a weakly typed language such as VBScript) leads to good programming habits.

Define variables to be used across all functions and subroutines as Public. In the following code, the variable g_strTemp contains the path and filename of the text file that is used by all of the program's subroutines to store temporary data.

The g_ prefix denotes the variable as global and helps remind you (and others that read your code) of its special status.

```
<script type="text/vbscript">
Option Explicit

public g_strTemp
```

Window_Onload

The subroutine Window_Onload is called automatically when the HTA is run (or more specifically, when the _Onload event is fired). This process sets the initial parameters of the HTA, including its size as well as instructions to initially disable the Save button (because there is no data to save yet). The window property references the current window, and cmdRun and cmdSave are element IDs. Element IDs make the elements programmatically available to any of the procedures within the script.

```
'-------------------------------------------------------------------
' Initialize the Application
'-------------------------------------------------------------------
Sub Window_Onload
    Dim iReturn
    Dim objFSO
    Dim objShell
    Dim objLocator
    Dim strHTML
    window.ResizeTo 640,480
    cmdRun.disabled = False
    cmdSave.disabled = True
```

The next segment of code provides basic error handling. The application turns on error handling and then attempts to create a new WMI object. A failed attempt automatically sets the Err variable to the error code (which will be a non-zero value). The script looks for a non-zero value and displays a bit of helpful information to assist the user with correcting the problem.

TIP

While you are debugging, remember to disable the On Error... statement, because it hides true errors from you (by simply passing right over the offending statements, as it is supposed to do). Although this sounds obvious, you'd be surprised how often this might slow down your hunt for software bugs.

```
On Error Resume Next
Set objLocator = CreateObject("WbemScripting.SWbemLocator")
If Err <> 0 Then
    strHTML=Heading("The WMI Core Components do not " &_
```

```
         "appear to be installed.","Red") & " As a result, both " &_
         "the PC Check and Process Check have been disabled. <br><br>"
      strHTML=strHTML & "<i>To run diagnostics, please download " &_
         "and install the core WMI services for Windows.<br>" &_
         "Download these files from Microsoft <a href=" & chr(34) &_
         "http://msdn.microsoft.com/library/default.asp?url" &_
         "=/downloads/list/wmi.asp" & chr(34) & ">here</a>." &_
         "</i><br><br>"
```

If the system does not support WMI, the script disables the diagnostic tests that rely on WMI, Check Process and Check PC. Each HTML input element within the <body> portion of the program has a unique ID. The Check Process and Check PC checkboxes have IDs of chkProcess and chkPC, respectively. These IDs make these elements available to any of the procedures within the script, allowing us to disable them in this Window_Onload procedure by setting disabled to True and checked to False. Likewise, any other subroutine can set a value, check or uncheck a checkbox, get or set a textbox value, or disable a control.

```
         output.innerHTML= strHTML
         chkPC.checked=False
         chkPC.disabled=True
         chkProcess.checked=False
         chkProcess.disabled=True
      End If
```

In the following code, the Public variable g_strTemp is defined as a file residing in the %temp% folder of the user's computer. This file is used to redirect output from the execution of external binaries (for example, ping, tracert, and ipconfig) called from this program. Using the %temp% folder is a good idea, especially if you do not have control or knowledge of the computer on which the script is running. This special folder generally permits read and write access to normal users, whereas other folders may be restricted and cause the program to halt.

```
         On Error Goto 0
         Set objShell=CreateObject ("WScript.Shell")
         g_strTemp=objShell.ExpandEnvironmentStrings("%temp%") &_
          "/tempdiagnostics.txt"
      End Sub
```

PerformDiagnostics

The PerformDiagnostics subroutine is called from the OnClick event of the Run button. This effectively is the launching point of the bulk of the program. The application may take a bit of time to complete its processing. For this reason, all of the buttons are temporarily disabled while it is running. Disabling the buttons prevents the user from trying to click a button while the program is running.

The large area at the bottom of the application is defined by `id="output"`, and that area is used by the application to display all of the results of the diagnostics performed by the program. The property `innerHTML` is used to set or retrieve the HTML contents of the element. For example, the code `output.innerHTML="Client Diagnostics Tool."` sets the HTML of the output element to that text. This process is used throughout the remainder of this script to update the program display of both the program updates as well as program output.

Depending on the `checked` state of the three diagnostic checkboxes, the program calls each of the three subroutines that actually perform the individual tests.

```
'-------------------------------------------------------------------
' Run the diagnostics
'-------------------------------------------------------------------
Sub PerformDiagnostics
    cmdQuit.disabled=True
    cmdSave.disabled=True
    cmdRun.disabled=True
    output.innerHTML="Client Diagnostics Tool."
    If chkPC.checked = true Then DoPCCheck()
    If chkProcess.checked = true Then DoProcessCheck()
    If chkNet.checked = true Then DoNetworkCheck()
    cmdSave.disabled = False
    update.innerHTML="Checks complete. Click Save to save results " &_
      "to a text file."
    cmdQuit.disabled=False
    cmdSave.disabled=False
    cmdRun.disabled=False

End Sub
```

DoPCCheck

The `PC Check` subroutine uses WMI to demonstrate how to retrieve a variety of information from a local computer. The results are formatted and displayed in the HTA output `<div>`, and the status is displayed in the update `<div>`.

```
'-------------------------------------------------------------------
' Run PC Check
'-------------------------------------------------------------------
Sub DoPCCheck()
    Dim objLocator, objServices
    Dim colService, colQueryResult
    Dim objItem, objShell
```

```
Dim strWQL, strHTML
Dim iFreeSpace
update.innerHTML="Running PC Check.<hr/>"
```

As used before, the Heading function converts a basic unformatted string and color into a predefined HTML formatted string. The formatted string includes the appropriate style tags to set the string color, plus spacing and horizontal rule tags. The DoPCCheck subroutine returns data from five different WMI classes: Win32_OperatingSystem, Win32_BIOS, Win32_ComputerSystem, Win32_Processor, and Win32_LogicalDisk. The procedure calls the ExecQuery to return a collection of objects for each of these classes and displays specific properties. Although the script must include individual calls of the ExecQuery method, the initial WMI setup of the Locator object and server connection can be set once and reused for the subsequent queries. This is a space-saving measure, and shorter code usually equates to fewer errors.

```
strHTML=Heading("Local Computer Information","Green")
On Error Resume Next
Set objLocator = CreateObject("WbemScripting.SWbemLocator")
Set objServices = objLocator.ConnectServer(".")
strWQL="Select * from Win32_OperatingSystem"
Set colQueryResult = objServices.ExecQuery(strWQL)
```

The five classes are each enumerated in a similar fashion using a For…Each loop to iterate the returned collection. The data from these five classes supports the design requirement of collecting the processor, disk, and memory information about the client computer. The individual properties are collected, formatted, and displayed. The output is formatted using HTML syntax. Rather than updating output.innerHTML, the procedure builds a single string named strHTML. At the end of specific tests, strHTML is appended to the existing output.innerHTML and rendered by the browser.

```
strHTML=strHTML & "<br><b>Operating System</b><br>"
For Each objItem In colQueryResult
    strHTML=strHTML &_
    "Operating System: <i>" & objItem.Caption & "</i><br>" &_
    "OS Version: <i>" &  objItem.Version & "</i><br>" &_
    "Service Pack Installed: SP <i>" &_
    objItem.ServicePackMajorVersion & "." &_
    objItem.ServicePackMinorVersion & "</i><br>"
Next

strWQL="Select * from Win32_BIOS"
Set colQueryResult = objServices.ExecQuery(strWQL)
strHtml=strHTML & "<br><b>BIOS Information</b><br>"
```

```
For Each objItem In colQueryResult
    strHTML=strHTML &_
     "Bios: <i>" & objItem.Description & "</i><br>"
Next
```

The total physical memory is converted from bytes to megabytes and rounded to the next whole number using the FormatNumber method.

```
strWQL="Select * from Win32_ComputerSystem"
Set colQueryResult = objServices.ExecQuery(strWQL)
strHtml=strHTML & "<br><b>Computer System Information</b><br>"
For Each objItem In colQueryResult
    strHTML=strHTML &_
     "Current User: <i>" & objItem.UserName & "</i><br>" &_
     "Computer Name: <i>" & objItem.Name & "</i><br>" &_
     "Make and Model: <i>" & objItem.Manufacturer & " " &_
     objItem.Model & "</i><br> Total Physical Memory <i>" &_
     FormatNumber((objItem.TotalPhysicalMemory/1024)/1024,0)&_
     " </i>MB<br>"
Next

strWQL="Select * from Win32_Processor"
Set colQueryResult = objServices.ExecQuery(strWQL)
strHtml=strHTML & "<br><b>Processor Information</b><br>"
For Each objItem In colQueryResult
    strHTML=strHTML &_
     "Processor: <i>" & objItem.Name & "</i><br>" &_
     "ClockSpeed: <i>" & objItem.MaxClockSpeed & " </i>Mhz<br>" &_
     "Percent Loaded: <i>" & objItem.LoadPercentage & "</i><br>"
Next
```

Until this point, the properties have been generally unique (unless you have run this on a multiprocessor system, which would have returned multiple Win32_Processor objects). A notable exception is with the enumeration of the Win32_LogicalDisk collection, which likely includes more than one drive. The script expects a larger data set and builds an HTML table to display the results. The table and first row containing the header information is first constructed.

```
strWQL="Select * from Win32_LogicalDisk"
Set colQueryResult = objServices.ExecQuery(strWQL)
strHTML=strHTML &_
 Heading("Storage Information","Green") &_
 "<table Border=1 class=table width=600px>" &_
 "<tr style='FONT-WEIGHT: bold'><td>Drive</td><td>" &_
 "Free Space</td><td>Description</td></tr>"
```

Next, a loop iterates through each object in the collection and creates a new table row for every logical drive. Notice that the following code includes an additional test to verify that objItem.FreeSpace is a number. If the function IsNull returns True, the objItem.FreeSpace is set to 0. This setting is necessary to account for drives with removable media, such as a floppy or a CD-ROM drive. If a disk is not inserted, the FreeSpace of that drive is reported as Null. The Null value returns an error in the subsequent method, FormatNumber, which expects an input parameter of a number. Lastly, the string strHTML is appended to the existing HTML contained within the <div> with an element id="output".

```
For Each objItem In colQueryResult
    If Not IsNull(objItem.FreeSpace) Then
      iFreeSpace=(objItem.FreeSpace / 1024)/1024
    Else iFreeSpace=0
    End If
    strHTML=strHTML &  "<tr>" &_
      "<td>" & objItem.Caption & "</td>" &_
      "<td>" & FormatNumber(iFreeSpace,2) & " MB free</td>" &_
      "<td>" & objItem.Description & "</td>" &_
      "</tr>"
Next
strHTML=strHTML & "</table><br>"

output.innerHTML=output.innerHTML & strHTML
On Error Goto 0
End Sub
```

DoProcessCheck

The DoProcessCheck subroutine uses WMI to retrieve all of the active processes. The process for connecting to the WMI provider and returning a collection from the Win32_Process is similar to the aforementioned WMI calls made in the DoPCCheck function. This subroutine includes additional tests and formatting to display the active processes and a few key properties in an HTML table.

```
'-------------------------------------------------------------------
' Run Process Check
'-------------------------------------------------------------------
Sub DoProcessCheck()
    Dim strHTML, strCmdLine, strWQL
    Dim objLocator, objServices, objItem
    Dim colQueryResult, strFlagApp
    Dim bFlagApp
    Dim iPageFaults, iWSS, iPWSS
```

```
On Error Resume Next
Set objLocator = CreateObject("WbemScripting.SWbemLocator")
Set objServices = objLocator.ConnectServer(".")
update.innerHTML="Processes.<hr/>"
strHTML=Heading("Current Processes","Green")
```

The variable `strFlagApp` supports a larger design to flag suspicious processes. These processes are identified beforehand by the script administrator. If any are found running on the system, they are flagged and highlighted in bold red type. The flagged applications are also presented below the table of active applications in a separate list. This design is meant to help recipients of the diagnostics information quickly identify previously known compatibility issues with other applications.

```
strFlagApp="Flagged Applications:<br>"
bFlagApp=FALSE
```

The WMI class `Win32_Process` contains a lot of information about active processes, including the process name, the command line that was used to start it, how much memory the process has consumed, the number of page faults caused by the process, and others. Because we want to list all of the processes, we call on the `ExecQuery` method with a basic query string.

An HTML table is used to format the results. The first row contains the table header information for the data that the script displays.

```
strWQL="Select * from Win32_Process"
Set colQueryResult = objServices.ExecQuery(strWQL)
strHTML=strHTML & "<table class=table width=600px border=1>" &_
 "<tr style=" & chr(34) & "FONT-WEIGHT: bold" & chr(34) &_
 "><td>Process</td><td>Command Line</td><td>Memory (KB)</td>" &_
 "<td>Peak Mem(KB)</td><td>Page Faults</td></tr>"
```

The objects and properties of the returned process collection also contain different types of data or none at all. The following code checks that each of the properties contain properly typed and formatted data. This verification also presents an opportunity to add more legible meanings to the data. For example, if a command line is not present, the script replaces the value with `n/a`.

Similarly, the script confirms that each of the values of the current and peak memory utilization for each process is a number and then converts that value to megabytes. If the value is not a number, the script replaces the value with -1 to denote an error.

```
For Each objItem In colQueryResult
    If IsNull(objItem.CommandLine) Then
        strCmdLine="n/a"
```

```
Else strCmdLine=objItem.CommandLine
End If
If IsNumeric(objItem.WorkingSetSize) Then
    iWSS=(objItem.WorkingSetSize / 1024)
Else iWSS=-1
End If
If IsNumeric(objItem.PeakWorkingSetSize) Then
    iPWSS=(objItem.PeakWorkingSetSize / 1024)
Else iPWSS=-1
End If
If IsNumeric(objItem.PageFaults) Then
iPageFaults=objItem.PageFaults
Else iPageFaults=-1
End If
```

Flagged applications are detected using a `Select...Case` statement. This allows for multiple applications to be quickly added to the list. The default flagged application, mshta.exe, is actually the HTA application itself to demonstrate on any system how the flagged application operates and looks. Add a comma-delimited list of strings to flag additional applications.

```
Select Case objItem.Description
    Case "mshta.exe"
        bFlagApp = TRUE
        strHTML=strHTML & "<tr style='COLOR: RED'>"
        strFlagApp=strFlagApp+objItem.Description & "<br>"
    Case Else
        strHTML=strHTML & "<tr>"
End Select
```

A separate HTML table row is built for every process, and the numeric data is rounded to the next whole number.

```
strHTML=strHTML & "<td>" & objItem.Description & "</td>" &_
 "<td>" & strCmdLine & "</td>" &_
 "<td align=right>" & FormatNumber(iWSS,0) & "</td>" &_
 "<td align=right>" & FormatNumber(iPWSS,0) &"</td>" &_
 "<td align=right>" & FormatNumber(iPageFaults,0) & "</td>" &_
 "</tr>"
Next
strHTML=strHTML & "</table><br>"
```

After the presentation of the active processes, the flagged applications are simply listed, and the entire strHTML string (containing all of the processes in formatted HTML) is displayed directly in the HTA.

```
        If bFlagApp Then strHTML=strHTML & strFlagApp
        output.innerHTML=output.innerHTML & strHTML
    End Sub
```

DoNetworkCheck

The `DoNetworkCheck` subroutine uses WSH `Run` methods and `FileSystemObject` methods to drive external applications such as `ping`, `ipconfig`, and `tracert`. These external applications run in a separate process, and their output is redirected to a temporary text file, `g_strTemp`. The design requirements of this script include measuring network latency and the consistency of the network connection. The script uses the network tools `ping` and `tracert`, which are installed on most Windows platforms.

```
'-----------------------------------------------------------------
' DoNetworkCheck
'-----------------------------------------------------------------
Sub DoNetworkCheck()
    Dim bNetworkUp
    Dim iReturn
    Dim objShell, objFSO
    Dim strHTML, strPingAddress, strTRAddress
```

Before running the test, the user can specify the IP address of a remote computer to measure the quality of the connection between the two computers. By default, this value is set to the `127.0.0.1`, or `localhost`. This application uses independent boxes that are tested for valid IP octets. If the IP address is invalid, the subroutine exits. Otherwise, the script concatenates the four octets into a properly formatted IP address and sets the address strings `strPingAddress` and `strTRAddress` to this value.

Alternatively, a single field could be used that accepts either an IP address or a fully qualified domain name (FQDN) of a system. The IP address could be validated for proper form by using a regular expression.

```
    If txtIP1.Value > 223 or txtIP2.Value > 255 or txtIP3.Value > 255_
    or txtIP4.Value > 254 or txtIP1.Value < 1 or txtIP2.Value < 0_
    or txtIP3.Value < 0 or txtIP4.Value < 0 Then
        update.innerHTML="IP address is invalid. Please correct."
        Exit Sub
    End If

    strPingAddress=txtIP1.Value & "." & txtIP2.Value & "." &_
     txtIP3.Value & "." & txtIP4.Value
    strTRAddress=txtIP1.Value & "." & txtIP2.Value & "." &_
     txtIP3.Value & "." & txtIP4.Value
```

The first process gathers the network configuration of the machine from the command ipconfig.

NOTE

A design assumption is that all of the client machines that will run this program will have ipconfig (as well as the other external applications) already present on the system. However, any time you run an external application, be sure that you are absolutely certain of how it will run. For example, the reg.exe command, which can be used to read and write registry settings from the console, underwent a major overhaul on Windows XP, and most of the parameters changed. If you can't deliver the version of the external application that you tested on (such as by copying it locally to the machine and calling the executable with the exact path to the known application), then be aware of what parameters you use in case they change in future versions.

The script uses the WScript.Shell.Run method, as opposed to the Exec method, which supports standard streams (StdIn, StdOut, StdErr). This design decision was made because this program may be run on a number of Windows computers. The Exec method was introduced in later WSH versions, and it would fail on earlier versions of Windows. The Run method works on Windows platforms with older version of WSH. The Run method calls the command ipconfig /all and redirects the output (using the greater than sign >) to the temporary data file. The remaining parameters in the Run method instruct it to not display a window (0) and to pause the script until the external command is finished running (true).

```
update.innerHTML="Collecting IP and Network Information.<hr/>"
    Set objShell = CreateObject("WScript.Shell")
    iReturn=objShell.Run("%comspec% /C ipconfig /all > " &_
      g_strTemp,0,true)
```

Immediately following the execution of ipconfig, the script calls a function ProcessResults to do just that. ProcessResults reads the contents of the text file that contains the ipconfig output data and returns the results in a string. Any specialized processing to the results (such as displaying only the IP address) can be done in this function after the string has been returned from the ProcessResults function. ProcessResults is a generic function that simply reads a file and returns its contents to the calling function. Its mechanics are covered in the next pages where the function is called out.

```
strHTML=ProcessResults(g_strTemp)
output.innerHTML=output.innerHTML &_
  Heading("IP Configuration Settings","Green") & strHTML
```

Measuring network connections with `tracert` or multiple `ping`s or connecting to a computer using any of the `net` commands (such as `net use`) can sometimes take quite a bit of time—especially if the destination computer is down and not responding. To detect this status before initiating one of these longer-timeout network commands, this script includes a preliminary test to check the status of the network connection. A single `ping` request is sent, and if a reply is returned, a flag, `bNetworkUp`, is set and the more comprehensive tests are run. `Ping` is a good solution, because it can be configured to be small and quick to complete (for example, a single `ping`).

```
update.innerHTML="Checking Network Connection.<hr/>"
    bNetworkUp = False
    iReturn=objShell.Run("%comspec% /C ping " & strPingAddress &_
    " -n 1 > " & g_strTemp,0,true)

    strHTML=ProcessResults(g_strTemp)
    If (InStr(strHTML, "Reply")) Then bNetworkUp = True
    If bNetworkUp Then
        strHTML= Heading("Network appears to be online.","Green")
        output.innerHTML= output.innerHTML & strHTML
```

The requirements specify that the script measure the consistency of the network connection. The next test sends 10 `ping`s to the remote machine and displays the results. For a problematic connection, this test can identify either high latency or dropped packets. Obviously, if the problem is intermittent, this test might not catch it. However, sending out a number of `ping`s gives a general sense of the quality and speed of the connection. This `ping` test is similar to the previous reconnaissance `ping` test; however, sending out 10 `ping`s takes longer (especially if the machine is down or filtered by a firewall), and waiting for the return of all of these `ping`s might frustrate the user of this tool.

```
update.innerHTML="Performing Ping.<hr/>"
        iReturn=objShell.Run("%comspec% /C ping " & strPingAddress &_
        " -n 10 > " & g_strTemp,0,true)
    strHTML= Heading("Ping Test to " & strPingAddress &_
    " : Succeeded","Green")

    strHTML= strHTML & ProcessResults(g_strTemp)
```

The `traceroute` command measures the latency between every hop between the client and its destination. This test takes the most amount of time to complete, but it pinpoints sources of latency (such as if a particular ISP or backbone was having problems). The script calls the command in a similar fashion to the aforementioned `ping` command.

```
update.innerHTML="Performing Traceroute.<hr/>"
    iReturn=objShell.Run("%comspec% /C tracert " & strTRAddress &_
      " > " & g_strTemp,0,true)

      If Err Then alert("An error occurred")
    strHTML=strHTML & Heading("Trace Route Results","Green") &_
     ProcessResults(g_strTemp)
```

If the network is down (for example, bNetworkUp = False), then the script displays a message informing the user. Lastly, all of the results of the network checks are displayed to the user.

```
Else
    strHTML=Heading("Network appears to be offline.","Red") &
strHTML
    End If
    output.innerHTML= output.innerHTML & strHTML
    On Error Goto 0
End Sub
```

Heading

Previously, the Heading function was briefly described as it was first called out. The following code contains the actual function. This function takes a text string and a color as input and returns a formatted HTML string.

```
'------------------------------------------------------------------
' Format the Heading
'------------------------------------------------------------------
Function Heading(strHTML, strColor)
Heading = "<br><br><div style=" & chr(34) & "color=" & strColor &_
  chr(34) & ">" & strHTML & "</div><hr/><br>"

End Function
```

ProcessResults

The function ProcessResults reads a text file and returns the contents to the calling procedure as a string. This function does not perform any text parsing (such as blank line stripping). Error checking is limited to a call to the method FileExists. Because the text file read by this function has been created by the script, additional error checking is not required. The method uses the OpenTextFile method to open the file. OpenTextFile parameters are covered in Chapter 3 under the discussion of the FileSystemObject and in Chapter 4 in the discussion of the ReadFile function.

After the file has been read, the function deletes the file. Again, this is possible and safe to do because it is this script that generated the file initially. Deleting the file when finished with it is a good thing to do because it keeps the system clean and does not leave any residual files on it after the script has exited.

```
'-----------------------------------------------------------
' Read a text file and iReturn the contents
'-----------------------------------------------------------
Function ProcessResults(strFileName)
Dim objFSO, objTextFile
Dim strHTML, strReadLine

Set objFSO = CreateObject("Scripting.FileSystemObject")
If (objFSO.FileExists(strFileName)) Then
   Set objTextFile = objFSO.OpenTextFile(strFileName,1)
   strHTML=""
   Do While objTextFile.AtEndOfStream <> True
       strReadLine=objTextFile.ReadLine
       strHTML=strHTML & strReadLine & "<br>"
     Loop
     objTextFile.Close
   objFSO.DeleteFile(strFileName)
   Else    strHTML="Error reading file."
End If

ProcessResults=strHTML

End Function
```

SaveResults

The SaveResults function writes the output generated by the script diagnostics to a text file specified by the user. The VBScript function InputBox provides an easy way to ask the user for the desired filename, as shown in Figure 9.3. The parameters of the InputBox function follow:

```
strAnswer = InputBox(strPrompt, strTitle, strDefault, xPosition,
yPosition, helpfile, context)
```

strAnswer is a string containing the value entered by the user. The only required parameter, strPrompt, is the text to display to the user asking for his input. The strTitle labels the title bar of the InputBox, and the strDefault is a string that prepopulates the text field. Optionally, specify the location of the dialog box by setting the xPosition and yPosition. The InputBox also supports a linked help file and context, which could provide additional information to the user.

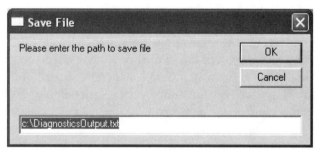

FIGURE 9.3 The VBScript function `InputBox` provides an easy method for getting user-supplied information.

The `SaveResults` subroutine exits if the user entered a blank entry but does not otherwise check for the use of illegal characters in the filename. Next, the script creates a new text file using the method `CreateTextFile` and writes, line-by-line, the content of the `output.innerText`.

The `innerText` property is similar to the `innerHTML` property that this script uses frequently to post diagnostic data to the `<div id="output">`. However, whereas the `innerHTML` supports HTML, the `innerText` supports only the plain text (no formatting). This characteristic is useful because we want only the text to be written to the save file. Otherwise, had the script used `innerHTML`, the save file would contain actual HTML, which would be more difficult for the user to read.

Alternatively, we could have written a savefile.html *that is formatted as an HTML document, in which case we could have preserved the formatting and written out* innerHTML.

```
'--------------------------------------------------------------------
' save dialog
'--------------------------------------------------------------------
Sub SaveResults
   Dim objFSO
   Dim strFileName
   Dim objTextFile

   Set objFSO = CreateObject("Scripting.FileSystemObject")
   strFileName = InputBox("Please enter the path to save file"_
     , "Save File","c:\DiagnosticsOutput.txt")
   If strFileName = "" Then Exit Sub

   Set objTextFile = objFSO.CreateTextFile(strFileName)
   objTextFile.WriteLine output.innerText
   objTextFile.Close

End Sub
```

QuitScript

The `QuitScript` subroutine cleans up any leftover files and exits the function. The previous `ProcessResults` function should have adequately cleaned up each of the text files, and this extra step here is generally superfluous. However, providing a shutdown, cleanup type function on exit is generally a useful and proper step to take. The method `window.close()` is an HTA method that closes the current window.

```
'-------------------------------------------------------------------
'Quit
'-------------------------------------------------------------------
Sub QuitScript
    On Error Resume Next
    Set objShell = CreateObject("WScript.Shell")
    Set objFSO = CreateObject("Scripting.FileSystemObject")
    objFSO.DeleteFile g_strTemp
    Set objFSO = Nothing
    window.close()
End Sub
```

Creating the Frontend with HTML

One of the advantages to using an HTA as the wrapper or shell of your script is the ease of formatting a frontend. This ease is the result of the fact that the HTA is a basic Web page and supports HTML formatting.

When the HTA is opened, the browser parses and renders all of the HTML, and then it calls `Window_onload`. This function could immediately call other subroutines. The next code uses standard HTML to define the look and interactivity of the HTA frontend. The applications buttons, checkboxes, and text fields are defined using HTML `input` elements. Subroutines are initiated by defining `onClick` events. The script references element properties by their assigned `id`. `Classes` define the style to be used in presenting the elements.

Also notice that the comments are denoted using the VBScript syntax of an apostrophe (`'`) *inside* the script tag. Outside the script tag, use the comment convention `<!--- comment --->`, as seen in the opening lines of this script.

```
'-------------------------------------------------------------------
'Format the HTA using standard HTML
'-------------------------------------------------------------------
</script>
</head>

<body>
```

The script is displayed in a single table cell comprising three rows and four columns, as illustrated in Figure 9.4. Diagramming the construction of the table layout on a piece of paper is helpful in programming the table elements. In this script, the bottom two rows span all four columns contained within the top row. The size of the table elements is also important in ensuring that the HTA looks good and that if the user resizes the HTA, it scales appropriately. Figure 9.4 also shows the initial sizes of the elements. Two of the sizes are set to *, which means that the browser will float that dimension to make up the total size of the defined container width or height. The rest of the elements remain a fixed size. This results in HTA window-resizing behavior that mimics a typical application.

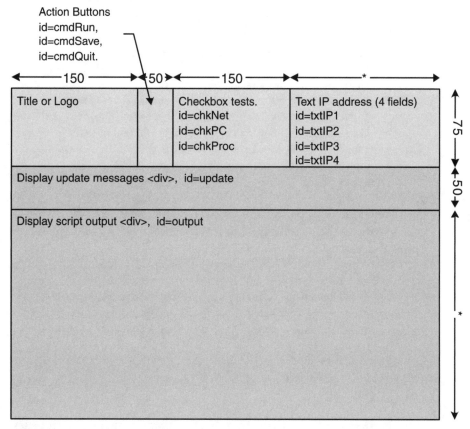

FIGURE 9.4 The *ClientCheck* HTA frontend is constructed using a single HTML table broken into three rows and four columns.

```
<table width="100%" height="100%" class="table">
    <tr height = 75px>
    <td align="center" width=150px><b><i>Client Diagnostic
Tool</i></b></td>
    <td width=50px
```

The HTML input elements provide the user interaction with the script. The elements are referenced in the script by their id attribute. The elements support a number of properties. The Style button was defined earlier in the <style> section of this script and is associated with the elements by the defined class (for example, class="button"). The value sets the text label of the button. Many HTML elements also support events (such as onClick) that fire when certain actions are recorded. For example, in the following code, when the user clicks the Run button, the input element onClick event calls the subroutine PerformDiagnostics, which initiates the diagnostics tests. Similarly, clicking the Save and Quit buttons starts the SaveResults and QuitScript procedures, respectively.

```
<input id=cmdRun  class="button" type="button" value="Run"
 name="cmdRun" onClick="PerformDiagnostics()"><br>
<input id=cmdSave class="button" type="button" value="Save"
 name="cmdSave" onClick="SaveResults()"><br>
<input id=cmdQuit class="button" type="button" value="Quit"
 name="cmdQuit" onClick="QuitScript()"><br>
</td>
```

The checkboxes permit the users to define what tests they wish to perform. Specify whether the checkbox elements are initially checked by including the definition CHECKED.

```
<td width=150px>
<input id=chkNet class="button" type="checkbox" CHECKED >
 Do Network Check<br>
<input id=chkPC class="button" type="checkbox" CHECKED >
 Do PC Check<br>
<input id=chkProcess class="button" type="checkbox" CHECKED >
 Do Process Check<br>
</td>
```

The input element text defines the text fields, which this script uses to collect the four octets of the IP address. The maximum length and initial values are set here as well.

```
<td width=* valign=top>
Check Network Connection to:<br>
IP Address
<input id=txtIP1 class="tbox" type="text" MAXLENGTH=3 Value="127">.
<input id=txtIP2 class="tbox" type="text" MAXLENGTH=3 Value="0">.
<input id=txtIP3 class="tbox" type="text" MAXLENGTH=3 Value="0">.
<input id=txtIP4 class="tbox" type="text" MAXLENGTH=3 Value="1">
</td>
</tr>
```

The second row consists of a single column that spans all four columns of the previous row. In order to control the text in this area, the table data cell is assigned an id="update". Many of the functions that perform the diagnostic tests update their status by setting the property update.innerHTML to the update message.

```
<tr>
    <td  height = 50 colspan=4 id="update">
    Select tests and click Run to begin.<hr/></td>
</tr>
```

The output of the text is displayed in a similar fashion to that of the previous update text. However, the next code uses a slightly different approach. Rather than assigning the element id="output" to the table data cell, a new division element <div> is created and that is assigned the id="output". The <div> (and <span>) elements permit inline text to be independently referenced. Because this text is the only item in this table cell, the id="output" should be assigned to the table data cell, but this alternate method demonstrates associating a body of text with an id, which can be used if table elements are not used.

```
<tr>
    <td height = * colspan=4><div id="output"
    style= "width:100%; height:100%; overflow:scroll"></div>
    </td>
</tr>

</table>
</body>
</html>
```

SUMMARY

The *ClientCheck* HTA script shows how easy it is to incorporate a basic frontend to a script by writing your script as an HTA. As opposed to running within an Office application, this script relies on a browser to run, which increases the number of machines that can run it. Be careful, however, because these scripts can pack a powerful punch by running in the local security context of the currently logged-in user. But it is this flexibility and power that help make HTAs a powerhouse in your scripting arsenal.

KEY POINTS

- Consider using hypertext applications (HTA) to provide an easy-to-design user interface to your scripts.
- An HTA provides user interactivity to your scripts, such as checkboxes, radio buttons, text fields, and other controls.
- Use the WMI method `ExecQuery` to efficiently return WMI class data from a local or remote computer.
- Be sure to understand the object type returned from a method or function—for example, does a method return an object or a collection of objects? What is the type of the object that it returns? Use the software development kit (SDK) to look up specifics on a new method that you might be unsure of. This script uses a number of `ExecQuery` method calls, which return a collection of objects that must be enumerated before they can be listed.
- If you can't (or don't need to) script a solution to a problem, consider shelling out and harnessing an existing tool and then parsing the results. Although perhaps not as elegant as coding your own solution, many IT-based tools are purposefully quick and dirty, and shelling out works just fine.

10 Worm Vulnerability Detector

OVERVIEW

Worms, virus attacks, and other malware that indiscriminately probe and attack any server pose an increasing threat to our IT infrastructure. Many of the more successful worms like *Code Red* and *Nimda* actually exploited known vulnerabilities patched by the software vendor months prior to the attacks. As a result of the increasing frequency and distribution of these worms, antivirus and security software vendors continue to strengthen their product offerings. However, even with this heightened sensitivity to new threats, IT administrators still face the inevitable and daunting task of checking every computer for vulnerable services and applied updates.

Consider a thin, rapid-response script that could assist you in your reconnaissance of vulnerable servers. Using some of the technologies that we have explored thus far—such as running command-line utilities against each of your computers, looking for specific services running, and checking for applied patches—we can create a script that assesses your vulnerability to a particular worm or exploit.

Although this script assists in discovering vulnerable machines, it should not replace dedicated security infrastructure such as intrusion-detection systems, firewalls, or antivirus software. This script provides a framework to automate some of the steps that you might manually take to assess these vulnerabilities.

SCENARIO

Last night a new worm made the six o'clock news. This worm exploits vulnerability in unpatched Microsoft Internet Information Server (IIS) installations. As the security administrator of a software company, you want to scan your network for potentially vulnerable systems. Microsoft released a patch for this particular exploit a while back, and because you follow a good patch triage and deployment plan for your corporate servers, you are not too worried about those systems. But you do want to ensure that developers who might install IIS on their own work systems patch their systems as well. You search the Internet for available network scanning tools that scan a domain for vulnerable machines but to no avail; these scanners have not yet been updated for this particular worm, and you want to get a jump on assessing your environments exposure now.

After reading the security bulletin describing the worm, you understand that the worm only attacks Microsoft IIS 5.0 (Windows 2000) on TCP Port 80 (the standard Web port), and the update that patches the system and closes the vulnerability is Microsoft MS03-007 (Q815021).

With this information, you want to quickly identify those machines that may be vulnerable and then take additional action. For all the systems in your domain, you want to scan running instances of IIS, look for open ports TCP 80 and TCP 443, and check whether Q815021 is installed. By taking a multipronged approach such as this, you hope to discover a reliable list of systems from which you can initiate remediation efforts.

ANALYSIS

An easy-to-configure scanner such as this improves the response rate by quickly identifying possibly vulnerable machines via a central console. This script automates and centralizes the scanning of computers in your domain using fairly cursory checks—but, nonetheless, checks that can help focus your remediation energy during the early hours of a possible attack.

As such, your needs map to the technical requirements listed in Table 10.1.

TABLE 10.1   Translating business objectives into technical requirements

Business objectives	Technical requirements
Scan computers in your domain.	Use the *DumpComputerList* script presented in Chapter 8 to generate a list of computers in the specified domain. Ping each machine to see if it is responding on the network, and then for each active computer in this list, perform each of the selected scans.
Scan for open Windows services (such as Microsoft World Wide Web server named *w3svc*).	Use WMI `Win32_Service` class to connect to the remote computer and interrogate the `State` of the selected service. Highlight services that are `Running` or `Stopped`.
Scan for open network ports.	Harness a third-party command-line-based port-scanning tool to scan the specified port of the target computer.
Scan for installed Microsoft QFE updates.	Use WMI `Win32_QuickFixEngineering` class to determine if the specified QFE update is installed.

This script consists of several functions that scan for services or systems vulnerable to a specific worm. The script checks for specific services running, harnesses the popular *nmap* network scanner to look for specific open ports, and looks to see whether a specific Quick Fix Engineering (QFE) hotfix has been applied.

NOTE

The script harnesses the features of the popular command-line port scanner nmap, *written by Fyodor. Download the Windows version of the* nmap *scanner and detailed installation instructions from* www.insecure.org/nmap/nmap_ download.html. *To use* nmap, *you will also need the* WinPcap *packet capture* architecture for Windows, which you can download from http://winpcap.polito.it/ install/default.htm. This software needs to be installed only on the computer performing the actual scan. If you choose not to use the nmap scanner, you can still perform the other tests by selecting those you wish to run from the setup worksheet. You can use your own command-line port scanner by editing the function PortCheck to handle the syntax arguments and output of another port scanner.

NOTE

This script uses the Win32_QuickFixEngineering WMI class to collect a list of all installed updates on a remote computer. A bug exists in calls to the Win32_QuickFix-Engineering class on earlier versions of Windows 2000 computers (pre-SP3), which may result in the script hanging and causing the client computer's Winmgmt.exe process to increase in both memory and CPU utilization. Microsoft fixed this bug in both a hotfix for pre-SP3 computers and in Windows 2000 SP3. More information about this problem and its resolution can be found at the Microsoft Knowledge Base (http://support.microsoft.com/default.aspx?scid=kb;en-us;279225).

For this reason, you should specify scanning only for QFE updates if you know that your target machines have SP3 or the hotfix Q279225.exe applied. A link to the hotfix is also included in the previous Knowledge Base article. If you do scan a computer and Microsoft Excel reports an error that Excel is waiting for another application to complete an OLE action, you have probably encountered this bug. You will see the name of the current computer and should terminate Excel, check the Winmgmt.exe process on the remote computer, and remove the computer from future scanning of QFE updates (or patch the computer as outlined in the KB article).

This script builds from the Excel-based *DumpComputerList* script presented in Chapter 8 and uses this tool to build its list of computers to scan and to record output. The functions and subroutines used by this script follow:

SetupScan: Parses the *Setup* worksheet for desired scans.

■ Updates the *Output* worksheet with the scans identified on the *Setup* worksheet.

SystemScan: Iterates through the list of computers and calls the individual scanning functions for scans to perform.

■ Supports single or multiple selected computers.
■ Supports single or multiple selected scans.

Pingable: Determines if a system responds to network pings.

■ Uses the Shell object Run method to execute a ping command.
■ Parses the output using a FileSystemObject.

ServiceCheck: Checks for specific running services.

■ Returns the state of a service running on a remote computer using WMI Win32_Services class.

PortCheck: Uses a command-line port scanner to scan a machine for an open port.

■ Uses the Shell object Run method to execute a custom *nmap* run against a remote computer. *Nmap* is a popular, open source port scanner.

■ Parses the output using a FileSystemObject.

QFECheck: Checks for specific QFE hotfix.

■ Performs a WMI ExecQuery method to determine if a QFE hotfix has been applied.

DisplayResults: Checks for specific hotfix or update.

■ Displays formatted text in a specified cell in Excel.

Distribution and Installation

Use this evolved script in much the same fashion as its predecessor, *DumpComputerList,* presented in Chapter 8. Define the parameters used to drive the script on the enhanced *Setup* worksheet. First, like the *DumpComputerList* script in Chapter 8, enter the domain name and the name of the *Output* worksheet and click the first button (Pull Computers from Directory) to initiate the directory query of a specified domain. The script creates and populates the *Output* worksheet using the same script presented in Chapter 8. Next, specify the scan information in the Service Scan, Port Scan, and QFE Scan boxes shown under the second button, as seen in Figure 10.1. Place an x next to the scans you wish to include on the *Output* worksheet. With this selectivity, you can build a substantial list of scans on this worksheet and then toggle only those you wish to copy to the *Output* worksheet. (Before you run the scan on the *Output* worksheet, you can further specify the scans to actually run.) The Column parameter specifies to which column number in the *Output* worksheet that scan will be added. Update the path to the *nmap* executable (or you can clear the cell if *nmap* is in your Path environment variable).

In Excel, columns can be referenced by letter (A, B, or C) or by number (1, 2, or 3). Scripting generally uses the numerical column reference because it is easier to iterate using numbers than letters.

When you have added and configured your scans and are ready to update the *Output* worksheet, click the second button (Format Output Worksheet for System Scan). This initiates the subroutine SetupScan, which iterates through each of the three scan lists and adds the scan type and definition into the appropriate columns on the *Output* worksheet. The scan types are service scan (s), port scan (p), and QFE scan (q).

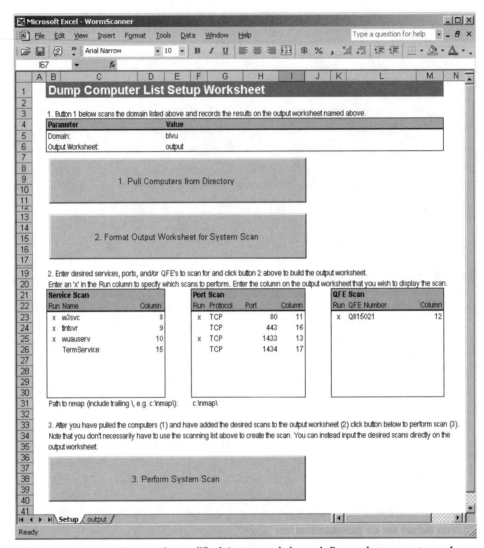

FIGURE 10.1 *WormScanner's* modified *Setup* worksheet defines what scans to perform.

The second button formats only the output worksheet given the parameters listed in the setup worksheet. The actual code that performs the scans (SystemScan) uses data only from the output worksheet to determine what to scan. You could actually skip the second step and manually format the output worksheet if you choose. This second step offers convenience and ensures that scan syntax on the output worksheet is correctly formatted. This is the beauty of designing your own software—you can customize the look, feel, and operation to exactly what you want or need.

After you have clicked the second button to run the script SetupScan, click the *Output* worksheet and confirm that your selected scans have been transposed onto this worksheet, as seen in Figure 10.2. Let's look at some of the differences between this script's *Output* worksheet and the *Output* worksheet of *DumpComputerList* in Chapter 8.

Microsoft Excel - WormScanner

File Edit View Insert Format Tools Data Window Help

	A	B	C	D	E	G	H	I	J	K	L
1	**Computer Listing**										
2	Last Scan: 6/14/2003					Run Scan (x)	x	x	x	x	x
3						(s or p)	s	s	s	p	q
4	Process? (x)	Name	Status	Discovered	Domain	name/port	w3svc	tlntsvr	wuauserv	TCP-80	Q815021
5						Ping Status					
6	x	dc1	Active	6/14/2003	blvu						
7	x	nt	Active	6/14/2003	blvu						
8	x	svr	Active	6/14/2003	blvu						
9	x	wkstn	Active	6/14/2003	blvu						
10		workstn004	Active	6/14/2003	blvu						
11		workstn005	Active	6/14/2003	blvu						
12		workstn007	Active	6/14/2003	blvu						
13		workstn009	Active	6/14/2003	blvu						

Setup output   Ready

FIGURE 10.2 Excel worksheets provide a great canvas to list computers and the scan data on one worksheet.

Column 1 now contains x marks. When run, the scanning script processes only those computers with an x next to their name. This allows you to rescan one or several computers without having to scan all computers again—such as to check if a computer was successfully patched or a service state disabled. Similarly, notice the x marks along row 2. These xs denote which scans should be performed or skipped. Row 3 lists the scan type as either s, p, or q. The script looks for this value to determine whether to perform a service scan, port scan, or QFE scan, respectively. Row 4 contains details about the specific scan. For a service scan, row 4 contains the name of the service for which to look. For a port scan, the script looks for protocol-port of the port scan to run (for example, TCP-80 or UDP-43). For a QFE scan, it looks for the QFE HotFixID. The QFE HotFixID is usually a well-known reference to the update provided by Microsoft.

Return to the *Setup* worksheet and execute the actual scan by clicking the last button (Perform System Scan). The script activates the *Output* worksheet and begins to march across the worksheet, performing scans for every selected computer while displaying its progress along the way. If you do not have the port scanner *nmap* (or other command-line port scanner) installed and configured, the script simply passes by those scans and notes the results with a dash (-). Figure 10.3 shows the results of a basic scan.

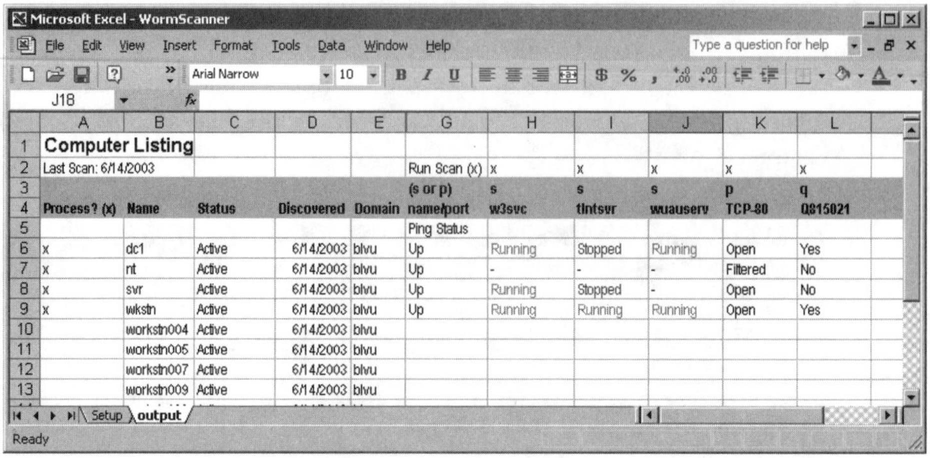

FIGURE 10.3 The script generates a matrix of information to help assess your susceptibility to worms or viruses that might exploit a vulnerable service or system.

The script uses the VBA type library reference to the Microsoft WMI Scripting V 1.2 Library. You may need to downgrade this library to V 1.1 if the newer version is not available on your computer. (From the Visual Basic Editor, select Tools, References. Deselect the reference marked as *Missing*, and look for and add an older version of that library. More about how Excel uses references is covered in Chapter 8.)

Output

The script records its output on the *Output* worksheet named on the *Setup* worksheet as seen in Figure 10.3.

Summary of Solution

This Excel-based script extends the *DumpComputerList* script presented in Chapter 8 by adding functionality to scan selected computers for sources of possible vulnerabilities. The tool scans for running services, looks for open network ports, and checks for installed QFE updates.

OBJECTIVE

An Excel-based tool that scans a directory of computers looking for specific services, open network ports, or installed hotfixes to identify possible worm vulnerabilities.

Excel Object Model

- Worksheet manipulation
- Cell/range formatting

RecordSet

- EOF
- Fields.Item(x).Value

Scripting.FileSystemObject

- DeleteFile
- OpenTextFile
- ReadLine

VBA Options

- Option Explicit
- Option Compare Text

VBA Error Handling

- On Error Resume Next
- On Error Goto 0

User Data Types

- Using Type to declare user-defined data types

WbemScripting.SwbemObject

- ConnectServer
- ExecQuery

WshShell

- Run
- ExpandEnvironmentStrings

THE SCRIPT

This script continues where the *DumpComputerList* script of Chapter 8 left off. Although some basic cell formatting changes were made to the Chapter 8 script, the rest of the script remains intact as presented in Chapter 8. This script is presented in the module named *DumpComputerList* of the *WormScanner.xls* workbook. The *WormScanner* script discussed in the rest of this chapter is presented in the module named *WormScanner*. To access the code, open the workbook *WormScanner.xls* and the Visual Basic Editor (from the workbook, press Alt-F11), and then click the module named *WormScanner*, as seen in Figure 10.4.

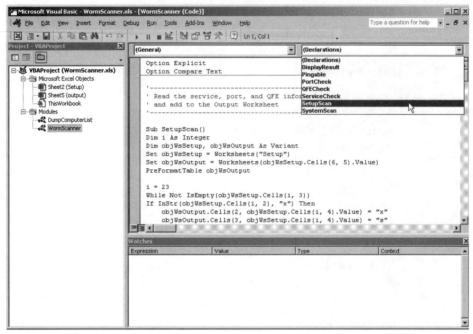

FIGURE 10.4   The WormScanner-specific code is found in module named *WormScanner*.

Notice in Figure 10.4 that the Declarations drop-down list contains all of the subroutines and functions for this particular module. This is a handy method to find a particular subroutine or function in the Visual Basic Editor.

The module named *DumpComputerList* contains the *ComputerList* code presented in Chapter 8. The Declarations drop-down box provides easy access to subroutines and functions within the module.

The script is presented on the CD-ROM as /chapter 10/scripts/wormscanner.xls. This script is contained within an Excel document that requires Excel 2000 or later as well as Windows Script Host 5.6. As with other scripts in this text, remember to test this script in a test environment before deploying to any production systems.

Module-level Code

`Option` statements must be specified for each module, and we have added both the `Option Explicit` and `Option Compare Text` statements to this module. Remember,

`Option Explicit` forces declaration of all variables, and `Option Compare Text` sets the default string comparison method to `Text`.

```
Option Explicit
Option Compare Text
```

SetupScan

The subroutine `SetupScan` iterates through the three scan lists located on the *Setup* worksheet and transposes the scans marked with an x to the specified columns in the *Output* worksheet.

First, the script calls the subroutine `PreFormatTable` to clear any headings that might have existed on the *Output* worksheet, such as any from a previous run. Next, it calls the subroutine `TransposeColumns` with several arguments. `TransposeColumns` is named as such to illustrate the concept of *factoring*. Factoring is the process of boiling down a repetitive procedure of code to the point that minimizes repetition. For example, instead of creating three blocks of code, each looping through various data and displaying a similar output, the `SetupScan` function simply makes three calls to one function. When and if this code ever needs to be maintained, the maintainer need only change the code in one place instead of three.

The scan configuration information consists of the scan type and scan properties. The first set of scan data is the services scan, and it is specified on the *Output* worksheet by a scan type of s. The only property of the service scan is the name of the service (such as w3svc).

The second scan type is a port scan denoted by a p. The port scan properties consist of the protocol (TCP or UDP) and the port number (1-65534). For example, a standard Web server is usually at TCP port 80. The script summarizes this information on the *Output* worksheet as protocol-port (for example, TCP-80).

The last scan performed by this script is to look for installed QFE updates, denoted by a scan type of q. The property for the QFE scan consists of the QFE HotFixID.

In this script, `TransposeColumns` requires three arguments plus an optional fourth. You can deem an argument optional using the keyword `Optional` when declaring the function or subroutine. The arguments in this script represent the operating columns. The first argument, iColName, represents the column number that contains the name of the scan. The second argument, iColColumn, represents the column number that contains the destination column number in which to insert the scan data on the *Output* worksheet. The third argument represents the name of the scan, either s, p, or q for service scan, port scan, or QFE scan, respectively. The fourth argument, iColOptName, is optional and represents the number of a second column that can contain a second name for the scan. It is used only with the port scan, which concatenates the two names such as TCP and 80 to display TCP 80.

```
' ------------------------------------------------------------------
' Read the service, port, and QFE information from setup worksheet
' and adds to the Output Worksheet
' ------------------------------------------------------------------

Sub SetupScan()
Dim i As Integer
Dim objWsSetup, objWsOutput As Variant
Set objWsSetup = Worksheets("Setup")
Set objWsOutput = Worksheets(objWsSetup.Cells(6, 5).Value)
PreFormatTable objWsOutput

TransposeColumns 3, 4, "s"
TransposeColumns 7, 9, "p", 8
TransposeColumns 12, 13, "q"

End Sub
```

The subroutine `TransposeColumns` reads the specified scan configuration data from the *Setup* worksheet and transposes the scans marked with an x to the specified columns in the *Output* worksheet. The subroutine begins a loop on row 23 (which corresponds to the first row of each of the scans on the *Setup* worksheet) and steps through consecutive rows until the loop encounters an empty cell. For each scan item, the loop first looks for an x, and, if one is present, it reads the scan configuration information and inserts it into the specified column on the *Output* worksheet.

In addition to the scan properties, `TransposeColumns` places an x in row 2 of the *Output* worksheet. The actual scanning subroutine `SystemScan` looks for this x to determine whether it should perform that scan. In this manner, the `Transpose-Columns` subroutine enables all scans by default. Conversely, you can disable any scan by simply removing the x from above it on the *Output* worksheet.

```
Sub TransposeColumns(iColName, iColColumn, strScanType, Optional
iColOptName)
Dim i
Dim strName
Dim objWsSetup, objWsOutput As Variant
Set objWsSetup = Worksheets("Setup")
Set objWsOutput = Worksheets(objWsSetup.Cells(6, 5).Value)
i = 23
While Not IsEmpty(objWsSetup.Cells(i, iColName))
If InStr(objWsSetup.Cells(i, iColName - 1), "x") Then
    objWsOutput.Cells(2, objWsSetup.Cells(i, iColColumn).Value) = "x"
    objWsOutput.Cells(3, objWsSetup.Cells(i, iColColumn).Value) = _
     strScanType
```

```
    If Not IsEmpty(objWsSetup.Cells(i, iColOptName)) Then
        strName = objWsSetup.Cells(i, iColName).Value & "-" & _
         objWsSetup.Cells(i, iColOptName).Value
    Else
        strName = objWsSetup.Cells(i, iColName).Value
    End If
    objWsOutput.Cells(4, objWsSetup.Cells(i, iColColumn).Value) = _
     strName
End If
i = i + 1
Wend

End Sub
```

SystemScan

The subroutine SystemScan loops through all of the computers listed on the specified *Output* worksheet, and for every row marked with an x in column 1, the script performs the selected scans. As we saw in the subroutine SetupScan, the scans are defined and in rows 2 through 4 of the specified *Output* worksheet. If a scan is marked in row 2 with an x, the script performs the scan. Otherwise, it increments the column counter j to look for the next scan. It continues processing the columns until it finds five consecutive blank columns, and then it moves on to the next computer. By placing or removing the xs, you can easily configure the script to scan all computers for all scans, or only scan one computer for a particular scan.

SystemScan uses an array of the custom data type aComputer described in the *DumpComputerList* script of Chapter 8 (and defined in the module *DumpComputerList* of this workbook) to store the list of computers used in the scanning. The script declares a number of strings used to store the input and result values from each of the different scans types. For example, strServiceName contains the name of the service to scan for, and strServiceState contains the result of that scan, such as if the service is Stopped or Running. The script uses the variable iBlankColumns to count the number of blank columns to indicate the end of the scans to perform. The script also declares the *Setup* (objWsSetup) and *Output* (objWsOutput) worksheet objects and a string (strDomain) defining the current domain.

The script populates the array aComputer from the computers listed in the *Output* worksheet using the function LoadComputerList, also previously discussed in the *DumpComputerList* script.

```
'----------------------------------------------------------------
' Main program to conduct service and port scans
'----------------------------------------------------------------
```

```
Sub SystemScan()
Dim aComputer() As ComputerData
Dim strDomain, strWsOutput, strServiceName, strServiceState As String
Dim strProtocol, strPort, strPortState As String
Dim strQFE, strQFEState As String
Dim objWsSetup, objWsOutput As Worksheet
Dim IFinalRowNum As Integer
Dim i, j, iBlankColumns As Integer
Set objWsSetup = Worksheets("Setup")
strWsOutput = objWsSetup.Cells(6, 5).Value
strDomain = objWsSetup.Cells(5, 5).Value
Worksheets(strWsOutput).Activate
IFinalRowNum = LoadComputerList(aComputer, objWsOutput)
```

SystemScan begins three loops and confirms several conditions before actually initiating a scan. In the first loop, the script cycles through all of the computers contained in the array aComputer. The scanning script gets the name and row number of the current computer via the user-defined data type elements aComputer(i).Name and aComputer(i).Row, where the counter variable i increments from 1 to the upper bounds of the array. To keep the current computer displayed on the screen, the script selects a cell on the current row—this keeps the data from scrolling off the screen.

Next, the script checks whether column 1 of the current computer row contains an x, and if it does, the script continues to process that computer. If an x does not exist, the script proceeds to the next computer.

As the next prescan check, the script calls the Pingable function to ping the computer to confirm that it is up and responding on the network. Some of the Windows network functions such as getting objects from a remote computer have a significant timeout if the computer does not exist, whereas a ping has a much shorter timeout. If the computer replies to the ping, then the script sets the state to Up and proceeds to set up the loops that cycle through each of the desired scans.

```
For i = 1 To UBound(aComputer)
    Cells(aComputer(i).Row, 7).Select
    If InStr(1, Cells(aComputer(i).Row, 1).Value, "x") Then
        If Pingable(aComputer(i).Name) Then
            Cells(aComputer(i).Row, COL_OPTION1) = "Up"
            j = 8
            iBlankColumns = 0
```

The script iterates through each of the user-desired scans using two separate loops. The first loop counts up to five blank columns. The second loop looks for an empty cell at row 4 of the current column. Using these two loops, the script scans for the presence of a listed scan while allowing for up to four blank cells between scans. Adding blank columns to the *Output* worksheet between categories of scans might be useful, and the script permits up to four consecutive blank columns.

As if looking to see if it should process a computer, the script looks at row 2 of the current column for an x to see if it should perform that scan. If an x is present, the script performs the scan; otherwise, it skips the scan and moves on to the next one.

```
While iBlankColumns < 5
While Not IsEmpty(Cells(4, j))
If InStr(Cells(2, j), "x") Then
```

At this point the script is committed to performing the specified scan. The script sets iBlankColumns to 0 (remember, it is looking for consecutive blank columns, not a total number of blank columns), and calls the DisplayResult subroutine to highlight the current cell that is being processed as working. The script opens a Select Case statement to determine which scan to perform based on the value for the scan type in row 3 of the current column. If this cell contains an s, it performs a service scan.

```
iBlankColumns = 0
DisplayResult aComputer(i).Row, j, "Working", 3
Select Case Cells(3, j)
```

The script gets the name of the service to look for from row 4 of the current column. Next, it calls the ServiceCheck function to determine the state of that service. It sends the current computer name and the service name to ServiceCheck and assigns the returned service state to strServiceState.

The script processes strServiceState through a second Select Case to provide more granular reporting of the results. If the service state is Running, it displays the result in green. It displays a Stopped service in red. The ServiceCheck function returns a 0 if it encounters an error (such as if the service is not installed), which this Select Case statement displays as a dash (-). Any other value of strServiceState is simply displayed.

```
Case "s"
    strServiceName = Cells(4, j)
    strServiceState = _
     ServiceCheck(aComputer(i).Name,
strServiceName)

    Select Case strServiceState
      Case "Running"
```

```
                    DisplayResult aComputer(i).Row, j, _
                        strServiceState, 4
                Case "Stopped"
                    DisplayResult aComputer(i).Row, j, _
                        strServiceState, 3
                Case 0
                    DisplayResult aComputer(i).Row, j, _
                        "-", 0
                Case Else
                    DisplayResult aComputer(i).Row, j, _
                        strServiceState, 0
            End Select
```

The script defines a port scan as a scan type of p. The port scanner requires three parameters to run: the name of the computer, the protocol to scan (TCP or UDP), and the port to scan (an integer between 1 and 65535). Row 4 of the current column contains both the protocol and the port in the format *protocol-port* (similar to TCP-80). The script splits the string and assigns the variable strProtocol based on the protocol of T for TCP, and U for UDP. The case-insensitive comparison of InStr uses a text comparison because of the Option Compare Text statement defined at the beginning of this module. strPort is defined as the second part of the protocol-port pair.

Although the syntax of the VBA Split method differs from the JScript split method used in previous scripts, the method achieves similar results. Both methods return an array of strings split at the defined delimiter. For example, passing the string TCP-80 through the split function with a delimiter of a dash (-) results in an array of two elements, which can be directly accessed as follows: Split("TCP-80","-")(0) = TCP and Split("TCP-80","-")(1) = 80.

After the script has populated the strings strProtocol and strPort, it calls the function PortCheck to port scan the selected computer. The scan returns the result to the variable strPortState, which the script then displays. (We also could have further parsed the results and color-coded the output like we did in the service scan code previously.)

```
        Case "p"
            On Error Resume Next
            strProtocol = Split(Cells(4, j), "-")(0)
            If InStr(strProtocol, "u") Then
                strProtocol = "U"
            Else: strProtocol = "T"
            End If
```

```
strPort = Split(Cells(4, j), "-")(1)
strPortState = PortCheck( _
 aComputer(i).Name, strProtocol, strPort)
On Error Goto 0
DisplayResult aComputer(i).Row, j, _
 strPortState, 0
```

The final scan looks to see if a specified QFE is installed on the current computer. The script pulls the name of the QFE from row 4 of the current column and calls the QFECheck function to determine if it is installed and displays the results.

 Due to the previously mentioned bug in the Win32_QuickFixEngineering class, if you intend to run this script against a group of computers for which you do not know what service pack version is installed, you could add a function to this script that checks whether SP3 is installed. If SP3 is installed, the script calls QFECheck, otherwise, it simply skips the QFE scan.

```
Case "q"
    On Error Resume Next
    strQFE = Cells(4, j)
    strQFEState = QFECheck(aComputer(i).Name, _
     strQFE)
    On Error Goto 0
    DisplayResult aComputer(i).Row, j, _
     strQFEState, 0
    Case Else
End Select
```

As we've seen, the SystemScan subroutine includes a number of loops and conditionals, which may become confusing, especially when remembering which End If or Wend applies. One method to keep these statements straight in your mind is by inserting loop or conditional termination comments, as seen in the following code. These are snippet comments designed to tie an ending statement with its mate. Another approach to streamlining complex nested loops or conditionals is by factoring out the inner branches of code. Using this method, a loop within a loop becomes a loop that simply calls a function, and this function contains another loop. Factoring the loops in this manner increases the number of functions you need, but it can help untangle more complicated code.

```
End If  ' Scan marked to be processed with an x
j = j + 1
Wend ' Loop looking for an Empty Cell
```

```
                    j = j + 1
                    iBlankColumns = iBlankColumns + 1
                    Wend ' Loop looking for iBlankColumns
```

If the function `Pingable` returns `false`, indicating that the computer does not respond to a `ping`, then the script displays the computer as `Down` and moves on to the next computer. The completion of the loop through the array `aComputer` completes this subroutine and *WormScanner* as a whole.

```
                Else ' else pingable if
                        Cells(aComputer(i).Row, COL_OPTION1) = "Down"
                End If  'end pingable if
            End If  ' Computer marked to be processed with an xNext i
        Next i
    End Sub
```

The remaining subroutines and functions presented in this chapter support the subroutine `SystemScan`.

Pingable

The `Pingable` function uses a WSH shell object to call the command shell and execute a `ping` command. The `ping` command redirects its output to a temporary file, which the function reads and parses.

TIP

`Pingable` calls the `Run` method instead of the `Exec` method because of its ability to suppress the command shell dialog box. If we used the `Exec` method, we would see multiple command shell boxes pop up all over the screen. It's too bad that the `Exec` method, with its support of the standard streams, doesn't offer a property to suppress the display of the called function.

The function instantiates a `FileSystemObject` to read this redirected text file line-by-line and parses it for the text `Reply`, which indicates whether the `ping` was successful. (A successful `ping` returns a line similar to: *Reply from 192.168.0.3: bytes=32 time<1ms ttl=128*—this script looks for that first word *Reply*.) This `ping` test is very similar to the `ping` test presented in the *ClientCheck* HTA presented in Chapter 9.

```
'-----------------------------------------------------------------
'Sends single ping to a machine to check if it is up.
'-----------------------------------------------------------------
```

```
Function Pingable(strComputer) As Boolean
Dim objShell As WshShell
Dim strTemp As String
Dim objFSO As FileSystemObject
Dim iReturn As Integer
Dim objTextFile As TextStream
Pingable = False
Set objShell = CreateObject("WScript.Shell")
strTemp = objShell.ExpandEnvironmentStrings("%temp%") & _
 "/tempping.txt"
iReturn = objShell.Run("%comspec% /C ping " & strComputer & _
        " -n 1 > " & strTemp, 0, True)
Set objFSO = CreateObject("Scripting.FileSystemObject")
Set objTextFile = objFSO.OpenTextFile(strTemp, 1)
While Not objTextFile.AtEndOfStream
    If InStr(objTextFile.ReadLine, "Reply") Then Pingable = True
Wend
objTextFile.Close
objFSO.DeleteFile (strTemp)

End Function
```

ServiceCheck

The function ServiceCheck queries a remote computer and returns the state of a specified service (such as whether the service is Stopped or Running). To do this, the script makes a WMI connection to the remote computer and returns an object to the specific Win32_Service. The function returns the WMI State property of this object, which represents the service state. If the service does not exist on this computer, WMI will throw an error, but the script handles this error and returns a 0, indicating the service is not installed. Also, notice the absence of the ImpersonationLevel in the WMI moniker. Although this makes the WMI call shorter and easier to read (and remember), also recall from Chapter 3 that if you plan to run this script on an operating system earlier than Windows 2000, you need to set the Impersonation-Level to 3 (wbemImpersonationLevelImpersonate), which instructs WMI objects to run under the context of the caller. This script requires WMI Core Components 1.5 or later (available on Windows 2000 and later). If you plan to run this script on an earlier version of WMI Core Components, refer to Chapter 3 for how to specify the ImpersonationLevel.

Early versus Late Binding—In this function, note the declaration of objService *as a* WbemScripting.SWbemObject *data type. In Visual Basic (including VBA), when a variable is declared as a data type, it is known as early binding. Early binding causes the object to be verified (syntax and type checking) when it is compiled as opposed to at runtime. This method has performance advantages over late binding, which does not declare an object as a specific type. To make an early binding to an object, we must have added an Excel reference to the WMI Scripting Library. Early binding has other advantages. With a reference established to this type library, Excel recognizes this object and offers additional functionality to the developer— such as auto-completing the syntax when using this object. For example, if we type objService., after we type the period (.), Excel provides a list showing all of the different properties and methods that can be called with a* WbemScripting.SWBemObject *object. In VBScript, we cannot declare objects like this and are limited to the Variant data type. Early binding is not available in VBScript or JScript.*

```
'------------------------------------------------------------
'Uses WMI to check the state of a specified service
'------------------------------------------------------------

Function ServiceCheck(strComputer, strServiceName) As String
Dim strWMI As String
Dim objService As WbemScripting.SWbemObject
On Error Resume Next
strWMI = "WinMgmts://" & strComputer & _
  "/root/cimv2:Win32_Service='" & strServiceName & "'"
Set objService = GetObject(strWMI)
ServiceCheck = objService.State
If Err > 1 Then ServiceCheck = 0
On Error GoTo 0
End Function
```

PortCheck

The function PortCheck essentially provides a script harness for an external command-line port scanner. This script is written for the syntax of the popular *nmap* scanner; however, it would be fairly easy to modify this code to support your other favorite command-line port scanner.

Much like the previous Pingable function, PortCheck creates a Shell object to Run an external command and redirects the output to a temporary file. It then uses the FileSystemObject to read and parse the text-based output and look for keywords designating the port status as either Open, Closed, or Filtered. The function returns the state of the port to the calling function.

The variable `strNmapCommand` contains the *nmap* command syntax to execute the scan. Although *nmap* supports a huge variety of command-line scanning options, this script uses a basic syntax to scan a single port of one computer, as follows:

```
nmap -s[U|T] -p n computername
```

The `-s[U|T]` parameter specifies the type of scan; `-sT` instructs *nmap* to use a TCP scan, and `-sU` specifies a UDP scan. The argument `-p n` specifies the port number, such as `-p 80` to scan port 80. The last argument specifies the name of the computer to scan. For example, to scan for a Web server listening on TCP port 80, you might use:

```
nmap -sT -p 80 webserver
```

and to look for a DNS server that typically listens on UDP port 53, this might do the trick:

```
nmap -sU -p 53 dnsserver
```

The script assembles these commands for the current computer and port scan and stores them in the variable `strNmapCommand`. Recall when the script transformed the protocol stored to be stored `strProtocol` from TCP to `T` and UDP to `U`. It's because *nmap* uses the argument `-sT` to specify a TCP scan, and `-sU` for UDP scans. By transforming the values early on, we can simply concatenate the value to `-s` to assemble the argument. Though not very fancy, it demonstrates a method to harness a proprietary, external application within your scripts.

```
'-----------------------------------------------------------------
'Uses nmap scanner to determine remote port state
'-----------------------------------------------------------------
Function PortCheck(strComputer, strProtocol, strPort)
Dim objShell As WshShell
Dim strTemp, strNmapCommand, strLine, strNmapPath As String
Dim iReturn As Integer
Dim objFSO As FileSystemObject
Dim objTextFile As TextStream
strNmapPath = Worksheets("Setup").Cells(39, 6).Value
PortCheck = "-"
Set objShell = CreateObject("WScript.Shell")
strTemp = objShell.ExpandEnvironmentStrings("%temp%") & _
 "\tempportscan.txt"
strNmapCommand = strNmapPath & "nmap -s" & strProtocol & " -p " & _
```

```
    strPort & " " & strComputer
    iReturn = objShell.Run("%comspec% /C " & strNmapCommand _
    & " > " & strTemp, 0, True)
    Set objFSO = CreateObject("Scripting.FileSystemObject")
    Set objTextFile = objFSO.OpenTextFile(strTemp, 1)
    While Not objTextFile.AtEndOfStream
        strLine = objTextFile.ReadLine
        If InStr(strLine, "open") Then PortCheck = "Open"
        If InStr(strLine, "close") Then PortCheck = "Closed"
        If InStr(strLine, "filter") Then PortCheck = "Filtered"

    Wend
    objTextFile.Close
    objFSO.DeleteFile (strTemp)
    End Function
```

QFECheck

The function QFECheck looks for the installation of a particular QFE update by making a WMI call to the remote computer and looking for that particular update. This function uses the WMI ExecQuery method to execute a small WQL query of the Win32_QuickFixEngineering class that searches for the specific QFE HotFixID. If the query returns any results, QFECheck is set to Yes; otherwise, it remains at No. For any errors, QFECheck is set to No.

TIP

Microsoft recently changed their naming convention for QFE updates. Now instead of Q123456, some are called KB123456.

```
'-----------------------------------------------------------------
'Checks a remote machine if the QFE is installed
'-----------------------------------------------------------------

Function QFECheck(strComputer, strQFE) As String
Dim colQFE As Variant
Dim objQFE As Variant
QFECheck = "No"
On Error Resume Next
Set colQFE = GetObject("WinMgmts://" & strComputer & "/root/cimv2") _

  .ExecQuery("SELECT * FROM Win32_QuickFixEngineering " & _
   "WHERE HotFixID='" & strQFE & "'")
```

```
For Each objQFE In colQFE
    QFECheck = "Yes"
    Exit For
Next
If Err > 1 Then QFECheck = "No"
On Error GoTo 0
End Function
```

DisplayResult

The subroutine `DisplayResult` provides ancillary support to this script to display text in a specified cell and format it to a specific color.

```
'----------------------------------------------------------------
' Display formatted output at the specified location
'----------------------------------------------------------------

Sub DisplayResult(iRow, iColumn, strText, iColorIndex)

Cells(iRow, iColumn) = strText
Cells(iRow, iColumn).Font.ColorIndex = iColorIndex

End Sub
```

SUMMARY

WormScanner extends the *DumpComputerList* script from Chapter 8 to include a number of useful scanning functions. Although many of the features and functionality presented in this script were built from techniques covered in previous chapters, uniting them in Excel provides insight into creating richer applications using individual script components as building blocks. Plus, by now you are probably noticing these reoccurring themes and techniques and how easy it is to extend a script (or set of scripts) to add your own functionality and flexibility. This script has been designed to be tolerant to extra spaces (remember `iBlankLines` and `iBlankColumns`) and does not rely on any specific computer order. Because of this, you can rearrange and classify computers and the results any way you see fit—and even use some of Excel's built-in functions and features to further customize your *Output* worksheet. For example, Figure 10.5 shows an *Output* worksheet that has been substantially customized; but even with these modifications, the scanning script still processes the computers and scans correctly. Just remember what cells you can edit and what cells the script depends on, and you should leave the same.

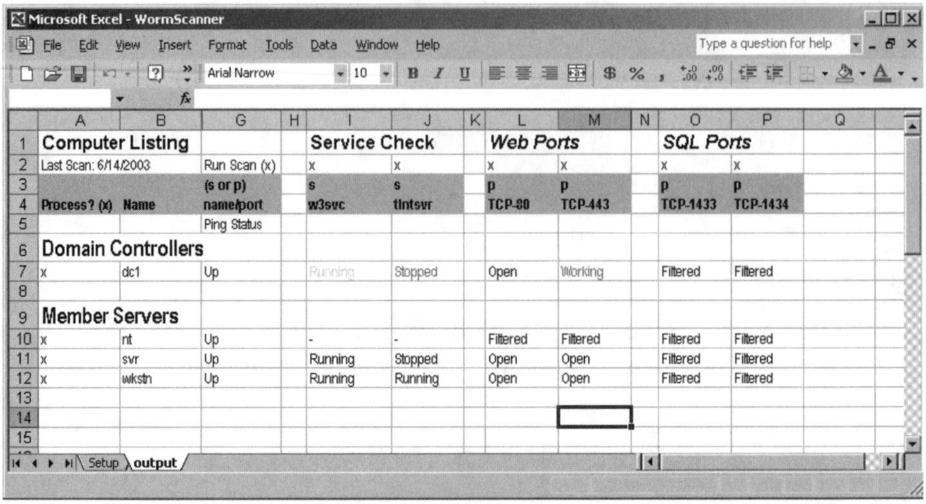

FIGURE 10.5 Using Excel as your script canvas allows you to fully customize your output. Even with these modifications, the *WormScanner* script still runs and updates this *Output* worksheet and preserves the formatting.

KEY POINTS

- *Wormscanner* demonstrates some of the many possibilities of bolting on scanning or system-query functions to the *DumpComputerList* script.
- Use the WScript.Run method over the Exec method if you need more control over the format of the resulting program (for example, hiding a command shell terminal window).
- Excel supports WMI objects, methods, and properties, enabling it as a capable inventory or scanning tool—such as scanning for services and their state or installed Quick Fix Engineering updates.
- Use Excel worksheets to configure script setup options, which make your script more accessible to others and quicker to reconfigure and run.
- Consider harnessing the ping command to quickly assess whether a computer is on the network before engaging a more lengthy remote Windows network call.

Next, we'll look at another extension of the *DumpComputerList* script that enumerates and displays all of the local accounts and groups on computers in your domain. If you have ever survived an IT audit, you'll likely appreciate the convenience this next script affords.

Remotely Enumerate Local Computer Accounts

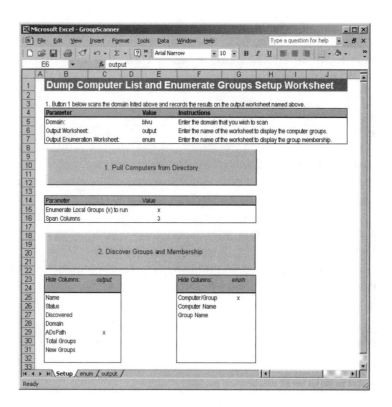

OVERVIEW

As we've come to see, creating scripts that report the configuration of accounts in Active Directory (AD) is relatively easy. However, AD is not the only repository for user accounts. Every Windows NT technology computer (from Windows NT 4.0 to Windows Server 2003) also maintains a local account database called the security accounts manager (SAM) database. Computers that belong to workgroups (not a domain) use the SAM exclusively to store user and password information for local account authentication. But even computers that belong to a domain use the SAM and contain legitimate logon accounts. Some of these accounts, such as the local administrator user account or any member of the administrator's local group, provide super-user access to that computer. Plus, because these accounts are distributed across your organization (one per computer), managing these SAMs becomes more difficult, and consequently, these accounts become prime targets for attackers looking for back doors into your network.

Many IT audits that assess an organization's security examine the membership of local computer groups to determine who actually has management or data access to that computer. The audits typically consist of an enumeration of all local accounts and group membership as well as domain groups that are members of local groups.

Tools help enumerate these groups. Microsoft provides a few tools such as Microsoft Baseline Security Analyzer (MBSA) to inspect computers for which accounts have administrator-level access. The ubiquitous NET command provides both NET USER and NET GROUP to list group information (although NET GROUP is restricted to run only on domain controllers). Plus, a number of resource kit tools and other tools are available on the Internet to enumerate these accounts.

Invariably, you may find yourself wanting to display a custom or specific list of accounts for which you find these tools too rigid (or simple) to produce. Using the *DumpComputerList* script from Chapter 8, this chapter explores how to use Excel and Windows scripting technologies to create a robust, easy-to-modify tool that lists local accounts and groups as well as their memberships.

SCENARIO

Your company is a member of multiple regional and national regulatory agencies and consortiums that require stringent operating procedures safeguarding your customers' data. Compliance with these organizations gives peace of mind to your current and potential customers. Additionally, the marketing department places a predominant logo on your Web site boasting of your compliance with these industry-respected guidelines.

These agencies measure compliance through audits of your IT systems every six months. These audits inspect all aspects of your IT operations—from disaster recovery to physical security to topology reviews. A particularly time-consuming aspect of the audit is a thorough review of the user account access to your companies' more than 500 servers. For each server you must provide a report of all local users and groups, including the membership of these local groups. Additionally, you must expand the domain groups that are members of the local groups to provide a user-level access map to these computers. The auditors compare this access map to that of a previous audit, looking for changes that you then must explain. For previous audits, you have written batch files to dump this data to a text file, but the reconciliation process comparing the current and past access lists takes a lot of time.

What you really want is a tool that pulls, enumerates, and documents the users and groups of a list of systems and flags additions and deletions between runs. You decide to use Microsoft Excel as the basis for this tool because of its ability to quickly manage lists of data. Plus, it's easy to archive audits simply by copying and renaming the worksheets.

ANALYSIS

Like the *WormScanner* script presented in Chapter 10, the *GroupScanner* script builds directly on the *DumpComputerList* tool presented in Chapter 8 and includes techniques for user and group enumeration covered in Chapters 4 and 5. Whereas previous scripts targeted AD using the ADSI LDAP provider, the scripts in *GroupScanner* target the local SAM using the ADSI WinNT provider. Additionally, the script demonstrates one approach for tackling the more difficult problem of reconciling multiple runs of the tool and displaying the large amounts of collected data in a print-friendly format.

The business objectives map into the technical requirements listed in Table 11.1.

TABLE 11.1 Translating business objectives into technical requirements

Business objectives	Technical requirements
Generate a list of all computers in your specified domains.	Begin with the *DumpComputerList* script presented in Chapter 8 to generate the list of computers from the specified domain.
List all local groups for each specified computer in the list.	Using the ADSI WinNT provider, attach to each selected remote computer and get a list of all group objects. Display these objects on the *Output* worksheet. Subsequent runs of the script should highlight new group additions and mark group deletions.

TABLE 11.1   Translating business objectives into technical requirements *(continued)*

Business objectives	*Technical requirements*
For each local group discovered, enumerate and display its membership.	Use the ADSI Members method to return a collection of objects representing the membership of a specific local group. Display this output on the *enum* worksheet. Like the previous requirement, additions or deletions from a group's membership should be highlighted in subsequent runs.
Highlight new local group additions and new memberships to the local groups compared to a previous run.	Load all previously run data from the *Output* worksheet, and compare to a current scan. Highlight new additions, and mark missing groups that may have been deleted.
Format the output to facilitate printout and distribution to external parties (the auditors).	Programmatically use Excel formatting features to distinguish the output using borders, shading, and colors. Also, allow the user to specify the number of columns in which to display the enumeration output.

ON THE CD

This script is perhaps the most complex in this text. The concepts are not necessarily advanced; however, the algorithms used to manipulate and format the data can be difficult to understand by simply reading the code that defines them. For this reason, open the script contained in the file GroupScanner.xls and walk through the code execution using the Excel debugger. By walking through the script line-by-line, you get a better appreciation of how the algorithms work and the flow of the script. Getting the local groups of a computer and their memberships is fairly easy; in fact, the first script, *ADSIQuery*, uses similar methods to extract groups and their membership from AD. However, this script maintains the list of groups and their membership. Multiple runs of the script highlight new group additions, cross out deleted groups, and format the data for easy printout in Excel. It is the formatting and auditing features of this script that increase its complexity.

A majority of this script consists of manipulating and formatting the data—both the first time it is collected and displayed and during subsequent runs that compare current results against a previously created baseline. As previously mentioned, the script mechanics of actually connecting to the computer and getting the names of the local groups and then their memberships make up only a fraction of the code. However, the script gives insight to some of the challenges and design questions of creating a user interface and manipulating and storing collected data.

Assess each situation when considering your design approach (and tool selection), and decide how much effort should be placed on usability versus strict functionality. For example, as previously demonstrated, it is easy to generate a list of computers, then generate a list of local groups of that computer, and finally generate a list of the membership of those groups, but it is much more difficult to store and highlight changes across multiple runs. In fact, although previous scripts focused on the technical aspects of development, such as actually gathering the data, a majority of the code in this script manages the comparison and presentation of data. In this case, it is the management and presentation of the data that fulfills many of the key requirements of the script.

A description of the functions and subroutines used in *GroupScanner* follows:

`GroupScan`: Scans computer list for eligible computers from which to get local groups.

- Starting point for the script.
- Checks if *Output* worksheet exists and cycles through previously generated list of computers looking for candidates to enumerate local groups.

`GetGroups`: Adds new groups if they exist.

- For a specific computer, gets list of current groups and compares against previous runs.
- Adds new groups to the *Output* worksheet.
- Calls function to get group membership, if selected by user.
- Displays new groups in a matrix of user-specified number of columns.

`GetMembership`: Collects and displays the memberships of a specific group.

- Records the membership information on the *enum* worksheet.
- Adds new groups to the *enum* worksheet.
- Adds new memberships to the *enum* worksheet.

`LoadRow`: Reads group information into user-defined `GroupData` data type.

- Scans row by row to read previously discovered information.
- Adds information into `GroupData` user-defined data type.

`AddElement`: Compares and adds an object to the `GroupData` data type.

- Adds an account name to a previously constructed list of accounts, if it does not already exist.

`MarkElementDeleted`: Highlights groups no longer present.

- Marks the cells of accounts that are no longer present.

`InitializeEnumWs`: Builds the *enum* worksheet.

- Creates and names the worksheet used to store the enumeration data.
- Formats the worksheet.

CheckIfGroup: Determines if an object is a group.

- Checks the account's `objectClass` to see if an object is a group object.
- Formats the cells of group objects to blue italic to distinguish them from a user or computer account.

FindLastRow: Returns the last row used on a worksheet.

- Used to determine the next row to use for inserting new data.

HideColumns: Hides user-selected columns on a worksheet.

- Automates hiding user-selected columns to make the data easier to read.

Distribution and Installation

Install this script in much the same fashion as its predecessor, *DumpComputerList*. When you open the *GroupScanner.xls* workbook, select the *Setup* worksheet. The *Setup* worksheet provides preliminary configuration of the script, as seen in Figure 11.1. As you did for the *WormScanner* configuration, specify the domain to scan as well as the name of the *Output* worksheet. New to this script, you must also specify the name of the *Output Enumeration* worksheet (*enum*). The *enum* worksheet stores the memberships of each of the local groups found on a computer. After you have specified these three options, click the button 1. Pull Computers from Directory to run the *DumpComputerList* script and built the *Output* worksheet.

NOTE

If the script fails because of missing references, switch to the Visual Basic Editor, deselect the reference marked as missing, and look for an older version of that library to add. (More about how Excel uses references is covered in Chapter 8.)

As you did for the previous scripts in Chapters 8 and 10, review the *Output* worksheet and select which computers you want to scan by placing or removing xs in column 1.

Next, return to the *Setup* worksheet, and specify whether you want to enumerate the local group membership by putting an x in the value column for Enumerate Local Groups (x) to run. Also, specify the number of columns that you wish the enumeration data to span. This step is useful for restricting horizontal output of the script to allow it to easily print its output on standard-size paper. Notice the addition of the `HideColumns` configuration from the previous *Setup* worksheets. This script lets you specify columns to hide by placing an x next to its name, which is useful for reducing the amount of data that you display or print in a report.

When complete and ready to run, click 2. Discover Groups and Membership to start the scan.

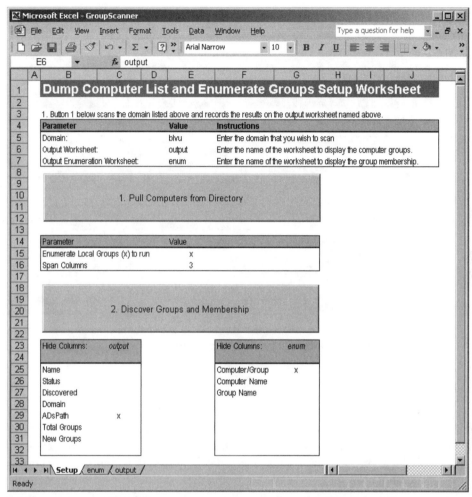

FIGURE 11.1 Specify the domain to scan, the names of the data worksheets, and other formatting options on the *Setup* worksheet.

Output

The *Output* worksheet displays the local groups for every computer listed, as seen in Figure 11.2. Notice that the list of groups spans only three columns and that the *ObjADs* (column 6) column is hidden. Both of these configurations ensure the data on the worksheet fits in view without scrolling to the right. Also, notice the formatting differences of the group named *aNewGroup*. *aNewGroup* was added to the computer named *wkstn* between two runs of the script. The script highlights the addition, making it easy to pick out additions to your groups between subsequent runs.

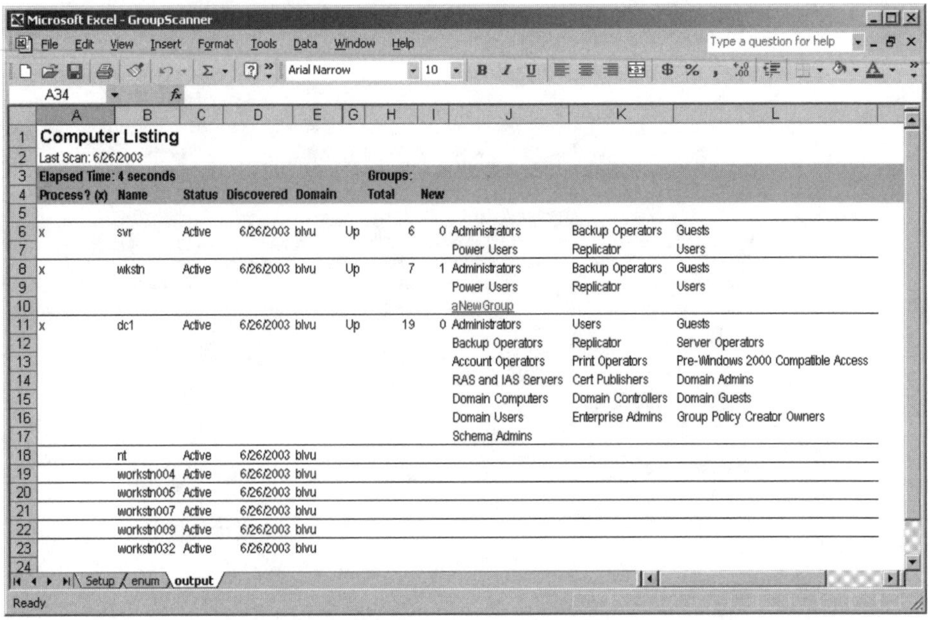

FIGURE 11.2   The *Output* worksheet shows a list of computers and their local groups. New groups are highlighted by color and underlined.

Let's now look at the group enumeration detail on the *enum* worksheet, as seen in Figure 11.3. The *enum* worksheet provides a display of the membership of each computer's local group. New additions are underlined and color-coded. The script italicizes groups that are members of other groups as a reminder to check the membership of those groups as well. Using these formatting differences, it's easy to spot the new additions of the group *Blvu/Domain Admins* and the user account *Trooper* to the *Administrators* group on the computer named *wkstn*. In this example, let's hope *Trooper* isn't an attacker.

Summary of Solution

This Excel-based script extends the *DumpComputerList* tool presented in Chapter 8 by adding functionality that enumerates the local computer groups and their membership. The worksheets created by the script are designed to persist across multiple runs of the script to identify and highlight changes to a previous run. Highlighting the changes makes auditing the membership from run to run easier.

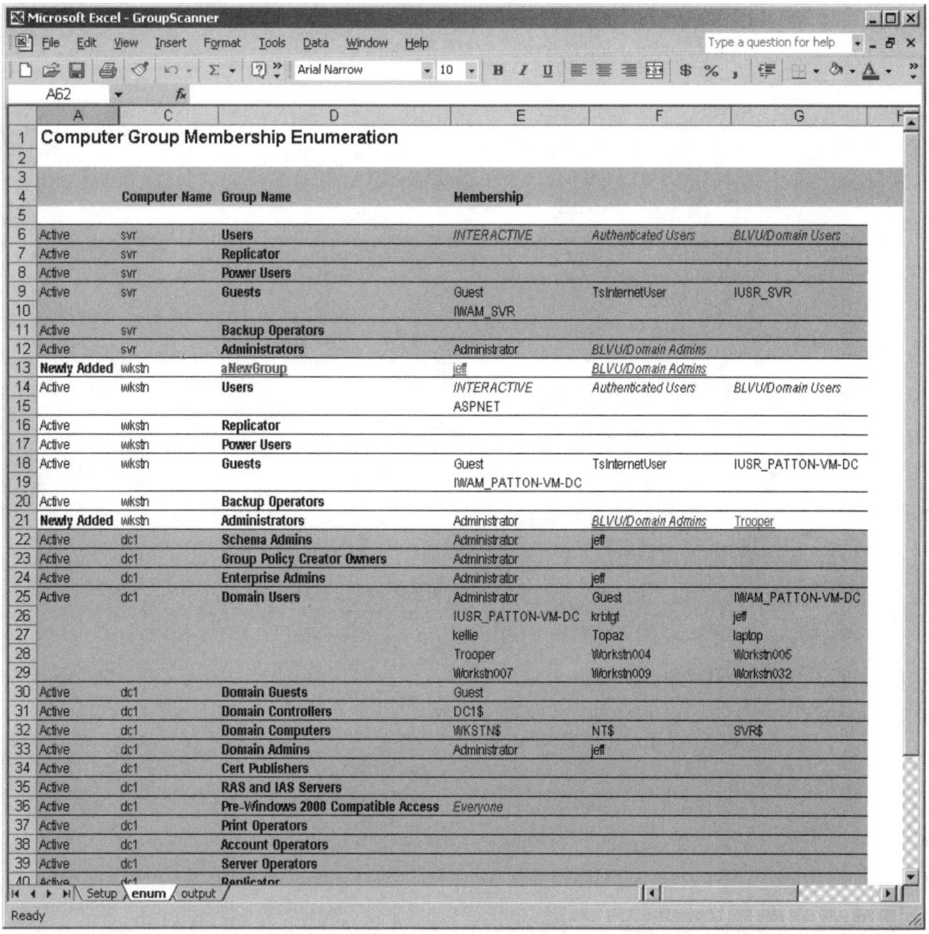

FIGURE 11.3   The *enum* worksheet displays the account membership of each of a computer's local groups, highlighting groups as members as well as any new additions.

OBJECTIVE

An Excel-based tool that scans a directory of computers and pulls their local groups and enumerates their membership. The script compares any new run against the previous run to highlight changes, making it easier to audit the list.

ADSI

- WinNT Provider
- `Members`
- `Filter`
- `Parent`

Excel Object Model

- Worksheet manipulation
- Cell/range formatting

VBA Options

- `Option Explicit`
- `Option Compare Text`

VBA Error Handling

- `On Error Resume Next`
- `On Error Goto 0`

User Data Types

- Using `Type` to declare user-defined data types

THE SCRIPT

This script leverages the code from *DumpComputerList.xls* presented in Chapter 8 and a couple of functions from *WormScanner* in Chapter 10. A new module named *GroupScanner* contains the code exclusive to this script.

ON THE CD

The script is presented on the CD-ROM as /chapter 11/scripts/groupscanner.xls. This script is contained within an Excel document that requires Excel 2000 or later, as well as Windows Script Host 5.6.

Module-level Code

Like *DumpComputerList* and *WormScanner*, this VBA script sets the `Option Explicit` and `Option Compare Text` statements. Additionally, the script creates a new user-defined data type called `GroupData` in which to store information about the groups read into the system. `GroupData` contains the same properties as `ComputerData` used in *DumpComputerList* and behaves in much the same way. `GroupData` contains the name of the group, the row, and the column in which the group is stored on the local worksheet and whether it is confirmed to exist.

GroupScanner also defines a new user-defined data type named `Position`. `Position` contains two properties, `row` and `column`. The data type `Position` is used to return both a `row` and `column` property from the function `LoadRow`. `LoadRow` returns the row and column number of the next cell to which to add any new data.

The function also declares several public variables representing the *Setup*, *Output*, and *enum* worksheets used by this script. Declaring these as public variables allows any of the code to access these common worksheets. The public variable `g_iSpanCol` represents the number of columns that the user chooses to span to display the group output data. Allowing the users to control the width of the output helps them fit the output in a printed report or on a screen without needing to scroll.

```
Option Explicit
Option Compare Text
Type GroupData
    Name As String
    row As Integer
    Column As Integer
    Exists As Integer
    Class As String
End Type
Type Position
    row As Integer
    col As Integer
End Type
Public g_objWsSetup As Worksheet
Public g_objWsOutput As Worksheet
Public g_objWsEnum As Worksheet
Public g_iSpanCol As Integer
```

GroupScan

The function `GroupScan` begins the script. `GroupScan` iterates through each computer on the *Output* worksheet, and, if the computer is selected by an x in column 1 and responds to a `ping` command, the function sends the name of the computer to the `GetGroups` function to determine the groups for that computer.

The function declares the user-defined data type variables `aComputer` and `aGroup`, which contain the list of computers and groups discovered and managed throughout the script. Also, the script declares the variables `dtStart` and `dtEnd` and uses them to time the execution of the script.

```
'-------------------------------------------------------------------
' Main program to conduct group scan
'-------------------------------------------------------------------
```

```
Sub GroupScan()
Dim aComputer() As ComputerData
Dim aGroup() As GroupData
Dim i, iRow As Integer
Dim result As Variant
Dim dtStart As Date
Dim dtEnd As Date
dtStart = Now
```

GroupScan uses the *Setup* worksheet to gather script configuration information. The script sets the previously declared public variable g_objWsSetup to this worksheet. The script also defines the variable g_iSpanCol, which represents the number of columns that the script user wants the output to span. For example, if a computer has eight local groups, and the user set the columns to span to three, then the script would display the groups in three rows of three columns, and the last row would contain only two groups.

```
Set g_objWsSetup = Worksheets("Setup")
g_iSpanCol = g_objWsSetup.Cells(16, 5).Value
```

Whereas the *Setup* worksheet always exists, the name of the *Output* worksheet may be specified by the user and will not always exist. It must first be created by clicking the first button on the *Setup* worksheet, which calls the *DumpComputerList* script. If the user has not created the *Output* worksheet in this manner (hence, it does not exist), the script notifies the user of this and then exits. At this point, the user can click the first button to run the *DumpComputerList* script and then run GroupScan a second time after the *Output* worksheet has been created.

The script calls the MsgBox method to display a message to the user along with an exclamation point icon (identified by the constant vbExclamation). VBA recognizes the constants from type libraries, and they do not need to be explicitly declared as before when we were coding VBScript WSH scripts.

```
On Error Resume Next
    Set g_objWsOutput = Worksheets(g_objWsSetup.Cells(6, 5).Value)
If Err <> 0 Then
    MsgBox "Can not find Output worksheet. Have you run '1. " & _
    "Pull Computers from Directory'?", vbExclamation, _
    "Check Configuration"
    End
End If
On Error GoTo 0
```

The script next activates the *Output* worksheet and calls the *DumpComputerList* function `LoadComputerList` to scan and read the *Output* worksheet for selected computer names. These computer names are stored in the `aComputer` array.

```
objWsOutput.Activate
i = LoadComputerList(aComputer, objWsOutput)
```

On the *Setup* worksheet, the user can choose whether to enumerate the group memberships. This choice is selected on the *Setup* worksheet by placing an x in `Cell(15,5)`. If the user chooses to enable this functionality, the script calls the subroutine `InitializeEnumWs`, which creates and formats a new worksheet to store this membership data.

```
If g_objWsSetup.Cells(15, 5).Value = "x" Then InitializeEnumWs
```

Next, the script initiates a loop through each of the items in the `aComputer` array to determine if it should process the local groups of that computer. This script relies on distinctive formatting to help identify new or changed groups or memberships. The script uses the Excel formatting methods and properties to visually format the data with lines and shading. For every computer in `aComputer`, the script turns on a top row border from the first cell to the last spanned cell of the current row.

```
For i = 1 To UBound(aComputer)
    Set result = Range("B5:B65536").
     Find(aComputer(i).Name, , ,xlWhole)
    iRow = result.row
    Cells(iRow, 7).Select
    Range(Cells(iRow, 1), Cells(iRow, 10 + g_iSpanCol - 1)). _
     Borders(xlEdgeTop).LineStyle = xlContinuous
```

Next, the script inspects the first column of the current row to see if it should process that computer. If this cell contains an x, then the script displays a status message to the user and calls the `pingable` function to see if the computer is up or down. (This is the same `pingable` function used in the *WormScanner* script in Chapter 10.) If the computer is up (or at least if it replies to a `ping`), then the script calls the `GetGroups` function to begin the processing of the local groups of the computer. If the computer does not respond to the `ping`, then the script displays the computer as down. If the computer does not respond to a `ping` or was not selected to check the groups, the script proceeds to the computer.

```
If InStr(1, Cells(iRow, 1).Value, "x") Then
    DisplayResult iRow, 8, "Working", 3
```

```
            If Pingable(aComputer(i).Name) Then
                Cells(iRow, 7) = "Up"
                GetGroups aComputer(i).Name, iRow
            Else
                Cells(iRow, 7) = "Down"
                DisplayResult iRow, 8, "", 0
            End If
        End If
    Next i
```

The script calls several functions at the conclusion of this script. `PostFormatTable` formats the *Output* worksheet so that all data is visible, and `HideColumns` turns on the hidden property for selected columns identified in the *Setup* worksheet. Lastly, the function displays the elapsed time taken to run the function using the method `DateDiff`, which returns the number of seconds (s) between the start date (`dtStart`) and end date (`dtEnd`).

```
    PostFormatTable
    HideColumns 3
    HideColumns 7
    dtEnd = Now
    objWsOutput.Cells(3, 1) = "Elapsed Time: " & DateDiff("s", dtStart,
    dtEnd) & " seconds"
    End Sub
```

GetGroups

`GetGroups` displays the local groups of a computer on the *Output* worksheet. Most of the function's complexity stems from its designed ability to look for any previously identified groups and then add to that any new groups found. Similarly, `GetGroups` highlights any groups that no longer can be found (for example, they may have been deleted).

The array `aGroup` contains the group data of the current computer and will be dynamically filled with group data that either already exists on the *Output* worksheet or is discovered by the ADSI call to the subject computer. The variable `pos` is a user-defined variable type `Position` and contains the row and column number of the next blank cell into which a new group could be added.

The variables `iTopRow` and `iStartCol` represent the top-left starting point for the list of groups it will display. `GetGroups` displays the groups on the *Output* worksheet beginning in column 10, and so the subroutine sets `iStartCol` to `10`. Similarly, the subroutine sets `iTopRow` to `iRow`, which was sent to the function as an argument and represents the row number of the current computer being processed. The function uses these references to track where it should display data when building the matrix of the groups.

The variables iRow and iCol represent the current row and column of a particular group object. The variable iNewGroup provides a counter for new groups added to the computer and is displayed at the completion of this function on the *Output* worksheet.

```
'-------------------------------------------------------------------
' List the groups of a computer- operates on Output worksheet
'-------------------------------------------------------------------
Sub GetGroups(strComputer, iRow)
Dim aGroup() As GroupData
Dim pos As Position
Dim iTopRow, iStartCol, iCol As Integer
Dim iNewGroup As Integer
Dim objADs, objGroup As Variant
iStartCol = 10
iTopRow = iRow
ReDim aGroup(0)
```

GetGroups's first phase is to check and load any previously discovered groups discovered during a prior run of this script. If the cell iTopRow, iStartCol is empty, then the script prepares to add new data and sets the current row (iRow) and column (iCol) to iTopRow, iStartCol, respectively. If, however, data exists in the cell, the script calls the LoadRow function to read in this previously collected data. After this conditional, we are sure that the variables iRow, iCol have been set and represent the next available cell to add any new group name.

```
If IsEmpty(Cells(iTopRow, iStartCol)) Then
    iRow = iTopRow
    iCol = iStartCol
Else
    pos = LoadRow(iTopRow, iStartCol, aGroup())
    iRow = pos.row
    iCol = pos.col
End If
```

The script makes a connection using the ADSI WinNT provider directly to the subject computer. The script sets the Filter property to Group, which limits returned objects to only those that are groups. The script sets the variable iNewGroup to 0 in preparation to count each new group that is discovered during this run of the script.

```
Set objADs = GetObject("WinNT://" & strComputer)
objADs.Filter = Array("Group")
iNewGroup = 0
```

With the ADSI connection made, the script begins to loop through the group objects on the remote computer. For each of these groups, the script calls the `AddElement` function to determine if the group has already been discovered during the previous run of the `LoadRow` function. The name of the group (`objGroup.Name`) is sent to the `AddElement` function to check if it already exists in the array `aGroup`. If it does not already exist in `aGroup`, then `AddElement` returns `true` indicating that the group must be added to the *Output* worksheet.

```
For Each objGroup In objADs
    If AddElement(aGroup(), objGroup.Name) Then
```

If a group does need to be added, the current column (`iCol`) is checked to see if it should be wrapped. If `iCol` is greater than the sum of the starting column (`iStartCol`) and the number of desired columns to span (`g_iSpanCol`), then the function wraps the line. It inserts a new row into the worksheet, increments the row number, and sets the column back to `iStartCol`.

```
If iCol >= iStartCol + g_iSpanCol Then
    iRow = iRow + 1
    Rows(iRow).Insert
    iCol = iStartCol
End If
```

Now, sure that the addition of a new group will not exceed the desired number of columns, the script adds the name of the group to the worksheet, formats the cell as a new addition, and increments the new group counter. Lastly, the column variable (`iCol`) is incremented to prepare for the next addition. This step completes the addition of a new group to the *Output* worksheet.

```
iNewGroup = iNewGroup + 1
Cells(iRow, iCol) = objGroup.Name
Cells(iRow, iCol).Font.ColorIndex = 14
Cells(iRow, iCol).Font.Underline = _
 xlUnderlineStyleSingle
iCol = iCol + 1
End If
```

If group membership enumeration is selected, then GetGroups calls the function GetMembership. Given a computer name and group name, the function GetMembership lists all of the members of a given computer's local group and displays the results on the *enum* worksheet specified by the user on the *Setup* worksheet. To prevent the screen from flashing between the *Output* and *enum* worksheets, the script turns off ScreenUpdating and then turns it on again when it returns to the *Output* worksheet. This permits updates to the *Output* worksheet to be displayed (such as when updating the Working status). This completes the loop through each of the current computer's group objects.

```
    Application.ScreenUpdating = False
    If Worksheets("Setup").Cells(15, 5).Value = "x" Then _
     GetMembership objGroup.Name, strComputer
    Application.ScreenUpdating = True
Next
```

After the enumeration of the groups, the script calls the function MarkElementDeleted, which iterates through aGroup and highlights any groups where the Exists property is set to False. (We have not yet covered this function; AddElement sets the user-defined data type aGroup.Exists property to True for every group found in the current scan.)

Lastly, to close out this function, the number of total groups and new groups (iNewGroup) is displayed. Because aGroup was dynamically dimensioned whenever a new group was added to it, the script uses the UBound method to determine its count. The script calls PostFormatTable to autofit the cells in the *Output* worksheet to make all data visible.

```
    MarkElementDeleted aGroup()
    If UBound(aGroup) = 0 Then
        DisplayResult iRow, 8, "", 0
    Else
        DisplayResult iTopRow, 8, UBound(aGroup), 0
        Cells(iTopRow, 9) = iNewGroup
    End If
    PostFormatTable
    End Sub
```

GetMembership

The function GetMembership operates exclusively on the *enum* worksheet whose existence was earlier confirmed or created by the function InitializeEnumWs. GetMembership strongly resembles the previously discussed GetGroups function.

However, whereas GetGroups enumerates groups from a specific computer, GetMembership enumerates memberships of a specific group. The membership accounts of a group are displayed in the same horizontal manner that the groups were displayed for a computer. This function even uses the same functions: LoadRow to load previously gathered data and AddElement to see if these previous runs match current data.

Some differences do exist between these two functions, however. Whereas the computer name is used by GetGroups to find a unique instance of a prior run, GetMemberships uses both the computer name *and* the group name. This duplication is because a computer may have more than one group, and the same group name may be used for more than one computer. To accommodate this, the *enum* worksheet includes a column representing this unique key titled ComputerName/GroupName.

GetMembership uses the same variables described in the previous function GetGroups with a few exceptions. The variable aTemp represents an array used as temporary storage of the results from the Split method. Also, the starting column (iStartCol) begins on the *enum* worksheet at column 5 instead of at column 10 used by the *Output* worksheet.

To get started, the function activates the *enum* worksheet and begins processing.

```
'-----------------------------------------------------------------
' List the Members of a group- operates on Enum worksheet
'-----------------------------------------------------------------
Sub GetMembership(strGroup, strComputer)
Dim aGroupMembership() As GroupData
Dim pos As Position
Dim iTopRow, iStartCol, iRow, iCol as Integer
Dim iNewGroup as Integer
Dim result, objADsGroup, objMember, colMembers As Variant
Dim aTemp() As String
iStartCol = 5
objWsEnum.Activate
```

The input to GetMembership is the current group and the computer name being processed by GetGroups. This computername/groupname is also how GetMembership searches for previously collected data. The function uses the Excel find method to search for any previous occurrences of computername/groupname. If no occurrences exist, we know that new data will need to be added. However, we do not yet know where to add the data, because the absence of computer/groupname could mean one of two things: either the computer has not yet been discovered or the group on that computer has not yet been discovered. To resolve this problem, the script calls a second find function to look only for the computer.

```
Set result = Range("B5:B65536"). _
 Find(strComputer & "/" & strGroup, , ,xlWhole)
If result Is Nothing Then
    Set result = Range("C5:C65536"). _
     Find(strComputer, , ,xlWhole)
```

If the computer name cannot be found, the results of the second find will result in Nothing, indicating that we indeed need to add the new computer to the *enum* worksheet. This new computer is added to the bottom of any previously existing data on the worksheet. The script calls FindLastRow to determine the next row to add new data and sets this value to the variable iTopRow. Much like iTopRow represented the top row of group data for a computer in GetGroups, iTopRow here represents the first row of membership data for the current group. Next, the script sets the fill color of this new row. (*GroupScanner* alternates the fill color between computers to better visually separate the data.) Because this is a new computer being added to the bottom of the worksheet, the script checks the cell fill color of the computer immediately above it (located at iTopRow-1). If the fill color isn't gray (ColorIndex = 15), then the script sets the color of the new computer row to gray. In this manner, the script alternates color between computer listings.

```
If result Is Nothing Then
    iTopRow = FindLastRow
    If Cells(iTopRow - 1, 3).Interior.ColorIndex <> 15 Then _
    Range(Cells(iTopRow, 1), Cells(iTopRow, _
    iStartCol + g_iSpanCol - 1)).Interior.ColorIndex = 15
```

If, however, the computer is found (but the group was not), then we only need to add a new group adjacent to the other groups of that computer. The find statement returns the row number of the first occurrence of the object, and so the new computer/group object is added to the top of the previously recorded computer/group objects. The row is set to iTopRow, and a new row is inserted. (This results in iTopRow as the new, empty row.) This time, because the new group is a member of a previously recorded computer, its cell shading must match that of its brethren. The script sets the Interior.ColorIndex to the same as the row containing the computer object (iTopRow+1) and turns off italic formatting for the row, which may have been inherited from the row above.

```
Else
    iTopRow = result.row
    Rows(iTopRow).Insert
    Range(Cells(iTopRow, 1), Cells(iTopRow, _
     iStartCol + g_iSpanCol - 1)).Interior.ColorIndex = _
```

```
        Cells(iTopRow + 1, 3).Interior.ColorIndex
        Rows(iTopRow).Font.Italic = False
    End If
```

Remember, at this point in the script, we are still in midst of the conditional branch in which computer/groupname does not exist, but now we have determined to which row number we should add the new data and have stored that value in variable iTopRow. Next, the script adds the computername/groupname, computer name, and group name to the *enum* worksheet and then formats the new row with a continuous border across the top of the new cells. The script also sets the font color to teal (ColorIndex=14) to highlight the new addition. The new group has been added to the spreadsheet, but its membership has not yet been enumerated.

```
Cells(iTopRow, 2) = strComputer & "/" & strGroup
Cells(iTopRow, 3) = strComputer
Cells(iTopRow, 4) = strGroup
Cells(iTopRow, 4).Font.ColorIndex = 14
Cells(iTopRow, 4).Font.Underline = xlUnderlineStyleSingle
Cells(iTopRow, 4).Font.Bold = True
Range(Cells(iTopRow, 1), Cells(iTopRow, _
 iStartCol + g_iSpanCol - 1)).Borders(xlEdgeTop).LineStyle = _
 xlContinuous
```

But first we must complete the conditional branch for when computername/groupname does exist. The script executes this branch if it did find computername/groupname, and it sets iTopRow to the row number of the find. The script notes that this computer/group combination previously existed by marking it as active and setting its color to black.

```
Else
    iTopRow = result.row
    Cells(iTopRow, 1) = "Active"
    If Cells(iTopRow, 4).Font.ColorIndex = 14 Then
        Cells(iTopRow, 4).Font.ColorIndex = 0
        Cells(iTopRow, 4).Font.Underline = xlUnderlineStyleNone
    End If
End If
```

At this stage in the script, we have set the computer/group and know its row number. The script redimensions the dynamic array aGroupMembership to 0 in preparation for loading it with any previously discovered data. It then checks whether the first cell to store the group membership is empty. If the cell is empty, then the script assumes that no groups previously existed for this computer/group combination, and it sets the current row and column to `iTopRow` and `iStartCol`.

If, however, the cell is not empty, then the script assumes that the membership for this group has already been discovered in a prior run, and the script calls the `LoadRow` function to read the data. This function is the same `LoadRow` function called by the previous function `GetGroups`. And like the previous function, the next available cell to store any new membership data is set to the coordinates `iRow`, `iCol`.

```
ReDim aGroupMembership(0)
If IsEmpty(Cells(iTopRow, iStartCol)) Then
    iRow = iTopRow
    iCol = iStartCol
Else
    pos = LoadRow(iTopRow, iStartCol, aGroupMembership())
    iCol = pos.col
    iRow = pos.row
End If
```

Whereas `GetGroups` established an ADSI connection to the subject computer to get all groups, this function establishes an ADSI connection to the current computer's specific group. For example, to get information about the administrators local group on the computer *wkstn*, you could create an ADSI connection string `WinNT://wkstn/Administrators`. The property method `Members` returns a collection of objects representing the membership of that specific group, which the script assigns to `colMembers`.

Some Windows NT 4.0 computer groups (namely the replicator group) do not support the ADSI `Members` enumeration, which causes an error. The script employs error handling to trap this error and displays a notification to the user that the error occurred. The script skips further membership enumeration if an error occurs (`Err <>0`).

```
Set objADsGroup = _
 GetObject("WinNT://" & strComputer & "/" & strGroup)
On Error Resume Next
Set colMembers = objADsGroup.Members
If Err <> 0 Then
    Cells(iRow, iCol) = "-Err-"
    Cells(iRow, iCol).Font.ColorIndex = 14
    Cells(iRow, iCol).Font.Underline = xlUnderlineStyleSingle
```

If no errors occurred, the script iterates through all of the members of the `colMembers` and builds and displays the grid of newly added group members, much like the previous function `GetGroup` built the grid of groups for a given computer. The border formatting should only be on the top row of a computername/group-name combination. If a second row is needed to store the group membership, after the row is added, the script turns off the inherited row border formatting and also turns off any italicized type.

```
Else
    For Each objMember In colMembers
        If AddElement(aGroupMembership(), objMember.Name) Then
            If iCol >= iStartCol + g_iSpanCol Then
                iRow = iRow + 1
                Rows(iRow).Insert
                Range(Cells(iRow, 1), Cells(iRow, _
                 iStartCol + g_iSpanCol - 1)).Borders(xlEdgeTop) _
                 .LineStyle = False
                Rows(iRow).Font.Italic = False
                iCol = iStartCol
            End If
```

The script then adds the text *Newly Added* to the first column of the row to highlight the addition of the new data to this particular computer/group combination. It also highlights the actual addition by underlining and changing the text to teal. A group, computer, or user account can be a member of a group. To differentiate user and computer accounts from groups, the script calls the function `CheckIfGroup` to determine if the new addition is a group and format it as blue italic font if it is a group.

```
            Cells(iRow, 1) = "Newly Added"
            Cells(iRow, 1).Font.Bold = True
            Cells(iRow, iCol) = objMember.Name
            Cells(iRow, iCol).Font.ColorIndex = 14
            Cells(iRow, iCol).Font.Underline = xlUnderlineStyleSingle
            CheckIfGroup objMember, iRow, iCol
            iCol = iCol + 1
        End If
    Next
End If ' end of error handling
On Error GoTo 0
```

Lastly, this function calls the `MarkElementDeleted` function to highlight any accounts previously listed that are no longer found as members to the group and calls `PostFormatTable` to format the *enum* worksheet.

```
MarkElementDeleted aGroupMembership()
PostFormatTable
Worksheets(Worksheets("Setup").Cells(6, 5).Value).Activate
End Sub
```

LoadRow

A major design element of this script is to report differences in group memberships over time. The function does this by highlighting differences between subsequent runs of the tool. The function LoadRow reads data from previous runs of the tool so that it can be compared to current data.

For example, from the function GetGroups, LoadRow steps through all of the cells in the group region of the *Output* worksheet starting at the top-left cell iTopRow, iStartCol. Through this iteration, LoadRow populates the array aElement and returns the position of the next available cell for the addition of a new group.

For example, if g_iSpanCol is set to 3, a previously run group list would look something like this:

	g_iSpanCol=3		
	10 (iStartCol)	11	12
iTopRow	administrators	backup operators	guests
iTopRow+1	power users	replicator	users
iTopRow+2	bangkoktesty	Web anonymous users	Web applications
iTopRow+3	WINS users	iRow,iCol	

On a subsequent run, LoadRow begins at Cells(iTopRow,iStartCol) and finds the group name administrators. This cell is not empty; the function reads the name into the first element of the array aElement. The function moves to the next row, backup operators, and adds that to the second element of aElement. After it reads the group guests, it encounters a blank cell. Like a typewriter, the function moves to the second row and begins again at the iStartCol and reads the power users group. It proceeds like this until it processes the remaining six groups. The function LoadRow returns the position of the next available cell—in this example, row iTopRow+3, column 11. In addition to the name, LoadRow records the row and column location of each of these group names, which is used by MarkElementDeleted to find groups needing to be highlighted as missing.

The function essentially consists of two loops with a number of conditions that must be met to break the loop. Starting at the origin `iRow`, `iCol`, the inner loop marches to the right across the row (by incrementing `iCol`) until it encounters a blank cell. This breaks out of the inner loop, and the script increments the `iRow` and sets `iCol` to the starting column (`iStartCol`) and checks the condition of the outer loop. If the conditions of the outer loop are met, the function marches across the second row, and so on.

The conditions of the outer loop determine how many rows of data actually make up the current data set. Remember that every new computer (or computer/group) data set contains a border along the top row. The script looks for this border, and if it sees a border, it knows that it has run into the next computer or group's data and should stop. However, the first row (which also contains a top border) must be processed, so the first row is excluded from this top-border check. Additionally, if the computer or group being checked is at the very bottom of the worksheet, then the script looks for an empty cell, which also denotes the end of the data set.

```
'---------------------------------------------------------------
' Build User Defined Data Type Group Data with Cell or Row data
'---------------------------------------------------------------
Function LoadRow(ByVal iRow, ByVal iCol, aElement() As GroupData) _
 As Position
Dim i, iColStart, iTopRow As Integer
Dim aTemp() As String
Dim bFirstRow As Boolean
iColStart = iCol
i = 1
```

The variable `bFirstRow` is set to `true` to ensure that the first row is processed (because it also has a top border set).

```
bFirstRow = True
```

The outer loop will run while the top border of the current cell does not exist or the routine is on the first row, *and* the current cell is not empty.

```
While (Not Cells(iRow, 2).Borders(xlEdgeTop).LineStyle = 1 _
 Or bFirstRow) And Not IsEmpty(Cells(iRow, iCol))
```

The inner loop will run while the current cell is not empty. The function also resets the formatting of the current cell to black, non-italic, not-underlined font.

```
While Not IsEmpty(Cells(iRow, iCol))
    If Cells(iRow, iCol).Font.ColorIndex = 14 Then
        Cells(iRow, iCol).Font.ColorIndex = 0
        Cells(iRow, iCol).Font.Italic = False
    End If
    If Cells(iRow, iCol).Font.Underline = _
     xlUnderlineStyleSingle Then Cells(iRow, iCol).Font _
      .Underline = xlUnderlineStyleNone
```

The function reads the data from the current cell into the array aElement, which is a user-defined data type of GroupData. LoadRow strips the domain name from any group (for example, *blvu/Domain Admins* is read as simply *Domain Admins*). The data's row and column position is also loaded into the aElement array. Because this script loads previously run data, it is still unknown whether the group data is current, so the script sets the property Exists to False.

```
ReDim Preserve aElement(i)
aTemp = Split(Cells(iRow, iCol), "/")
aElement(i).Name = aTemp(UBound(aTemp))
aElement(i).Column = iCol
aElement(i).row = iRow
aElement(i).Exists = False
```

The function increments i to reference the next aElement array member and increments iCol to move one cell to the right. The loop terminates when an empty cell is read on the far right side of the data.

```
        i = i + 1
        iCol = iCol + 1
    Wend
```

Now out of the inner loop, the routine increments iRow and sets iCol to the beginning (not unlike the action of a typewriter at the end of a line). Also, at this point in the script, at least one line has been processed so the script sets bFirstRow to false, and the outer loop is checked and, if the condition is still true, the inner loop is run again.

```
iCol = iColStart
iRow = iRow + 1
bFirstRow = False
Wend
```

LoadRow returns the row and column cell position of the next available cell in which to put data. This position is defined as the position to the right of the position of the last element read. It doesn't matter here if the next position extends beyond the number of spanned rows because the bounds are checked before data is actually added to the worksheet.

```
'Set the return row,col coords to last element
LoadRow.col = aElement(UBound(aElement)).Column + 1
LoadRow.row = aElement(UBound(aElement)).row
End Function
```

AddElement

The function AddElement searches the Name property of all of the members of the array aElement for the presence of a specific string corresponding to the name of an object. If a match is made, the Exists property is updated to True (confirming the object exists). The routine exits the loop, and the function returns False, indicating that the object does not need to be added.

If, however, a match is not made, then the function redimensions the array incrementally by one and adds the string as a new member of aElement. Variables, by default, are passed by reference to a function, which means that any updates to aElement are also reflected in the calling function.

```
'-------------------------------------------------------------------
' Search GroupData data type for a specific string and add it if new
'-------------------------------------------------------------------
Function AddElement(aElement() As GroupData, strName) As Boolean
Dim bAdded As Boolean
Dim i
bAdded = True
For i = 1 To UBound(aElement)
    If StrComp(strName, aElement(i).Name) = 0 Then
        bAdded = False
        aElement(i).Exists = True
        Exit For
    End If
Next i
If bAdded Then
    ReDim Preserve aElement(i)
    aElement(i).Name = strName
    aElement(i).Exists = True
End If
AddElement = bAdded
End Function
```

MarkElementDeleted

The function `MarkElementDeleted` is similar to the function `MarkMissingComputers` of Chapter 8, but this function is updated to work with `GroupData` user-defined data types. `MarkElementDeleted` iterates through all the members of the array `aElement`, and if the property `Exists` is `False`, the function highlights the cell by changing its color to red and turning on strikethrough.

```
'-------------------------------------------------------------------
' Highlight Elements that are no longer in the SAM
'-------------------------------------------------------------------
Sub MarkElementDeleted(aElement() As GroupData)
Dim i As Integer
For i = 1 To UBound(aElement)
    If Not aElement(i).Exists And aElement(i).Column > 0 Then
        With Cells(aElement(i).row, aElement(i).Column).Font
        .Strikethrough = True
        .ColorIndex = 3
        End With
    End If
Next
End Sub
```

InitializeEnumWs

The function `InitializeEnumWs` prepares the *enum* worksheet for use by the *Group-Scanner* script. The function sets the object `g_objWsEnum` to the worksheet named in the *Setup* worksheet. If an error results, the specified *enum* worksheet probably does not exist, and so the script creates a new worksheet, names it, and formats it for use as a new *enum* worksheet.

```
'-------------------------------------------------------------------
' Create new worksheet to hold group enumeration data
'-------------------------------------------------------------------
Sub InitializeEnumWs()
Dim strWsEnum As String
strWsEnum = Worksheets("Setup").Cells(7, 5)
On Error Resume Next
Set g_objWsEnum = Worksheets(strWsEnum)
If Err Then
    Set g_objWsEnum = Worksheets.Add(After:=Worksheets("Setup"))
    objWsEnum.Name = strWsEnum
    objWsEnum.Cells.Font.Name = "Arial Narrow"
    objWsEnum.Cells(1, 1) = "Computer Group Membership Enumeration"
```

```
    With objWsEnum.Range("A1").EntireRow.Font
        .Name = "Arial"
        .Size = 12
        .FontStyle = "Bold"
    End With
    ActiveWindow.DisplayGridlines = False
    objWsEnum.Cells(4, 2) = "Name/group"
    objWsEnum.Cells(4, 3) = "Computer Name"
    objWsEnum.Cells(4, 4) = "Group Name"
    objWsEnum.Cells(4, 5) = "Membership"
    objWsEnum.Range("A3:p4").Interior.ColorIndex = 15
    objWsEnum.Range("A3:p4").Font.FontStyle = "Bold"
End If
On Error GoTo 0
objWsEnum.Range("A5:A65536").ClearContents
objWsEnum.Range("A5:A65536").Font.Bold = False
objWsOutput.Activate
End Sub
```

CheckIfGroup

The CheckIfGroup function is called by the GetMembership function to specifically highlight groups that are members of other groups. The function takes an object returned from the Members property method and inspects its Class. If the Class is a group, then the function formats the cell as blue italic.

Additionally, if the object is a group, then the function also prefixes the name with the group domain, if it is not WinNT: or NT AUTHORITY. The domain is found from the object's Parent and then Split using the delimiter /. For example, if the computer *wkstn* had two groups—*wkstn/Administrators* and *blvu/domain admins*—then the script would display the groups as *Administrators* and *blvu/domain admins*.

```
' -------------------------------------------------------------------
' Determine if the object is a group and highlight cell if it is
' -------------------------------------------------------------------
Sub CheckIfGroup(objMember, iRow, iCol)
Dim aTemp() As String
Dim strPrefix, As String
If objMember.ObjectClass = "Group" Then
    With Cells(iRow, iCol).Font
        .Italic = True
        .ColorIndex = 5
    End With
```

```
        aTemp = Split(objMember.Parent, "/")
        strPrefix = aTemp(UBound(aTemp))
     If StrComp(strPrefix, "NT AUTHORITY") <> 0 And _
         StrComp(strPrefix, "WinNT:") <> 0 Then _
       Cells(iRow, iCol) = _
       strPrefix & "/" & Cells(iRow, iCol)
    End If
    End Sub
```

FindLastRow

The function `FindLastRow` returns the next available row number at the bottom of a worksheet. The function inspects three columns, -2, 5, and 10, when determining the last row. Multiple columns are checked because some data might extend below the computer or group's primary line (known as `iTopRow` for the computer). The function accepts up to 99 blank rows between legitimate data.

```
'----------------------------------------------------------------
' Find the last row on enum worksheet- this is where new computers
' will be added.
'----------------------------------------------------------------
Function FindLastRow()
Dim iRow, iBlankLines As Integer
iRow = 6
iBlankLines = 0
While iBlankLines < 99
    While Not IsEmpty(Cells(iRow, 2)) Or _
          Not IsEmpty(Cells(iRow, 5)) Or _
          Not IsEmpty(Cells(iRow, 10))
       iBlankLines = 0
       iRow = iRow + 1
       Wend
    iBlankLines = iBlankLines + 1
    iRow = iRow + 1
Wend
FindLastRow = iRow - 99
End Function
```

HideColumns

Mostly a vanity function, `HideColumns` reads the *Setup* worksheet and, depending on what cells contain an x, hides the corresponding column. The function is short because the counter variable `i` also corresponds to the column number being inspected. As the function walks across the columns, it increments the rows on the *Setup* worksheet and looks for xs. If an x is found, the function hides the column by setting the Hidden property to True.

```
'-------------------------------------------------------------------
' HideColumns- Hide Columns to aide in formatting
'-------------------------------------------------------------------
Sub HideColumns(iCol)
Dim iRow, i As Integer

iRow = 25
For i = 2 To 9
    If StrComp(Worksheets("Setup").Cells(iRow, iCol), "x") = 0 Then
        Worksheets(Worksheets("Setup").Cells(23, iCol).Value) _
        .Columns(i).Hidden = True
    End If
iRow = iRow + 1
Next i

End Sub
```

SUMMARY

GroupScanner focuses on more general development code to drive Excel to discover and display computer group membership information. The actual script to enumerate the groups of a computer and the membership of a group is short relative to the total length of this script. The bulk of the script, however, provides a framework in which the enumeration functions can run. This framework increases the utility of the script because it allows for the creation of a baseline against which future scans can be compared. For example, it could take hours to audit, review, and approve the membership of an organization's computers. A full second run and review could take just as long if you must compare the membership line-by-line. This script effectively automates this task by highlighting only rows that have changed. And although the example works with group enumeration, it could also be extended to track file ACLs, IIS configuration settings, shares, or any other large amount of data that could slowly change over time.

KEY POINTS

- Sometimes the presentation and formatting of your script will take more length and consideration than the code performing the actual work. For example, *GroupScanner* supports varying columns and multiple rows of data, which adds complexity to the functions.
- *Groupscanner* demonstrates reusing functions that handle similarly formatted but different data. For example, the functions `LoadRow` and `AddElement` work for both groups of computers and members of groups.
- Due to the likelihood of encountering a lot of data, this script includes additional functions to highlight new or deleted objects, such as if a group is added to a computer or a membership of a computer is changed.
- The script uses the ADSI WinNT provider to remotely connect to all specified computers and access their SAM for user and group information.
- Whereas *DumpComputerList* and *Wormscanner* iterate through arrays of objects to look for previously encountered data, *Groupscanner* taps into the proprietary Excel `find` function to achieve a similar result. Using the optimized `find` function executes searches much faster but limits the portability of the script to only Excel.

The next script is also the last Excel script that uses the *DumpComputerList* framework. It provides a method to rotate the local administrator passwords of computers in your domain and keep an archive record of all past successful password changes.

12 Local Password Audit and Change Tool

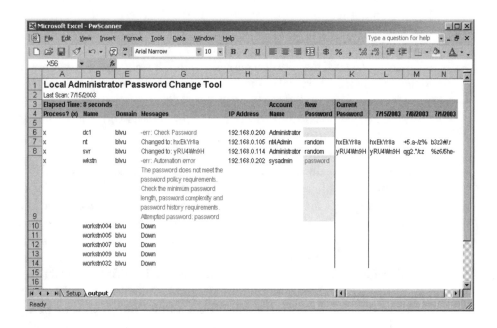

OVERVIEW

Enterprise IT generally frowns on using shared accounts for anything beyond anonymous services such as intranet access. Many company security policies specify guidelines surrounding the usage of shared accounts and restrict the usage of super-user or privileged shared accounts like the local administrator account. The local administrator can perform any action on its computer, such as adding new users and groups, installing any software, reconfiguring the system and network settings, and changing file access control lists. Additionally, changes made using a generic account like the local administrator account are difficult to audit, because the changes can be made by anyone who knows the username and password and can access the computer. Also, the changes are all recorded under the name of the local administrator (typically `computername\Administrator`). For this reason, many organizations usually add individual user accounts (either local or domain) to the administrators group, which provides a similar level of access to that of the local administrator but allows changes to be made (and audited) by unique user accounts.

Every Windows NT-based computer (including Windows XP, Windows 2000, and Windows Server 2003) configures a local administrator account at installation. When a computer joins a domain, the process adds certain domain accounts to the computer's local groups. For example, the domain group *Domain Admins* is added to the local group *Administrators*, and domain group *Domain Users* is added to the local group *Users*. This addition allows for the centrally managed domain accounts to access resources on local computers. Windows NT has operated in this manner since its inception, and it works well. Even with these centralized accounts, however, the original local administrator account persists and still provides high-level access to the computer.

The local administrator account provides a critical and special function. If the computer is physically off the network or its account has been deleted from the domain, often the only access to the computer might be through the local administrator account (or using a cached profile of a previously logged-on administrator).

Managing and rotating the password of the local administrator account for your computers is important. In many organizations, the local administrator password is the same for many (or all) computers, because they are built using automated scripts or processes that set the local administrator password the same. Also vulnerable are machines built with a configuration file in which the password is stored in the clear in a simple text file on a build floppy disk or network share. Sometimes, the builder of a computer sets the password to something like *password* and forgets to change it later to something more complex.

Many organizations change the name of the local administrator account from *administrator* to something more obscure as an extra level of protection. However, whereas you can change the name of the local administrator account using a centralized tool such as group policy, you cannot change the password from a central location for all computers in your domain. The script in this chapter examines one method and design of centrally managing the distributed local administrator passwords across your organization using an Excel-based VBA script. This script archives and rotates the local administrator passwords for the computers in your domain.

SCENARIO

Your company employs about 30 systems engineers and administrators to manage more than 1,500 servers across five data centers around the world. Your team repairs, rebuilds, and adds computers into your network constantly using a combination of Microsoft and third-party vendor tools to automate the build process. These tools set the local administrator account password to a common value known by the IT staff involved with the build and configuration process. Your company's information security officer maintains a list of critical infrastructure passwords and rotates them on a regular basis. However, rotating the local administrator accounts is difficult due to the number of computers and people involved in the process. When regularly rotating the passwords, it is important to keep them in sync and secured. In the hurried pace of building and rebuilding systems, sometimes the local administrator passwords for computers not built with the deployment tools are set to something other than the standard—or worse, the password has been set to *password*.

It is also important to be able to quickly pull up the local administrator password for a particular computer. On occasion, an administrator needs the password when a computer is off the network and no cached domain profiles exist on the computer to use to log on to it. You need a reliable password archive tool that sets the local administrator password and records it in a secure document. Users use their own domain accounts for routine maintenance of the computers, but you need quick and reliable access to the local administrator password for special uses or emergencies. Plus, you would like to use unique passwords for every system—if you ever give out the local administrator password for one system, you would prefer to rotate only the password for that system instead of having to rotate the password for all the systems. You want your tool to randomly set strong and unique passwords for every specified computer's local administrator account.

ANALYSIS

This script leverages the *DumpComputerList* tool presented in Chapter 8 and extends the functionality to determine the name of the local administrator account and set its password on selected computers. The script first determines the local administrator for the computer and then uses the SetPassword method to set the password for that account. The script traps and displays any errors to help inform you if (and why) setting a password to a particular value may have failed. Additionally, the script can randomly generate eight-character passwords for you using characters from six predefined character sets.

For similar reasons as in the *GroupScanner* scenario (Chapter 11), you decide to use Excel as your preferred storage document for these passwords. Excel provides an ideal tool for tracking a matrix of computer names and passwords that is easy to update and reference. Using the *DumpComputerList* tool, you already have a method for collecting all of the computer names in your domain, and the columns in Excel provide intuitive, easy-to-manage statistics surrounding this management of the local administrator password. For each selected computer you want to display the name of the local administrator account, messages informing the progress of the password-change efforts, the new proposed password, and the history of old passwords.

The characters used to create the password are defined by their ASCII decimal values (for example, the character A has an ASCII value of 65, and a has a value of 97). All characters that make up the six-character sets span the ASCII range from 33 to 122, as shown in Table 12.1.

TABLE 12.1   The password generator uses the ASCII values of characters to create the random passwords

Character set	ASCII start range	ASCII end range
0-9	48	57
A-Z	65	90
a-z	97	122
!"#$%&'()*+,-./	33	47
:;@=:?@	58	64
[\]^_`	91	96

To generate a password from any of these characters, the password generator simply randomly generates numbers between 33 and 122 and then converts each ASCII value to the corresponding character using the Chr method (e.g., Chr(65)=A). However, depending on which character sets the user chooses to include, gaps may exist in the ASCII range. For example, if a user selected only the alphanumeric character sets, then the included characters would span ASCII values from 48-57, 65-90, and 97-122. To accommodate these gaps, the script builds an array (g_aValidChar) that contains the ASCII values of all selected characters. In the previous example, g_aValidChar would contain 62 elements consisting of the ASCII values from each of the three alphanumeric character sets. Before the function considers a randomly generated character valid, it must be match the number in g_aValidChar.

A description of the functions and subroutines used in *PWScanner* follows. The script separates the identification of the local administrator account name from the actual changing of the passwords into two separate actions.

PwScan: Scans computer list for selected computers to process.

- Starting point for the script.
- Cycles through computer list, and calls the IDLocalAdmin and ChangePassword functions to identify the name of the local administrator and then change its password.

IDLocalAdmin: Discovers local administrator account.

- Returns the name of the local administrator account.

ChangePassword: Sets a new password for a specified user account.

- Changes the password for a user account.
- Confirms the successful change.

Distribution and Installation

You must run this script from a privileged account with administrator access to the target computers (such as a domain administrator). For example, to reach other domains you could use the RunAs utility to launch Excel under the context of a privileged user in the domain with which you wish to work. The entire script is contained within the Excel workbook.

The data contained within this script—your passwords—is extremely sensitive, and you should be sure to store this document in a protected location, such as offline and password protected. Doing so effectively keeps safe the proverbial keys to your castle.

TIP

Output

ON THE CD

To run the script, open the Excel workbook *pwscanner.xls* from the accompanying CD-ROM and review and set the configuration parameters on the *Setup* worksheet, as seen in Figure 12.1. Specify the name of the domain, and click the first button 1. Pull Computers from Directory to generate the list of computers from the specified domain. Next, on the *Output* worksheet, notice in column 1 that all computers are initially selected by x, as seen in Figure 12.2. This means that the script operates on all computers. At this point you can deselect and reselect computers, sort the list, or remove computers from the generated list until you have a list of computers that you wish to process.

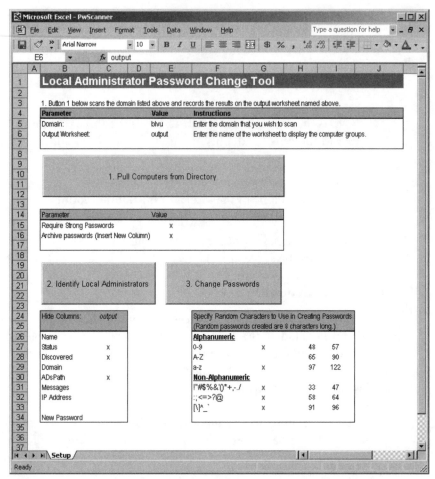

FIGURE 12.1 The *PWScanner Setup* worksheet lets you specify the domain, whether to enforce strong passwords, keep old passwords, or specify which characters to use when generating random passwords.

When satisfied with the computer list, return to the *Setup* worksheet and click the button called 2. Identify Local Administrators. Clicking this button calls the script that discovers the local administrator account for all selected computers. Return to the *Output* worksheet, and you see that the script has discovered the names of the local administrator account and listed them in the column titled Account Name, as seen in Figure 12.3. If the name of the local administrator account changes between subsequent runs, the script highlights the old name to draw your attention to the change. In Figure 12.3, two of the local administrator accounts have been changed to *nt4Admin* and *sysadmin*. Also, notice that the IP address is also listed, which may be helpful in further identifying or troubleshooting computer problems.

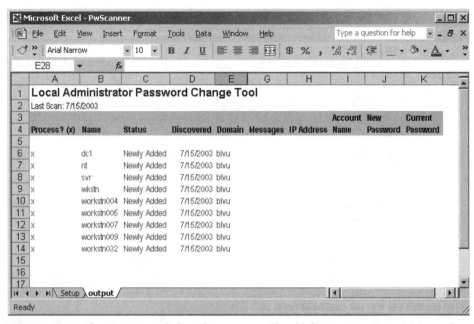

FIGURE 12.2 The *Output* worksheet lets you specify which computers to process and displays script progress and output results.

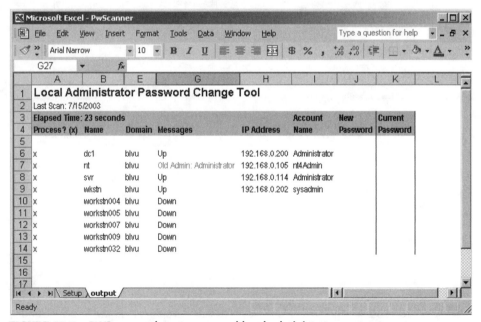

FIGURE 12.3 *PWScanner* detects renamed local administrator accounts.

The first step of the script is to identify the names of the local administrator accounts—they might not always be *Administrator*.

Changing the password is a two-step process. First, on the *Output* worksheet, enter the desired password in the column named New Password. If you want the computer to autogenerate a random password, enter the word random for the password. When you are ready to change the passwords, return to the *Setup* worksheet and click the button called 3. Change Passwords. The script iterates through the list of computers and attempts to change the passwords for all selected computers. When the process is complete, you see any messages highlighting progress and any errors that may have occurred. Figure 12.4 shows an example of the messages you see after changing the passwords. This figure actually contains quite a bit of data about how the script operates. In the first column, only four computers were selected to process. The workstnXXX computers (which were determined to be *Down* from the *Find local administrator* script) were not selected to process, and you can see this by the absence of x marks in the Process column. The script highlights any problems it encountered when trying to change the password. In this example, we selected to process the computer *dc*, but no new password was entered. As such, the script highlighted the cell needing attention and added a message informing the user of the problem.

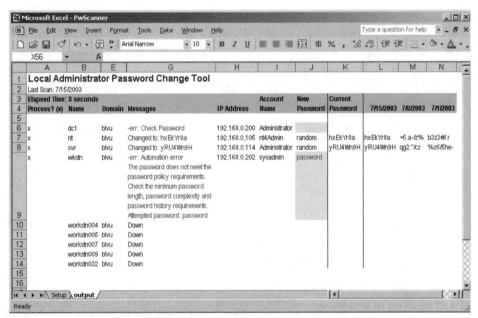

FIGURE 12.4 The Change Password function displays warning and status messages and archives current and past passwords.

In rows 7 and 8, we specified that the new password be random. The script randomly generated a new password and displayed a message informing us of the new password. The column Current Password always displays the most current password. Columns to the right of Current Password display old passwords—in this case, the script was previously run on July 1 and 8. Also, you may notice that the latest randomly generated passwords contain only alphanumeric characters. This is because, specifically for this latest run, only `0-9`, `A-Z`, and `a-z` character sets were enabled.

Lastly, in row 9, we tried to set the password to the word *password*. The script displays an error notifying us of the strong password policy and does not change the password.

Summary of Solution

Create an Excel-based script that displays a list of computers and discovers the names of the local administrator accounts on these computers. The primary purpose of the script is to provide a record of local administrator passwords for these computers and change them when needed.

OBJECTIVE

An Excel-based tool that scans a directory of computers, changes the password of the local administrator account, and maintains a record of the change.

ASCII
- ASCII character manipulation

ADSI
- WinNT Provider
- `SetPassword`
- `SetInfo`

Excel Object Model
- Worksheet manipulation
- Cell/range formatting

Random Number Functions
- `Randomize`
- `Rnd`

VBA Error Handling
- `On Error Resume Next`
- `On Error Goto 0`

User Data Types
- Using `Type` to declare user-defined data types

THE SCRIPT

When you open the Microsoft Visual Basic Editor with *PWScanner* loaded, you may notice modules named from previous scripts in this text; this script uses some of the functions from previous scripts. However, only the functions that this script uses are included (not all functions from previous scripts).

ON THE CD

The script is presented on the CD-ROM as /chapter 12/scripts/pwscanner.xls. This script is contained within an Excel document that requires Excel 2000 or later, as well as Windows Script Host 5.6. Because this script changes the passwords of your accounts on selected computers in your domain, be especially careful that you test this script in a test environment and become familiar and comfortable with its operation before using in production.

Module-level Code

As with the previous VBA scripts, *PWScanner* requires that variables be declared. It also sets the default global comparison property to `Text`. The script declares the worksheet objects for the *Setup* and *Output* worksheets as `Public` variables so that all other functions can access the worksheets without having to explicitly declare and reference them. The script also declares the array `g_aValidChar` as `Public`. This array stores the allowable valid characters, which the random-password generator uses to create the passwords.

```
Option Explicit
Option Compare Text
Public g_objWsSetup As Worksheet
Public g_objWsOutput As Worksheet
Public g_aValidChar()
```

PWScan

The main function in *PWScanner* is named `PWScan`. `PwScan` performs two different functions, depending on how it is called. If `PwScan` is called with a `1` (`iSelection = 1`), then it searches for the local administrator account on the specified computers. If `PwScan` is called with a `2` (`iSelection=2`), then it attempts to change the local administrator account password on specified computers. Because both of these functions require looping through the same list of computers and pinging them to see if they are responding, sharing this common code between the two functions keeps the script shorter and easier to read and encourages code reusability.

The aComputer array of user-defined data type ComputerData stores the list of computers generated from the *DumpComputerList* script of Chapter 8. The script uses the variables dtStart and dtEnd to time and display how long the entire script takes to execute. The variables iCol and iRow represent column and row values when stepping through worksheet data. The script uses the variables i and iCharSetCount as generic counter variables.

The variables bIDLocalAdmin and bChangePassword represent whether the script should find the name of the local administrator or change the local administrator password branch of code, respectively. Initially, both of these variables are set to False. If PwScan is called with a 1 (iSelection=1), then bIDLocalAdmin is set to True, and subsequent checks look to bIDLocalAdmin to check if the script should process this branch of code. Similarly, if PwScan is called with a 2 (iSelection=2), then bChangePassword is set to True.

To complete the basic script initialization, the function sets the Worksheet object g_objWsSetup to the *Setup* worksheet and redimensions the dynamic array g_aValidChar to 0 to prepare it for use. It also sets the start time of the script by setting dtStart to the current time via the Now method.

```
'-------------------------------------------------------------------
' Main program to conduct password scan
'-------------------------------------------------------------------
Sub PwScan(iSelection As Integer)
Dim aComputer() As ComputerData
Dim i, iRow, iCharSetCount, iCol  As Integer
Dim dtStart, dtEnd As Date
Dim computer As PingInfo
Dim bIDLocalAdmin, bChangePassword As Boolean
ReDim g_aValidChar(0)
dtStart = Now
Set g_objWsSetup = Worksheets("Setup")
If iSelection = 1 Then bIDLocalAdmin = True
If iSelection = 2 Then bChangePassword = True
```

The script allows you to specify what characters to use when creating random passwords. The script breaks these characters into six different sets. The first three sets represent the alphanumeric characters 0-9, A-Z, and a-z. The second three sets represent nonalphanumeric characters: !"#$%&'()*+,-./ ,   :;@=:?@ , and [\]^_` . You can select any or all of these sets of characters via the *Setup* worksheet by placing an x in the cell adjacent to the desired character set. The random passwords that the script generates comprise the characters from the sets you specify.

Note that your own network password policy may require the use of *strong* passwords enforced by local or group policy. The Windows NT (and later) strong password definition requires that passwords contain characters from three out of four of the following sets: 0-9, A-Z, a-z, and the nonalphanumerics. Regardless of your password policy, the script lets you select any number of character sets from which to create the random passwords. When the script attempts to set the password for a computer requiring a strong password, it fails if you have selected fewer than three character sets and notifies you that strong passwords are required. If you select *Require Strong Passwords* on the *Setup* worksheet (by placing an x in Cells(15,5)), then the script checks that at least three of the four character sets that define a strong password are checked. The script alerts you if fewer than three sets are selected.

The script uses a simple tally (iCharSetCount) to count the number of enabled character sets. The character sets are set on the *Setup* worksheet. For example, if the character set 0-9 is enabled, then the *Setup* worksheet has an x in Cells(27,7) and iCharSetCount is incremented by one. Similarly, whereas this script breaks up the nonalphanumeric characters into three sets, the strong password requirement considers them as one of the four possible sets of characters. Accordingly, the script considers an x in any of the three nonalphanumeric character set cells as a single tally. If the resulting tally is less than three, then the script displays a warning to increase the number of character sets used.

Some organizations may not require strong passwords for all domains and prefer to use fewer or specific character sets (for example, only alphanumeric characters) to create their passwords. In this example, do not specify *Require Strong Passwords*, and select only the character sets you desire. Using the strong password and custom character selection, this script accommodates both requirements.

```
iCharSetCount = 0
If g_objWsSetup.Cells(15, 5) = "x" Then
    If g_objWsSetup.Cells(27, 7).Value = "x" Then _
     iCharSetCount = iCharSetCount + 1
    If g_objWsSetup.Cells(28, 7).Value = "x" Then _
     iCharSetCount = iCharSetCount + 1
    If g_objWsSetup.Cells(29, 7).Value = "x" Then _
    iCharSetCount = iCharSetCount + 1
    If g_objWsSetup.Cells(31, 7) = "x" Or _
        g_objWsSetup.Cells(32, 7) = "x" Or _
        g_objWsSetup.Cells(33, 7) = "x" Then _
        iCharSetCount = iCharSetCount + 1
    If iCharSetCount < 3 Then
    MsgBox "You have selected to Require Strong Passwords. " & _
        "To satisfy Windows strong password requirements, please " & _
```

```
        "check at least 3 out of the 4 of the random character " & _
        "types to use in the strong password. 0-9, A-Z, a-z or " & _
        "non-alphanumerics. " & vbCrLf & vbCrLf & "Or, deselect " & _
        "Require Strong Passwords.", vbCritical, "Check Password " & _
        "Character Set"
        Exit Sub
        End If
    End If
End If
```

Next, the script looks for the existence of the *Output* worksheet and displays a warning if it does not exist; this functionality is described in more detail in Chapter 8, where it was first introduced.

```
On Error Resume Next
    Set g_objWsOutput = Worksheets(g_objWsSetup.Cells(6, 5).Value)
If Err <> 0 Then
    MsgBox "Can not find Output worksheet. Have you run '1. " & _
    "Pull Computers from Directory'?", vbExclamation, _
    "Check Configuration"
    End
End If
On Error GoTo 0
```

Next, PwScan activates the *Output* worksheet and begins its work. The script can also archive passwords. If PwScan is in the Change Password mode (bChangePassword is True), it checks whether Cells(16,5) on the *Setup* worksheet contains an x, which indicates that the Archive Passwords (Insert New Column) option is selected. If selected, the script inserts a new column and labels it with the current date. Password data from a previous run is maintained but is shifted one column to the right. In this manner, when the Archive Passwords feature is activated, all past password histories are stored.

```
objWsOutput.Activate
If bChangePassword Then
    If g_objWsSetup.Cells(16, 5) = "x" Then
        Columns(12).Insert
    End If
    Cells(4, 12) = Format(Now(), "m/d/yy")
End If
```

The script uses the LoadComputerList function from Chapter 8 to load the previously generated computer list from the *Output* worksheet into the array aComputer. The script uses this array for cycling through the list of computers. For each computer in the list, the script checks the first column of that computer's row number for an x to see if it should process that computer. If the script finds an x, it italicizes the row to highlight it as the current row of operation. The script also resets the font color and interior cell color to the default and displays the *Working* message.

```
i = LoadComputerList(aComputer, objWsOutput)
For i = 1 To UBound(aComputer)
    iRow = aComputer(i).row
    Cells(iRow, 7).Select
    If InStr(1, Cells(iRow, 1).Value, "x") Then
        Rows(iRow).Font.Italic = True
        Rows(iRow).Font.ColorIndex = 0
        Rows(iRow).Interior.ColorIndex = 0
        DisplayResult iRow, 7, "Working", 3
```

The first thing the script checks is whether the computer is up and running by pinging its name via the `Pingable` function. The `Pingable` function from Chapter 10 has been modified to return not only the response (via the property `responding`) but also the IP address of the computer, as resolved by DNS. The function `Pingable` returns the `ping` response and the IP address of the computer to the variable computer, which has been defined as a user-defined data type `PingInfo`.

View the changes to the `Pingable` function by opening the Visual Basic Editor from the PWScanner *workbook. Double-click the module named* WormScanner, *and you can review the function for the new changes.*

Sometimes, an incorrect DNS entry can cause problems when you are working remotely. Displaying the resolved DNS IP address alongside the computer's name can be helpful in troubleshooting. The script displays the IP address (`computer.ip`) of the computer (as returned by DNS) in `Cells(iRow,8)`.

If the computer responds to a `ping` (`computer.responding = True`), then the script updates the message cell to note that the computer is up. It then calls either the `IDLocalAdmin` or `ChangePassword` subroutine, depending on whether PwScan is in the Find Local Administrator or Change Password mode (as previously defined by the variable `iSelection`).

If the computer does not respond to a `ping` (`computer.responding= False`), the script updates the message cell to note that the computer is *Down*.

```
computer = Pingable(aComputer(i).Name)
Cells(iRow, 8) = computer.ip
```

```
If computer.responding Then
    DisplayResult iRow, 7, "Up: Working", 3
    If bIDLocalAdmin Then IDLocalAdmin iRow
    If bChangePassword Then ChangePassword iRow
Else
    DisplayResult iRow, 7, "Down", 0
End If
```

After processing the computer through either the `IDLocalAdmin` or `ChangePassword` function, the script turns off the italic formatting of the row to indicate that it is no longer the current row.

The script then updates the current password cell for the active computer. This process finds the last changed password for a computer and inserts this password into the current password column. In this manner, even if not all passwords have been changed with every run of the script, the current password column contains the last set password for a given computer. Figure 12.5 shows an example of how the script displays the current password of each computer in the current password column, regardless of when the password was last set.

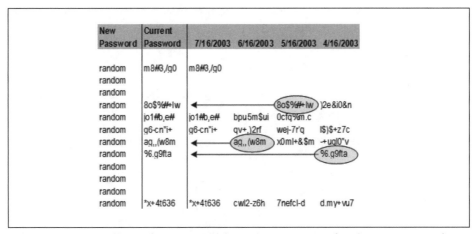

FIGURE 12.5   Even if you choose to run the *PWScanner* on only a few computers, the script rolls up all current passwords to the Current Password column.

The procedure that accomplishes this task uses a `Do..Loop..Until` statement to continue execution until `iCol` reaches a value or the end of the worksheet. (An Excel worksheet has 256 columns, which is why the upper bounds of this loop end at a value greater than 255.) Each cell, beginning at the column immediately to the right of the current column (12,) is checked for a value. If the value is found, then the current password column (11) is set to this value. If not, iCol is incremented and the next cell is checked.

```
            Rows(iRow).Font.Italic = False
        End If
    iCol = 12
    Do
        If Cells(iRow, iCol) <> "" Then
            Cells(iRow, 11).Value = Cells(iRow, iCol).Value
            Exit Do
        End If
        iCol = iCol + 1
    Loop Until iCol > 255
    Next i
```

Lastly, the script formats the results. It calls the PostFormatTable function (described in Chapter 8) and adds a right border to columns 10 and 11 to distinguish the different password columns (the new password, the current password, and the archived passwords). To close this subroutine, the script calls the HideColumns function (described in Chapter 11) to hide any columns the user specifies and displays the elapsed time taken to execute the script.

```
PostFormatTable
Range(Cells(3, 10), Cells(aComputer(UBound(aComputer)).row, 10)). _
  Borders(xlEdgeRight).LineStyle = xlContinuous
Range(Cells(3, 11), Cells(aComputer(UBound(aComputer)).row, 11)). _
  Borders(xlEdgeRight).LineStyle = xlContinuous
HideColumns 3
dtEnd = Now
objWsOutput.Cells(3, 1).Value = "Elapsed Time: " & _
  DateDiff("s", dtStart, dtEnd) & " seconds"

End Sub
```

IDLocalAdmin

The subroutine IDLocalAdmin remotely connects to a computer and enumerates its local users to find the name of the local administrator account. By default, this account is named *Administrator*. However, many organizations choose to change this name to something else in an effort to increase security. (Arguably, security is increased because in order to log on to an account, you must know both the account name and password, and if you change the well-known account name *Administrator* to something obscure, it makes logging on more difficult.) The subroutine `IDLocalAdmin` uses the row number of the current computer as input and from this row gets all the information it needs for processing. The computer name is stored in the variable `strComputer`. The objects `objADs` and `objUser` represent the ADSI connection to the remote computer and the reference to the individual user objects when a connection is made. The script uses the Boolean variables `bUser` and `bAdmin` to flag and track possible alerts. The script uses the temporary array `aTemp` to store the contents of the `objectSID` property, which is used to determine if a given account is the local administrator account.

```
'-----------------------------------------------------------------
' Scans a local computer user accounts looking for the local admin
'-----------------------------------------------------------------
Sub IDLocalAdmin(iRow)
Dim strComputer As String
Dim objADs, objUser As Variant
Dim bUser, bAdmin As Boolean
Dim g_aValidChar() As Integer
Dim aTemp() As Byte
```

First, the script sets `bUser` and `bAdmin` to `False`. As the script progresses, it updates these variables to `True` if and when the subroutine passes certain objectives. At the end, depending on the state of these variables, the script displays custom messages.

The script also sets the variable `strComputer` to the value of `Cells(iRow,2)`, which contains the name of the computer. `IDLocalAdmin` uses the ADSI WinNT provider to connect to the computer and sets the `filter` to all `User` objects. This ensures that all objects returned from this connection are only `Users` and not `computers` or `groups`.

```
bUser = False
bAdmin = False
strComputer = Cells(iRow, 2).Value
Set objADs = GetObject("WinNT://" & strComputer)
objADs.Filter = Array("User")
```

`objADs` references all user objects on the remote computer `strComputer`. The script cycles through each of the objects and attempts to identify the local administrator account, regardless of to what it may have been renamed.

But first, we reach our first checkpoint. Remember that at this point, the remote computer is up. (It responds to a `ping`.) However, if `objADs` contains no objects, then the account under which the script is running might not have sufficient rights to connect to this remote computer, or the computer could be a non-Windows-based computer. However, if the remote computer does have at least one user object that the script can read, then the bUser is set to True, which indicates that the script could successfully enumerate the local users of this remote computer.

```
For Each objUser In objADs
    bUser = True
```

Remember that the local administrator account may not always be named *Administrator*, which precludes us from simply looking for an account named as such. One method of determining the true local administrator is by examining the SID (security identifier) of the account. Every account in Windows has a unique SID. Unique accounts (such as local administrator or guest) have particular SID attributes that do not change. For example, the local administrator has a SID of S-1-5-21-x-y-z-500 and a guest account has a SID of S-1-5-21-x-y-z-501. Whereas the x, y, and z variables in the previous example change with each account, the Relative Identifier (RID) does not change. The RID value is the last value in the SID. The RID for a local administrator is 500, and the RID for a guest account is 501. When a SID is expressed as S-R-I-S-S (such as S-1-5-21-x-y-z-500), it is in a format known as the *string SID*.

TIP

To see the SID for yourself, use the resource kit utility getsid.exe. *This program displays the string SID for a given computer and account.*

The ADSI WinNT provider provides access to an attribute called ObjectSID, which contains the SID for a local account. For every user account the script gets the ObjectSID, which contains a byte representation of the SID stored in an *octet string*. The octet string format is not the same as the aforementioned SID string format. Unfortunately, whereas C++ and C# each include methods to convert one SID format to another, Visual Basic and JScript do not (although you can find third-party DLLs and external Microsoft DLLs that you can install to make these conversions for you). However, we can bypass this inconvenience by examining the construction of the ObjectSID in its native format. Each ObjectSID is an array of 28 elements. Each element is of a byte data type. Figure 12.6 shows an example of the ObjectSID elements for a number of accounts. Notice the elements numbered 24 and 25. They are the same for the administrator accounts. The 24 and 25 elements of a local administrator account have a byte value of 244 and 1, respectively.

In this manner, we do not need to convert the SID to a SID string to determine if an account is the local administrator account; we merely inspect if the values of the 24 element equals 244 and the 25 element equals 1.

Computer Name	Account Name	Element Number																										
		1	2	3	4	5	6	7	8	9	10	11	12	13	14	15	16	17	18	19	20	21	22	23	24	25	26	27
svr2	jeff	5	0	0	0	0	0	5	21	0	0	0	67	23	10	50	15	248	96	29	22	192	234	50	247	3	0	0
dc1	krbtgt	5	0	0	0	0	0	5	21	0	0	0	217	49	248	66	209	218	116	3	7	229	59	43	246	1	0	0
svr2	SQLDebugger	5	0	0	0	0	0	5	21	0	0	0	67	23	10	50	15	248	96	29	22	192	234	50	246	3	0	0
svr2	ACTUser	5	0	0	0	0	0	5	21	0	0	0	67	23	10	50	15	248	96	29	22	192	234	50	245	3	0	0
svr1	Guest	5	0	0	0	0	0	5	21	0	0	0	252	227	21	49	187	140	51	94	130	139	166	40	245	1	0	0
dc1	Guest	5	0	0	0	0	0	5	21	0	0	0	217	49	248	66	209	218	116	3	7	229	59	43	245	1	0	0
wkstn1	Guest	5	0	0	0	0	0	5	21	0	0	0	87	41	2	76	248	159	180	116	21	37	175	71	245	1	0	0
wkstn2	Guest	5	0	0	0	0	0	5	21	0	0	0	205	124	65	102	253	67	70	30	21	37	175	71	245	1	0	0
wkstn3	Guest	5	0	0	0	0	0	5	21	0	0	0	252	227	21	49	77	100	73	46	7	229	59	43	245	1	0	0
svr2	Guest	5	0	0	0	0	0	5	21	0	0	0	67	23	10	50	15	248	96	29	22	192	234	50	245	1	0	0
svr1	Administrator	5	0	0	0	0	0	5	21	0	0	0	252	227	21	49	187	140	51	94	130	139	166	40	244	1	0	0
dc1	Administrator	5	0	0	0	0	0	5	21	0	0	0	217	49	248	66	209	218	116	3	7	229	59	43	244	1	0	0
wkstn1	Administrator	5	0	0	0	0	0	5	21	0	0	0	87	41	2	76	248	159	180	116	21	37	175	71	244	1	0	0
wkstn2	Administrator	5	0	0	0	0	0	5	21	0	0	0	205	124	65	102	253	67	70	30	21	37	175	71	244	1	0	0
wkstn3	Administrator	5	0	0	0	0	0	5	21	0	0	0	252	227	21	49	77	100	73	46	7	229	59	43	244	1	0	0
svr2	Administrator	5	0	0	0	0	0	5	21	0	0	0	67	23	10	50	15	248	96	29	22	192	234	50	244	1	0	0
svr2	VUSR_svr2	5	0	0	0	0	0	5	21	0	0	0	67	23	10	50	15	248	96	29	22	192	234	50	244	3	0	0
svr2	ASPNET	5	0	0	0	0	0	5	21	0	0	0	67	23	10	50	15	248	96	29	22	192	234	50	242	3	0	0
svr2	IWAM_svr2	5	0	0	0	0	0	5	21	0	0	0	67	23	10	50	15	248	96	29	22	192	234	50	241	3	0	0
svr1	SWMAccount	5	0	0	0	0	0	5	21	0	0	0	252	227	21	49	187	140	51	94	130	139	166	40	241	3	0	0
svr2	IUSR_svr2	5	0	0	0	0	0	5	21	0	0	0	67	23	10	50	15	248	96	29	22	192	234	50	240	3	0	0

FIGURE 12.6 By examining the 28 elements making up the ObjectSID, we can identify different types of user accounts without converting the SID to a string SID format.

Through this process the script checks each of the user accounts for the current computer. If an account is found to be the local administrator account, the second checkpoint is reached, and the script sets the variable bAdmin to True. This variable tells the script that the local administrator account was successfully found. Later in this script, if bAdmin is still false, the script reports that it could not find the local administrator account.

```
aTemp = objUser.ObjectSID
If aTemp(24) = 244 And aTemp(25) = 1 Then
    bAdmin = True
```

Before the script records the name of the local administrator, it first looks to see if a local administrator has been found for this computer in a previous run. The script looks at the value of the cell containing the name of the local administrator account (Cells(iRow,9)). If the cell is not blank, and if the value of that cell is not the same as the name of the current local administrator, then the script displays a message of this discrepancy. (This might occur if someone changed the name of the local administrator account between subsequent runs of this script.)

Otherwise, the script updates the message cell that the computer is *Up* (it used to report the current status as *Up: Working*) and records the name of the local administrator on the *Output* worksheet.

This step completes the conditional that the local administrator has been found. The Next statement instructs the script to proceed to the next user account.

```
            If Cells(iRow, 9).Value <> objUser.Name And _
             Cells(iRow, 9).Value <> "" Then
                  DisplayResult iRow, 7, "Old Admin: " & _
                   Cells(iRow, 9).Value, 4
            Else
                  DisplayResult iRow, 7, "Up", 0
            End If
            Cells(iRow, 9).Value = objUser.Name
        End If
    Next
```

After all user accounts have been searched, the script reports its findings. If the local administrator account has not been found, bAdmin remains False, and the script updates the *Output* worksheet with the message that it could not find the local administrator account.

If the script could not find any user accounts (such as if the computer is running a non-Windows operating system or the permissions set on the remote computer prevent enumeration of the accounts), then it displays a message that it could not read the user data.

Lastly, to accommodate these lengthy messages, the script adjusts the width of the column containing these messages and turns on text wrapping. Text wrapping causes Excel to wrap text longer than the cell width to the next row in the cell.

```
    If Not bAdmin Then DisplayResult iRow, 7, _
     "-err: Can not find local administrator.", 3
    If Not bUser Then DisplayResult iRow, 7, _
     "-err: Can not read local system user data. Check permissions.", 3
    Columns(7).ColumnWidth = 25
    Cells(iRow, 7).WrapText = True
    End Sub
```

ChangePassword

The subroutine ChangePassword, as its name implies, actually changes the password of the local administrator account. Like IDLocalAdmin, ChangePassword obtains the computer and user account name as specified on a given row (iRow). The subroutine reads the values of the variables strComputer and strUser from the computer on the specified row. strPassword stores the new password for the account and is either read from the current row or randomly generated, depending on whether the script is configured to generate a random password.

The script uses the Boolean variable bRandom to indicate whether the password was randomly generated or explicitly specified. The variable objUser represents the connection to the remote computer-user object used to actually change the password. The script uses iCounter as a generic counter variable.

```
'------------------------------------------------------------------
' Change the password of the local administrator account
'------------------------------------------------------------------
Sub ChangePassword(iRow)
Dim strComputer, strPassword, strUser As String
Dim bRandom As Boolean
Dim objUser As Variant
Dim iCounter As Integer
strComputer = Cells(iRow, 2).Value
strPassword = ""
strUser = ""
```

Before attempting to change the password, the script checks that values of the cells containing the name of the local administrator and password are not blank. If the cells are blank, the script displays an error message and highlights the blank cell needing attention. The script does not attempt to change the password if either the name or password cell is blank.

First, the script checks that the user account name is not blank and displays an error if it is blank.

```
Select Case LCase(Cells(iRow, 9).Value)
    Case ""
        DisplayResult iRow, 7, "-err: Check Username", 3
        Cells(iRow, 9).Interior.ColorIndex = 35
    Case Else
        strUser = Cells(iRow, 9).Value
End Select
```

Next, the script checks that the password is not blank. If both the username and the password are blank, the script displays a custom message indicating such. Otherwise, it displays a message that the password is blank.

```
Select Case Cells(iRow, 10).Value
    Case ""
        If Cells(iRow, 9).Value = "" Then
            DisplayResult iRow, 7, _
                "-err: Check Username and Password", 3
        Else
            DisplayResult iRow, 7, "-err: Check Password", 3
        End If
        Cells(iRow, 10).Interior.ColorIndex = 35
```

In addition to checking whether the cell containing the password is blank, the script checks to see if the cell contains the value Random (or its case variants random or RANDOM). If it does, then the script generates a random password for the local administrator of the current computer. The script sets the variable bRandom to True and calls the function GetRandomPassword, which actually generates a random password comprising characters from the specified character sets on the *Setup* worksheet.

If the password cell is not blank and does not contain the word random, then it is assumed to contain a desired password for this account, and the variable strPassword is set to this value.

```
    Case "random"
        bRandom = True
        strPassword = GetRandomPassword
    Case Else
        strPassword = Cells(iRow, 10).Value
End Select
```

Before proceeding with the code to actually change the password, the script checks that strUser and strPassword are not blank. Using the ADSI WinNT provider, the script attempts to connect to the user object on the remote computer. An error occurs if the account under which this script is running does not have permissions to make this connection, or if the user account does not exist. To trap these errors, the script enables error handling and reports an inline message on the *Output* worksheet of the description of any encountered errors.

```
If strUser <> "" And strPassword <> "" Then
    On Error Resume Next
    Set objUser = GetObject("WinNT://" & strComputer & "/" & strUser)
    If Err <> 0 Then
        DisplayResult iRow, 7, "-err: " & Err.Description, 3
        Columns(7).ColumnWidth = 25
        Cells(iRow, 7).WrapText = True
        Exit Sub
    End If
```

The code to actually change the password is surprisingly simple. First, the script calls the SetPassword method to specify the new password. (Note that when using this method you do not need to know the existing password before you change it; however, you do need to have appropriate administrator-level permissions, such as being a member of *Domain Admins* or other account that is a member of the remote computer's *Administrators* group.) After the script calls the SetPassword method, it calls the SetInfo method to actually update the information on the remote computer. The change is not committed to the remote computer until the SetInfo method is called.

```
objUser.SetPassword strPassword
objUser.SetInfo
```

The actual changing of the password may fail for a variety of reasons. It fails, for example, if it is not long enough, if it is not a strong password when one is required, or if the user does not have permission to change the password. The script looks for any errors denoted when Err does not equal 0.

Assuming the account running the script has appropriate permissions to change the password on the remote computer, the most likely error you may encounter is trying to set a password that does not satisfy your password policy. (You can set a password policy via the local or group policy and control settings, such as the minimum length or whether the password must meet strong password requirements, among other settings.) For example, if your domain group policy requires strong passwords, and you specify a password of *apple*, an error (Err Number 2147022651) with the following description occurs:

Automation error
The password does not meet the password policy requirements. Check the minimum password length, password complexity and password history requirements.

A number of errors, which the script traps and displays, could occur while you try to change the password. However, the script prepares for three specific outcomes:

■ The random password generator creates a password that does not conform to the strong password settings.
■ The script user specifies a password that does not conform with the strong password settings (or another error occurs).
■ The password was successfully changed.

In the first case, it is possible for the random password generator to create a password that does not conform to the strong password settings, even if all character sets are selected. A strong password must contain characters from three out of four of the character sets. The password generator randomly creates a password from characters within the specified sets. So, once in a while, a password could be generated that is all uppercase, mixed case (no numbers), and so on. To check for this occurrence, the script determines if the specific error pertaining to the password policy is encountered (Err=-2147022651), that the user specified a random password to be generated (bRandom=True), and the user specified *Require Strong Passwords* on the *Setup* worksheet.

If the user did specify to require strong passwords, then the script definitely has the *possibility* of creating a truly strong password (and it was just random that the password generator created a "weak" password). When *Require Strong Passwords* is enabled, the script enforces that at least three out of four of the character sets are selected.

Given this design, if the previous criteria are satisfied and the *Require Strong Passwords* setting is enabled, then the script calls the ChangePassword function again to start over and create a new random password. This process continues until the password is successfully changed.

This script's password-generation automatic retry is geared toward changing the actual characters that make up the password. If your password policy restricts some other setting—such as requiring a password longer than eight characters—this script must be modified or it will loop indefinitely. For most organizations using the current version of Windows, the settings within this script are typical and work within customary password policies.

If, however, the random password generated fails, and the user has not specified *Require Strong Passwords*, then the script presents the user with a dialog box warning of this error. An example configuration that might trigger this alert is if the user selected to use only the uppercase character set for random passwords and tried to set an all-uppercase password on a computer that requires strong passwords.

```
If Err <> 0 Then
    If Err.Number = -2147022651 And bRandom Then
        If g_objWsSetup.Cells(15, 5) = "x" Then
            ChangePassword iRow
        Else
            MsgBox "You have selected to use Random Passwords" & _
            " , however this target computer's password " & _
            "settings require strong passwords. Enable " & _
            "Require Strong Passwords on the Setup worksheet.", _
            vbExclamation, "Check Setup"
        End If
    End If
End If
```

The second case that the script checks for is if the user specified a password that does not meet the password policy (or any other error). In this case, the script reports an error along with its description and also appends to the error message the attempted password. Note that this alert displays any error message caused by trying to set the password, not just errors related to the password policy.

```
        DisplayResult iRow, 7, "-err: " & Err.Description & _
        " Attempted password: " & strPassword, 3
        Columns(7).ColumnWidth = 25
        Cells(iRow, 7).WrapText = True
        Cells(iRow, 10).Font.ColorIndex = 3
        Cells(iRow, 10).Interior.ColorIndex = 35
```

The third case is if the password change is successful and does not result in any error. The current password cell is updated, and the script reports a message that the password has successfully been changed to the new password.

```
    Else
        Cells(iRow, 12).Value = strPassword
        Cells(iRow, 7).WrapText = False
        DisplayResult iRow, 7, "Changed to: " & strPassword, 5
    End If
  End If
End Sub
```

GetRandomPassword

The function GetRandomPassword generates a random eight-character password comprising the characters from the user-selected character sets. The script user specifies which character sets to use on the *Setup* worksheet. GetRandomPassword consists of two main sections. The first section determines whether the character sets have been loaded into the g_aValidChar array, and the second section actually generates the password.

The function uses the variable iRow to track the row on the *Setup* worksheet when reading the character sets. The variable randomASC2, as its name implies, is a randomly generated number between 33 and 122 representing the ASCII value of a proposed password character. If a randomly generated character exists within the g_aValidChar array of selected character sets, then the script sets the value of bValidChar to True. The variable strPassword stores the new password as the function concatenates characters to it.

```
'------------------------------------------------------------------
' Generate an 8 character random password
'------------------------------------------------------------------
Function GetRandomPassword()
Dim iCounter, iRow, i, randomASC2 As Integer
Dim bValidChar As Boolean
Dim strPassword As String
```

PWScanner declares g_aValidChar a Public variable, and the main function PwScan redimensions it to a length of 0. The first time the script calls GetRandom-Password, the upper bounds of g_aValidChar will be 0, and the function builds the array g_aValidChar.

The variable iCounter represents the current index of g_aValidChar, and the function increments iCounter for each character added to g_aValidChar.

The actual upper and lower ASCII boundaries for each character set are stored on the *Setup* worksheet in rows 26 to 31 in the random characters configuration section. The script steps through each of the rows and looks for an x in column 7 of that row. If an x exists, then the script creates a second loop beginning at the value in column 8 and ending at the value in column 9. For example, if Cells(26,8).Value=48 and Cells(26,9).Value =57, then the function would loop from 48 to 57 and populate g_aValidChar with the following values: 48, 49, 50, 51, 52, 53, 54, 55, 56, and 57. First, the script incrementally redimensions the array g_aValidChar to the index iCounter, then it adds the new value to this index, and then it increments iCounter. In this manner, the function builds g_aValidChar, and when complete, the upper bound of g_aValidChar equals the number of elements in the array.

```
If UBound(g_aValidChar) = 0 Then
    iCounter = 1
    For iRow = 26 To 31
        If g_objWsSetup.Cells(iRow, 7).Value = "x" Then
            For i = g_objWsSetup.Cells(iRow, 8).Value To _
            g_objWsSetup.Cells(iRow, 9).Value
                ReDim Preserve g_aValidChar(iCounter)
                g_aValidChar(iCounter) = i
                iCounter = iCounter + 1
            Next i
        End If
    Next
End If
```

The second section of GetRandomPassword actually creates the eight-character random password. The script creates the password by first randomly generating a character and checking that the character is a member of a valid character set (g_aValidChar). If the character is valid, the script adds the character to the string strPassword. This process repeats until the length (len) of the strPassword is eight characters long.

The function creates random characters until a valid character is generated. If a match is found, the Boolean variable bValidChar is set to True, and the loop terminates.

```
While Len(strPassword) < 8
    bValidChar = False
    While Not bValidChar
```

The function uses the Randomize statement to seed the built-in random-number generator with a value from the system timer. A seed value is required by a random-number generator to generate a random number. The more random the seed value, the more random the generated number. Many mathematicians and computer scientists have theorized how to practically generate a truly random number; however, for the purposes of this script, using the system timer to generate a random seed value produces satisfactory random numbers. The Rnd function actually generates the random number greater than 0 and less than 1 using the seed value generated from the Randomize statement.

The script uses the function (122-33+1)*Rnd+33 to convert a randomly generated number to a value between 33 and 122 and then rounds it to the nearest integer using the Round method. This new integer represents the ASCII value of a proposed character to be used in the new password. But first, the script loops through each of the elements in the g_aValidChar array and verifies that this new character is included in one of the selected character sets. If the character is valid, the function sets bValidChar to True, and the character is appended to strPassword.

```
Randomize
randomASC2 = Round((122 - 33 + 1) * Rnd + 33, 0)
For i = 1 To UBound(g_aValidChar)
    If randomASC2 = g_aValidChar(i) Then
        bValidChar = True
        Exit For
    End If
Next i
Wend
strPassword = strPassword + Chr(randomASC2)
```

One particular caveat when using Excel to store passwords is its interpretation of the characters = and ' as special characters denoting an action. When a cell value is prefixed with an equals sign (=) character, Excel regards that cell as a function, and a single quote (') denotes a text entry into the cell. To avoid misinterpreted passwords, the script discards any passwords that begin with an = or '.

```
Select Case Left(strPassword, 1)
    Case "=", "'"
        strPassword = ""
End Select
```

This process repeats until the length of the password is eight characters, at which time the script returns the randomly generated eight-character password to the calling function.

```
Wend
GetRandomPassword = strPassword
End Function
```

SUMMARY

Whereas the Active Directory Users and Computers MMC provides a centralized and fairly efficient method for changing passwords for accounts stored in AD, rotating privileged accounts passwords of individual local computers can be a laborious process. The *PWScanner* script demonstrates a method of tracking and changing the privileged local administrator account for all computers in your domain. The script connects to each remote computer you specify in your domain, identifies the (possibly renamed) local administrator account, and lets you specify or randomly generate a new password for that account. The script sets this password and keeps an archive of past password changes.

KEY POINTS

- *PWScanner* demonstrates how to find the local administrator account using basic, built-in ADSI calls.
- The script includes a modest level of error handling to notify the user of various problems that may result when working with remote computer accounts and passwords.
- The script reports its progress when changing the password and makes an attempt to always notify the user of what the current password is, or, in the event of a problem, what the attempted password was.
- The script uses random-number functions and ASCII character values to create random passwords that satisfy either user-selected character sets or strong password policies.

This script completes the Excel VBA-based scripts. The script presented in the next and final chapter provides additional functionality that you can plug into any of the previous scripts or into your own creations. The next script presented demonstrates how to use CDO to e-mail your script results to others.

13 E-mail Script Notification

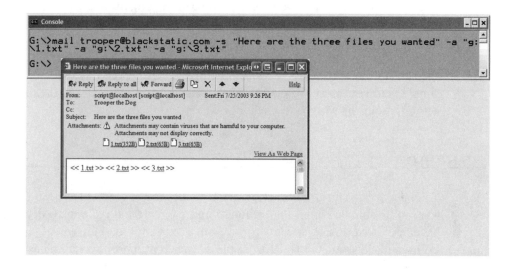

OVERVIEW

As we have seen so far, scripts come in many shapes and sizes. Some scripts provide real-time information via a command line or GUI interface. Other scripts can be scheduled to run at regular intervals and provide system administrators with a regular status of system health or configuration.

For example, if you remember, one of the requirements of the *DateQuery* script in Chapter 6 was to regularly query and display all accounts created in the previous two weeks. The solution included a new scheduled task that ran the script with the desired parameters and dumped the output to the console or a text file. The user could then review the data and e-mail the results to others. However, wouldn't it be much easier and more powerful to simply mail-enable the script to automatically e-mail the script results to any mail recipient?

As it turns out, e-mail-enabling your scripts is straightforward using a technology called Microsoft *Collaboration Data Objects*, or CDO. CDO provides an interface to the Messaging Application Programming Interface (MAPI) used by Microsoft messaging products such as Microsoft Exchange and Microsoft Outlook®. However, if your organization uses Microsoft Exchange, you can further leverage CDO to provide even more creative messaging functionality, such as posting script results to an Exchange public folder or using custom forms to collect and report script-generated data. Its more basic alternative, CDO for Windows 2000, allows you to harness CDO functionality on any Windows 2000 Server or Server 2003 (including Windows XP) without an Exchange Server and generate and send Simple Mail Transport Protocol (SMTP) e-mails using a local or remote mail server without requiring Microsoft Exchange Server.

SCENARIO

You have created quite a library of scripts that perform myriad tasks, and now you want to be able to e-mail the results or alerts to external recipients. Because you have been scripting for a while now, your script toolbox consists of batch files, HTAs, and elaborate Office VBA scripts, as well as numerous console-based scripts that you run from the command line or have set up as scheduled tasks. You want to develop an e-mail script that you can call from your existing scripts, as well as a function that you can include in new scripts to support e-mail from the start. Your scripts run from a variety of locations—some from your own computer that runs Windows XP Professional, others from utility servers located around your network.

Your company uses Microsoft Exchange as its primary messaging platform and supports relaying SMTP messages only for recipients who also have mailboxes in your mail system. This relaying restriction is typical for most corporate SMTP servers, but because you plan to configure your scripts to send only to recipients of your Exchange system, you do not expect this to be a problem.

NOTE

Mail relay restrictions have become more prominent recently due to the massive increase in Unsolicited Commercial Email (UCE, or Spam). When you generate and send an e-mail using CDO, you are sending it as an SMTP message to a remote SMTP server, which, depending on your To and From address, might act as a relay. For example, if your script sends an e-mail from script@localhost to someone@anexternaldomain.com via the SMTP server smtp.internal.com, *the internal SMTP server will likely reject the relay. However, an e-mail from script@ localhost to someone@internal.com sent to* smtp.internal.com *will most likely succeed.*

ANALYSIS

The solution to the scenario presented in this chapter consists of a JScript script harnessing a small CDO-based `Mail` function that can be reused in future scripts.

The script consists of a simple function that you can include in your existing scripts to enable them to e-mail their results to a mail recipient using SMTP. This standalone command-line-driven script accepts a number of messaging related parameters, such as the *To*, *From*, *CC*, and *BCC* recipients, custom *Subject* and *Body* fields, and attachment support. To support the e-mailing of external script output, the user can specify to mail the file as an attachment or e-mail the contents of the file as the body of the e-mail. The script also supports e-mailing binary files as well through standard e-mail attachments.

The CDO-based `Mail` function that drives the actual message processing detects whether a local SMTP service is running and uses that service or a specified remote SMTP server, if a local server cannot be found. This functionality uses CDO for Windows 2000, which uses SMTP and does not require Exchange server to create and send messages. This function is written in JScript but can be easily ported to VBScript. Table 13.1 shows how these functionality requests map into technical requirements.

The harness for the mail function provides a basic command-line-driven script that allows you to send e-mails in a variety of formats. This function interprets the command-line arguments passed to the script, parses them, and generates the e-mail. The script generally interprets the arguments in pairs such as `-arg` `[option]`. For example, `-s "This is the subject"` specifies the argument `-s`, which denotes the subject followed by the desired subject text. Because command-line arguments are delimited by spaces, you must enclose any text that contains spaces in quotations.

TABLE 13.1   Mapping functionality requests to technical requirements

Add mail functionality for new scripts.	Create new JScript function that can be inserted into new scripts and called as an integral part of the script. This functionality leverages CDO for Windows 2000.
Add mail functionality to old scripts.	Provide an auxiliary script that reads a text file or file attachment and mails the data independently of the script that created the data. This script also uses CDO and does not require an Exchange Server.

The main functions for these services follow:

Main Function: Provides a harness to the `Mail` function that parses command-line arguments to set up a new message.

- Supports specifying *To, From, CC,* and *BCC.*
- Specify custom subject and body lines.
- Supports multiple file attachments.
- Supports reading a text file as inserting the contents as the body of a new message.

`mail`: Provides a function to e-mail a message subject, body, and optional attachment to a recipient or recipients using SMTP.

- Supports both local and remote SMTP servers.
- Leverages Microsoft CDO technology.
- Does not require an Exchange server.

`ReadTextFile`: Opens and reads a text file and returns the result.

- Uses `FileSystemObject` to read a text file.
- Returns the results as a string to the calling function.
- Used by the mail harness script to insert the contents of a text file as the body of a message.

`ServiceCheck`: Searches a specified computer for a running service.

- Checks a local or remote computer for the `Win32_Service` state of a specified service.
- Used by the mail harness script to see if the SMTP service is running.

`ParamError`: Displays a commonly formatted error.

`Usage`: Displays help for the script.

The arguments supported by this script are shown in Table 13.2.

TABLE 13.2 *Mail* script command-line arguments and variables

Mail property	Argument	Variable in mail.js	CDO attribute (as referenced from objMsg)	Description
To	default	strTo	objMsg.To	The primary recipients of the e-mail.
Subject	-s	strSubject	objMsg.Subject	Subject of the e-mail.
Text body	-b	strTextBody	objMsg.TextBody	The main body of the e-mail.
From	-f	strFrom	objMsg.From	The sender of the e-mail.
CC	-cc	strCC	objMsg.CC	Secondary recipients of the e-mail.
BCC	-bcc	strBCC	objMsg.BCC	Recipients who will receive the e-mail but will not be visible to other recipients.
Attachment	-a	strAttachment, aAttachment	objMsg.AddAttachment	This method allows for multiple addition of attachments.

Distribution and Installation

Run the harness script *mail.js* from the command line or as a scheduled task. The script accepts command-line arguments, making it a convenient platform for scheduling the generation of a variety of e-mails.

Output

The *mail.js* function creates a custom message that is sent to either a local or remote SMTP server. Figure 13.1 shows an example of how to run this script to create a simple e-mail using the SMTP service installed on the local computer running the script. This simple e-mail consists of the subject *Some Subject Line* and the body taken from the file *body.txt* sent along with an attachment named *trooper.jpg*. This e-mail was processed by the local SMTP service, which looked up the correct destination of trooper@blackstatic.com and forwarded the message directly to the *blackstatic.com* SMTP mail server. The mail server then delivered the message to Trooper's inbox, and the results in Figure 13.2 show the message rendered by Trooper's e-mail program.

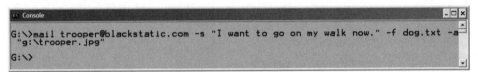

FIGURE 13.1   The script *mail.js* lets you send e-mails from the command line and supports multiple recipients, custom subject and body fields, and multiple file attachments.

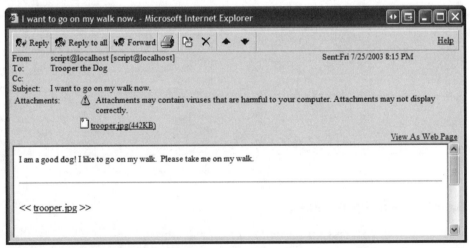

FIGURE 13.2   The script sends normal, standard SMTP messages, which are interpreted and rendered normally by e-mail client programs.

If an SMTP service is not installed and running on the computer running the *mail.js* script, you can specify an external SMTP server. Figures 13.3 and 13.4 show a second message sent to Trooper that the script forwards to the SMTP server located at *smtp.blackstatic.com*. Also, instead of including the contents of a text file as the body, the body is directly specified by using the -b argument.

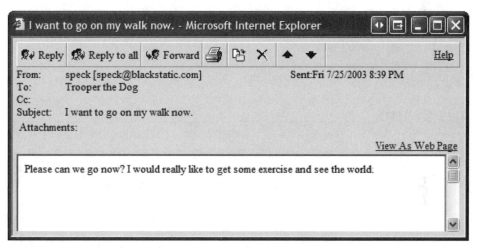

FIGURE 13.3   Specify a remote SMTP server using the -SMTP argument.

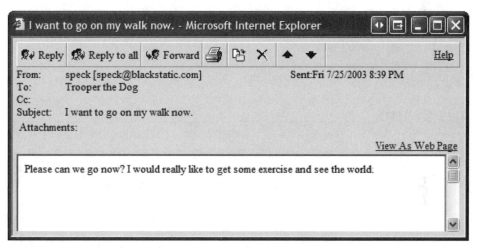

FIGURE 13.4   Using the *mail.js* script, customize the *To, From, CC, BCC, Subject,* and *Body* fields.

Summary of Solution

Create a harness and function to add e-mail support to your scripts. These options demonstrate a design to either incorporate messaging support directly into your script or as an independent method for mail-enabling other scripts or tools that otherwise do not support mail. Each of these solutions use the Windows CDO object model to create and send the mail messages.

OBJECTIVE

To create a command-line script to generate standard SMTP-based e-mails that can be used to e-mail the results of other task-specific scripts.

Array (JScript)

- push

CDO for Windows 2000

- CDO configuration object
- `sendusing`, `smtpserverport` and `smtpserver` Configuration fields
- `To`, `From`, `CC`, and `BCC` fields
- `Subject` and `TextBody` fields
- `AddAttachment`

FileSystemObject

- `OpenTextFile`
- `ReadLine`
- `Close`

THE SCRIPT

The script *mail.js* consists of a main function that acts as a harness to collect and parse arguments, which are then sent to a specific mail function. The mail function serves as the heart of the CDO mail routine and can be easily inserted into other scripts for which you wish to provide mail support. The harness demonstrates how to tap into the mail function as well as provides a simple method of generating command-line-based e-mail.

ON THE CD

The script is presented on the CD-ROM as /chapter 13/scripts/mail.js. Before continuing onto this next section, open the actual script from the CD-ROM and run it a few times in your test environment to get a feeling of how it works. Explore the entire code on your own, and then continue with the walkthrough, which will hopefully answer any questions you may have. Plus, seeing the code in its entirety in your script editor will give you a good sense of its scope and also allow you to tweak it as we walk through its function.

The script begins with the declaration of variables. This script is intended as a command-line mail utility and, as such, includes many extra parameters that your own individual unique deployments may not need—for example, you may not need to attach files or blind carbon-copy recipients.

The main string variables are defined in the previous Table 13.2. The script uses the variable colArgs to process the command-line arguments and bHelp to track whether the user specified (or requires) the script help commands. CDO adds e-mail attachments using the AddAttachment method, and the script keeps track of these attachments using the array aAttachment.

```
//------------------------------------------------------------------
// mail.js
//------------------------------------------------------------------
var strTo, strSubject, strTextBody, strFrom, strCC, strBCC;
var strAttachment;
var colArgs;
var bHelp = false;
var aAttachment = new Array();
```

The script initializes the common variables such as the subject and To and From addresses. This ensures that if the script is run missing one of these parameters, the script substitutes for generic values and continues to run. The domain @localhost refers to the computer that the user is using and in most cases is not deliverable on Windows systems. It is meant only to serve as a placeholder.

```
strTextBody="";
strSubject="Subject";
strSMTPServer="";
strFrom="script@localhost";
strTo="postmaster@localhost";
```

To begin the argument, parsing the script sets the variable colArgs to the collection of arguments and then begins looping through each of the members of the collection. The script calls the JScript toLowerCase method to convert the argument string to all lowercase and sets strArgs to this value. This setting permits a user to specify -From, -from, or -FrOm in a command-line argument, and the script processes all three the same as -from.

Next, the script checks whether the argument begins with a hyphen (-) using the regular expression /^-/. The script checks each argument against /^-/, which matches "if the beginning of a line followed by a hyphen." So, our arguments -s, -cc all match, but user@localhost does not match and does not satisfy the conditional. The argument parsing is constructed such that a nonhyphenated argument is considered the send to (To) address.

```
var colArgs = WScript.Arguments;
for(i=0;i<colArgs.length;i++) {
```

```
strArg=colArgs.Item(i).toLowerCase();
    if(/^-/.test(strArg))
        {
```

If the script encounters the help argument (-?), it simply sets the variable bHelp to true, which it later checks, displays the help, and then quits.

The script processes the arguments -s, -cc, -bcc, -from, -smtp, and -b in the same manner. When it finds these arguments, it sets the variable representing the parameter to the subsequent argument, then increments the argument counter to effectively skip over the value.

For example, if user calls the following mail function:

```
mail -s "Had a great day" trooper@blackstatic.com
```

the script interprets the first argument, -s, then immediately reads the next argument—in this case, the string "Had a great day"—into the variable strSubject. The script increments the argument counter i by one (setting it to the second argument), and then the for loop increments it again so the next argument interpreted is the string sunshine@somewhere.com, which the script parses as the recipient of the e-mail.

```
if(strArg=="-?") bHelp=true;
if(strArg=="-s") {
    if(i+1 > colArgs.length-1) ParamError();
    strSubject=colArgs.Item(i+1);
    i++;
    }
if(strArg=="-cc") {
    if(i+1 > colArgs.length-1) ParamError();
    strCC=colArgs.Item(i+1);
    i++;
    }
if(strArg=="-bcc") {
    if(i+1 > colArgs.length-1) ParamError();
    strBCC=colArgs.Item(i+1);
    i++;
    }
if(strArg=="-from") {
    if(i+1 > colArgs.length-1) ParamError();
    strFrom=colArgs.Item(i+1);
    i++;
    }
```

```
if(strArg=="-smtp") {
    if(i+1 > colArgs.length-1) ParamError();
    strSMTPServer=colArgs.Item(i+1);
    i++;
    }
if(strArg=="-b") {
    if(i+1 > colArgs.length-1) ParamError();
    strTextBody=strTextBody+colArgs.Item(i+1);
    i++;
    }
```

The script interprets the arguments -a, -f, and -p slightly differently from the previous arguments. The argument -a lets the user add an attachment to the e-mail and instructs the script to add the next argument to the end of the aAttachment array using the JScript push method. When preparing the actual message to be sent, the script iterates through this array and calls the AddAttachment method to add the attachment files represented by the members of the array. The script supports multiple attachments within a single e-mail by simply calling the script with multiple -a arguments (for example, -a [filename1] -a [filename2]), as seen in Figure 13.5. The filename must be specified as a drive letter and filename and not just the filename (such as "g:\1.txt").

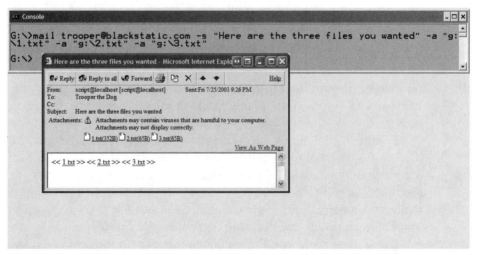

FIGURE 13.5 *Mail.js* script lets you attach a number of attachments to your command-prompt-script-generated e-mail.

Specify multiple attachments by repeated use of the -a argument.

```
if(strArg=="-a") {
    if(i+1 > colArgs.length-1) ParamError();
    aAttachment.push(colArgs.Item(i+1));
    i++;
    }
```

The script supports sending the contents of a text file as the body of the e-mail message by using the -f argument followed by the name of the file. This step is useful if you have an external function or program that regularly outputs its results to a text file, and you want a convenient and automated way to check this output. This script trims the ProcessResults function presented in the *ClientCheck.hta* script (Chapter 9) and ports it to JScript. This new function is called ReadTextFile and accepts the filename as an argument and returns the contents of the file.

```
if(strArg=="-f") {
    if(i+1 > colArgs.length-1) ParamError();
    strTextBody=strTextBody+ ReadTextFile(colArgs.Item(i+1));
    i++;
    }
```

Piping the output from one utility into another is a useful method of chaining programs into performing multiple sequential operations. For example, the following command:

```
DIR | FIND "My"
```

returns the directory results of all lines that contain the text "My". For example, a Windows XP system running this command from your home directory might return the lines *My Documents*, *My Pictures*, *My Music*, and so on.

Essentially, the output from the DIR command is sent as *Standard Out*. Standard Out is a part of the Standard Streams (also known as stdio), which includes Standard Out, Standard In, and Standard Error. The pipe character (|) pipes, or sends, this data into a subsequent utility—in this case, the FIND command. Normally, FIND requires a string to search for and a location to search. The utility (as do many command-line utilities) supports data input through stdio.

Using some tricks you can emulate piping data into your script by using the WScript.StdIn property and ReadAll method. WSH does not support robust streams handling. For example, when you get the AtEndOfStream property, WSH waits for a carriage return. If there is an input stream, the script processes the data normally. But if you read the property from a command line without any input stream, the script halts until you optionally enter some data at the console followed by a necessary carriage return. Whereas this behavior might be fine for a script that always expects an input stream, it is not ideal for a script such as this that supports an input stream as optional.

To overcome this behavior, this script looks for the -p argument, which indicates that the user wishes to pipe data to the script. Upon encountering this argument, the script reads the StdIn property and sets the contents to the strTextBody variable.

A second caveat of piping data to WSH scripts stems from how the command line must be constructed. When working with scripts, you must pipe the output from one program into the *cscript.exe* program with your script name as an argument, including the file extension. For example, to e-mail the results of a directory output to a user, use the following syntax:

```
DIR | cscript.exe mail.js someuser@somewhere.com -p
```

Figure 13.6 shows an example of how to pipe the output from a sample run of our first script, *ADSIQuery*, presented in Chapter 4, into this mail script. Checking your inbox for the results of a script is much more convenient than manually searching and opening a file share. In addition, storing your output in e-mail provides a record of historic output without renaming or archiving files.

```
if(strArg=="-p") {
    while(!WScript.StdIn.AtEndOfStream) {
        strTextBody=strTextBody + WScript.StdIn.ReadAll();
        }
    }
}
```

The script interprets any arguments not beginning with a hyphen as a potential e-mail recipient. But first, the script screens the argument to match a regular expression of the commercial at character (@). If matched, the script sets the variable strTo to this argument. If the argument does not contain an @, then the script sets bHelp to true to assist the user with a possibly erroneous argument.

```
else {
    if(/@/.test(strArg)) {
        strTo = strArg;
        }
    else bHelp=true;
    }
}
```

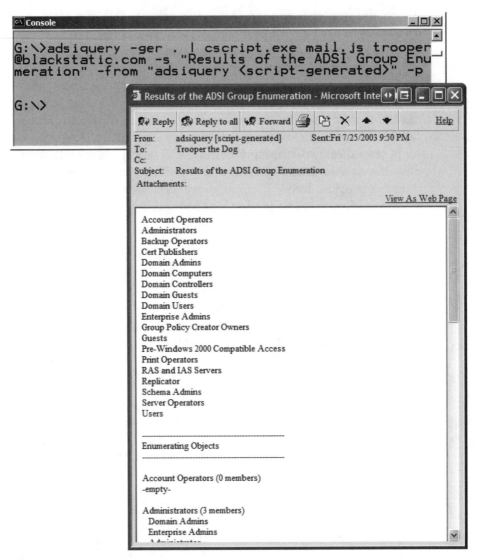

FIGURE 13.6 Piping data from one script or program to another script can be done using WSH and provides a terrific method of chaining the output from one program as input to another program.

CDO for Windows 2000 supports sending mail using either a locally installed SMTP service or sending the e-mail to a remote SMTP server. The code to redirect an e-mail to a remote SMTP server is a bit longer, but it permits you to send e-mails from any Windows 2000 (or later) computer without requiring installation of the SMTP service.

Coding your script to send an e-mail to a remote STMP server requires that you configure the CDO SMTP configuration properties. Specify a remote server by running the script with a -SMTP argument followed by the name of your SMTP server (for example, *smtp.yourcompany.com*). The script interprets your specification of an SMTP service to trigger the additional code necessary to send to a remote server.

If you do not specify to use a remote SMTP server and you do not have the SMTP service installed and running, then the script displays an error. The script calls the function ServiceCheck with the name of the local computer running the script (localhost) and the name of the SMTP service (SMTPSVC) to check the state of the service. If the service is running, ServiceCheck returns a true. Otherwise, it returns a false, and the script displays an error for the user.

```
if(!ServiceCheck("localhost", "SMTPSVC") & strSMTPServer=="") {
    WScript.Echo("It does not appear that the local SMTP service " +
    "is installed or else not running.\nPlease start this service " +
    "or specify an external SMTP server using the '-smtp server' " +
    "command.\nEnter mail -? for help.");
    WScript.Quit();
    }
```

Before passing the message parameters to the mail function, the script checks whether the user specified (or requires) help and calls the Usage function to display a help message to the user. Otherwise, the script calls the mail function with all of the user-specified parameters.

```
if(bHelp) Usage();
mail(strFrom, strSubject, strTextBody, strTo, strCC, strBCC,
 aAttachment, strSMTPServer);
```

Mail

The function mail processes a number of input parameters to configure and send an SMTP-based message using CDO. The two main CDO objects this function uses are Message and Configuration objects, represented in the script as objMsg and objConfiguration.

```
//--------------------------------------------------------------------
// Generates and sends an email using local or external SMTP server
//--------------------------------------------------------------------
function mail(strFrom, strSubject, strTextBody, strTo, strCC, strBCC,
 aAttachment, strSMTPServer)
{
var objMessage, objConfiguration, objConfigFields;
objConfiguration = new ActiveXObject("CDO.Configuration");
```

CDO uses the `Configuration` object to process the various configuration settings of the CDO entity. If you do not set a CDO `Configuration`, then CDO sets one for you prior to sending out each message using the default values. Alternatively, you can set the CDO `Configuration` once, which is then used for each message you send. In this script, first we configure the CDO `Configuration` object, then we tie this configuration to the `Message` that we send.

The `Configuration Fields` that we use are specified as name value pairs, and the names are stored as *Uniform Resource Identifiers* (URI). A URI is a string that represents a resource. The most common URI form is a Uniform Resource Locator (URL), such as *http://www.microsoft.com*. For example, the CDO configuration field `smtpserverport` is completely (and uniquely) identified by the URI `http://schemas.microsoft.com/cdo/smtpserver`.

You do not have to refer to the URI if you use a language that supports type libraries, such as VBA. In this case, after you have established the CDO reference, you can refer to the configuration simply as their name; for example, `smtpserver` *instead of* http://schemas.microsoft.com/cdo/smtpserver.

The three `Configuration` fields that this script uses are `sendusing`, `smtpserverport`, and `smtpserver`. The `Configuration` field `sendusing` instructs CDO whether to send the message (or other CDO entity such as a calendaring entry) using the local SMTP service (`sendusing = 1`), send to a remote SMTP server (`sendusing=2`), or send using an Exchange Server mail submission URI (`sendusing=3`). `sendusing` defaults to sending using a local SMTP service, and the script only configures this if a remote server is used.

The field `smtpserverport` specifies the desired TCP SMTP port to use, which is 25, by default. If your remote SMTP server listens on a different port, you can configure CDO to send to that port instead.

The last `Configuration` field that we specify is `smtpserver`. This field represents the name of the remote SMTP server with which we wish to forward the message. You can specify an IP address or resolvable name (for example, *smtp* or *smtp.yourcompany.com*). If you cannot communicate with your SMTP server, try pinging the name of the SMTP server to confirm that your name resolution is working correctly.

Lastly, we must update the `Configuration` by calling the `Update` method.

```
objConfigFields = objConfiguration.Fields;
if(strSMTPServer != "") {
    objConfigFields.Item("http://schemas.microsoft.com/cdo/" +
     "configuration/sendusing") = 2;
    objConfigFields.Item("http://schemas.microsoft.com/cdo/" +
     "configuration/smtpserverport") = 25;
```

```
objConfigFields.Item("http://schemas.microsoft.com/cdo/" +
  "configuration/smtpserver") = strSMTPServer;
objConfigFields.Update();
}
```

With the CDO configuration object defined, the script now defines a CDO.Message object that represents the e-mail to be sent. First, the script instantiates a new CDO.Message object and then sets its configuration to the previously defined configuration object objConfiguration. Then the script begins to set the message properties, such as the From address, the Subject, and TextBody.

This script demonstrates just a few of the many CDO properties, and there are many more. For example, CDO supports HTML e-mail, MIME, and a number of other fields, such as SMTP authentication, reply to addresses, and a raft of other configuration options. For more information on CDO technology, look up Messaging and Collaboration in the Platform SDK (see Chapter 14 for more information), or search for CDO on the MSDN Web site.

```
objMsg = new ActiveXObject("CDO.Message");
objMsg.Configuration = objConfiguration;
objMsg.From = strFrom;
objMsg.Subject = strSubject;
objMsg.TextBody = strTextBody;
```

As you are aware, you can usually specify multiple recipients in any of the message recipient fields (*To*, *CC*, and *BCC*). This script supports multiple recipients using the built-in CDO list support. Recipient lists in CDO are delimited by commas. For example, to send an e-mail to multiple people using CDO, you can set the *To* attribute to recipient1@someaddress.com, recipient2@someotheraddress.com, etc.

```
objMsg.To = strTo;
objMsg.CC = strCC;
objMsg.BCC = strBCC;
```

The script adds attachments to the message by looping through each of the members of the previously created aAttachment array and for each member it calls the AddAttachment method.

```
for(j=0;j<aAttachment.length;j++) {
    objMsg.AddAttachment(aAttachment[j]);
    }
```

Lastly, the script calls the `Send` method to actually send the message, which completes the `mail` function.

```
objMsg.Send();
}
```

ReadTextFile

The `ReadTextFile` function uses `Scripting.FileSystemObject` to open and read the contents of a text file. A more complete description of how this function works can be found in Chapter 9 in the description of the `ProcessResults` function. This function is used by the script to include the contents of a text file as the body of the message to send.

```
//------------------------------------------------------------------
// Read a text file and iReturn the contents
//------------------------------------------------------------------
function ReadTextFile(strFileName)
{
    var objFSO;
    var objTextFile;
    var strReadLine="";
    objFSO = new ActiveXObject("Scripting.FileSystemObject");
    if(objFSO.FileExists(strFileName)) {
       objTextFile = objFSO.OpenTextFile(strFileName,1);
        while(!objTextFile.AtEndOfStream) {
            strReadLine = strReadLine + objTextFile.ReadLine();
            }
          objTextFile.Close();
       }
    else {
       strReadLine="Error reading file: " + strFileName;
       }
    return(strReadLine);}
```

The `ServiceCheck` function uses WMI to scan a computer for whether a specified service is installed. If the `State` of the service is `Running`, the script returns `true`; otherwise, the function returns `false` (for example, it could be stopped or simply not installed).

```
//------------------------------------------------------------------
// ServiceCheck: Uses WMI to check if SMTP service is installed
//------------------------------------------------------------------
```

```
function ServiceCheck(strComputer, strServiceName)
{
    var strWMI;
    var objService;
try {
    strWMI = "WinMgmts://" + strComputer +
     "/root/cimv2:Win32_Service='" + strServiceName + "'";
    objService = GetObject(strWMI);
    if (objService.State=="Running") return(true);
    }
        return(false);
}
```

The ParamError function displays an error if the user did not input the argument parameters properly, and the Usage function displays a help message for the user on how to use this script.

```
//---------------------------------------------------------------------
// Display the help screen
//---------------------------------------------------------------------
function ParamError()
{
WScript.Echo("Parameter Error Received. Please check syntax.\n");
Usage();
}
//---------------------------------------------------------------------
// Display the help screen
//---------------------------------------------------------------------
function Usage()
{
WScript.Echo("Tool to send email from command prompt.");
WScript.Echo("Usage:\n");
WScript.Echo("mail [recipient list] -s \"subject\" -cc [list] " +
 "-bcc [list] -from sender@domain.com -a \"c:\\file.jpg\" " +
 "-f \"c:\\textfile.txt\" -b \"Body of Message\" " +
 "-smtp smtpserver.domain.com\n");
WScript.Echo("[list] denotes a comma delimited list of email " +
 "addresses with no spaces between the commas.");
WScript.Echo("-s \"subject\"  specify subject of message " +
 "(quote delimited).");
WScript.Echo("-cc [list] carbon copies recipients.");
WScript.Echo("-bcc [list] blind carbon copies recipients.");
WScript.Echo("-from [sender@somedomain.com] specify the sender " +
 "of the message.");
```

```
        WScript.Echo("-a [filename] attach a file to the message. Can be " +
          "used more than once.");
        WScript.Echo("-f [text filename] specify name of text file whose " +
          "contents you wish to include as the body. Can be used together " +
          "with -b body parameter.");
        WScript.Echo("-b \"body\" specify body of message (quote delimited)");
        WScript.Echo("-smtp [smtpserver.domain.com] specify name of " +
          "external SMTP server.");
        WScript.Echo("-p looks for piped input from another program " +
          "(specified with a | character).");
        WScript.Echo("* note- when piping input to a script, you must pipe " +
          "it to cscript.exe, " +
          "followed by the script.ext name.\n");
        WScript.Echo("-? Help- this screen.");
        WScript.Echo("");
        WScript.Echo("Example:");
        WScript.Echo("mail user@somedomain.com -s \"Subject of Message\" " +
          "-b \"Body of message\" -cc user2@domain.com,user3@domain.com");
        WScript.Echo("mail user@somedomain.com -a \"c:\attachment1.jpg\" " +
          "-a \"c:\attachment1.jpg");
        WScript.Echo("mail user@somedomain.com -smtp smtp.somedomain.com");
        WScript.Echo("dir c:\\ | cscript mail.js user@somedomain.com -p");
        WScript.Quit(0);
        }
```

SUMMARY

Even as a standalone utility, the ability to generate e-mails from your scripts to no-tify, alert, or report status can dramatically improve the information management of your environment. For example, instead of creating a reminder to check a file server for the results of a script that runs weekly, simply enable your script to mail the results to you when complete. In most organizations, e-mail provides a ubiqui-tous method of communication, and harnessing this communication medium likely can improve your scripts' usefulness even more.

KEY POINTS

- CDO provides a robust method for mail-enabling your scripts.
- Add support to your scripts for reading the standard in input streams to pipe data between your scripts.
- CDO does not require the SMTP service, Exchange Server, or another mail server to be installed on the computer that is running the script.

This script wraps up these 10 introductory scripts. We hope you have found them instructive and even perhaps a catalyst to ideas of your own. To wrap up this book, the next chapter lists and briefly discusses a few of the official Microsoft development resources to help you get more information with the creation of your scripts.

14 Where to Go from Here

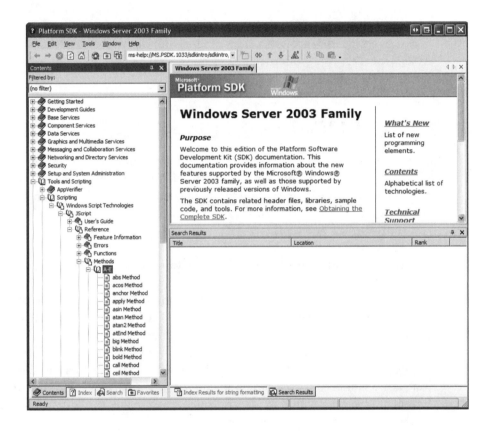

Throughout this text we've explored just a sampling of all of the scripting possibilities Windows offers. Using some of the larger and more accessible scripting technologies such as ADSI, WSH, WMI and some of the Excel objects, we've created a number of IT-centric scripts. More important than the applicability of these exact scripts is your ability to either morph these functions and subroutines to fit your own environment and needs, or, using these as a template (or to simply spark ideas), create your own scripts to solve your own puzzles and problems.

One thing for sure is that technology will never stop, and it's important to stay current with the online resources available to help keep you abreast of the latest scripting toolkits and SDKs. And who knows—the product you are installing today might in the future support its own SDK. For example, about two years after Microsoft released Internet Security Accelerator (ISA) Server, they released an SDK allowing developers to write scripts to help manage ISA Server deployments. Even from the examples included in this text, we've seen WMI classes with increasing support from Windows NT to Windows 2000 to Windows XP and Windows Server 2003.

The online resources include vendor-supported sites—such as Microsoft MSDN or TechNet—Internet newsgroups, and any of the hundreds of high-quality code sample sites.

MSDN: THE MICROSOFT DEVELOPER NETWORK

The Microsoft Developer Network (MSDN) provides the most development-related content for Windows products anywhere. Most of the research content is available free via the Web site, and for those needing more, Microsoft offers MSDN subscription services in which they provide access to developer editions of their software products and tools.

Access MSDN at: *http://msdn.microsoft.com*

MSDN provides interesting articles, announcements, and Web casts, as well as links to other communities, which are sometimes third-party Web sites that focus on or evangelize a particular Microsoft development technology. One of the most useful sections of the site is the MSDN library. The library contains articles, tutorials, *MSDN Magazine* reprints, case studies, white papers, and documentation on most of Microsoft technologies. Additionally, the MSDN library includes an online version of most SDKs, which is handy if you need quick information about an object or are wondering how development-friendly a particular product might be. Best of all, the MSDN library is searchable, which makes this a first stop when researching new scripting technologies.

SOFTWARE DEVELOPMENT KITS

Microsoft SDKs provide the instruction manual for how to access and use all of their products' development capabilities. Most of the SDKs can be downloaded to your computer and installed, which dramatically improves the speed and response of using the tool over accessing it online. The main SDK is called the Platform SDK, and you can download it from:

http://microsoft.com/msdownload/platformsdk/sdkupdate/

Don't get overwhelmed; the SDK includes development information about all aspects of Microsoft development tools that extend way beyond scripting. The SDK includes reference to Microsoft objects—for example, it shows how to properly use the ADSI `IADsGroup::Members` method, as well as references non-Microsoft technology, such as how to use the JScript `try...catch...finally` statement.

The SDK document viewer, shown in Figure 14.1, includes a Search dialog box, results, and the document window. In Figure 14.1, we searched for the word *member* and found more than 500 topics relating to this word. In this case, if we were looking for the ADSI method to return the members of a group, we are in luck because `IADsGroup::Members` came up first. But if you aren't quite sure what you are looking for, click the Index tab of the Search pane and try approaching it from that angle. For example, if we are looking for an ADSI interface, we can switch to the Index and begin typing *IAD....* As you can see in Figure 14.2, the Index results may help jog our memory.

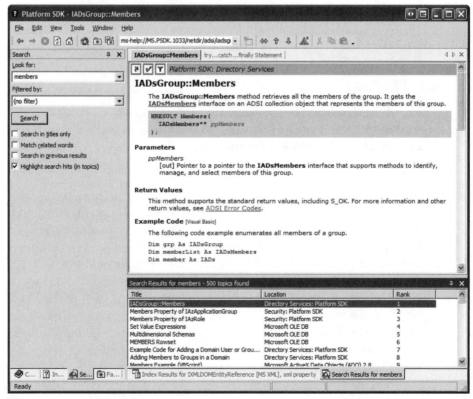

FIGURE 14.1 Search the Platform SDK for the names of objects, methods, and properties to access detailed usage syntax.

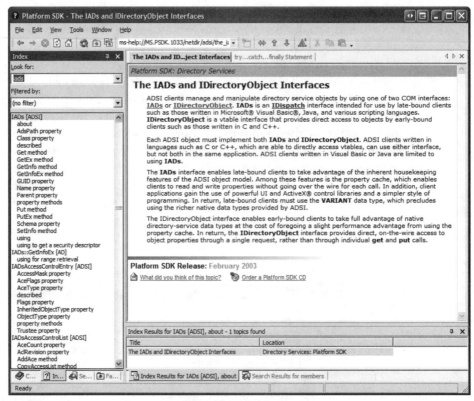

FIGURE 14.2 The Platform SDK Index lets you browse topics to help you find information if you don't know the exact name.

The documents in the SDK provide all of the technical information for how to explicitly use a particular object, method, property, or statement. For example, for a method, the SDK tells you about its arguments, return values, and version requirements and most of the time provides an example of how to use it. The information is good, but sometimes the consistency across technologies varies. For example, the ADSI and WMI technologies are both covered in the Platform SDK; however, the style and content of the documents covering these technologies feel dramatically different in their style and coverage. Plus, you may find some script-friendly technologies that provide examples in Visual Basic or JScript. Other technologies may provide example code only in C or C++.

Even for the casual scripter, the SDK is a terrific and essential resource.

OTHER MICROSOFT DOCUMENTATION

In addition to the SDK, Microsoft makes available independent documentation on various scripting technologies. For example, you can download the *JScript Language Reference*, which focuses solely on JScript—from how to use the language to short tutorials and the language reference. Whereas many of the reference material is contained in the SDK, the smaller and more focused JScript reference document is targeted at the scripter, and it can be easier to find specific information for how to do something JScript-related in this document. The guide also includes selected tutorials on accomplishing more common scripting tasks. For example, it has a nice tutorial on enumerating the local drives of a computer using `FileSystemObject`.

NEWSGROUPS

Microsoft hosts a number of development-centric public newsgroups, which are terrific places to get information. Access the public MSND newsgroups from the following URL: *http://msdn.microsoft.com/newsgroups/*

From here, you can expand the navigation tree to the categories that interest you. To reach scripting, expand Web Development and then Scripting. (In the probable event that Microsoft redesigns MSDN and changes the URLs, search on MSDN for *scripting newsgroups*, and the results should take you to the appropriate information.)

External to Microsoft, consider searching for and subscribing to globally available scripting newsgroups. Or, if you do not have a newsreader, use an HTTP newsgroup browser such as *http://groups.google.com* to search for answers to your questions.

SUMMARY

Microsoft offers a huge amount of online and downloadable reference materials that you can take advantage of to increase your programming knowledge and efficiency. Meandering through the MSDN Web site will show you not only new tools to add to your own repertoire but also all of the other programming offerings and capabilities used to make software of all shapes and sizes—from a short script to a full-blown enterprise application.

About the CD-ROM

The CD-ROM included with *IT Administrator's Top 10 Introductory Scripts for Windows* includes all the code and projects from the various examples found in the book.

CD-ROM FOLDERS

Each chapter is represented on the CD-ROM as a folder and contains two subfolders: *figures* and *scripts*.

Chapter #- Title\figures: Contains the figures of the chapter.

Chapter #- Title\scripts: Contains all the scripts and sample code listings for that chapter.

OVERALL SYSTEM REQUIREMENTS

■ Windows NT, Windows 2000, Windows XP Pro, or Windows Server 2003

Note that many of the example code and scripts in this text leverage and demonstrate scripting using Windows 2000 technologies. Some scripts will not run on or against a computer running Windows NT. If you use Windows NT, you must load the WMI and ADSI technologies for Windows NT, but some script functionality may be limited. Also, several of these scripts specifically target Active Directory, which requires a Windows 2000 Server or Windows Server 2003 Active Directory-based domain structure.

■ Pentium II processor or greater

■ CD-ROM drive

■ Hard drive

■ 128MB of RAM minimum, 256MB RAM recommended

■ Strongly recommended: WSH 5.6

Windows Script Host 5.6 provides the latest objects, methods, and properties for Windows Script Host. You can run all of the scripts in this text with WSH 5.6, and also ensure that your own scripting can leverage the latest development tools. Earlier versions of Windows include older versions of Windows Script Host, so you must download and install the latest version from the Microsoft Web site:

http://msdn.microsoft.com/library/default.asp?url=/downloads/list/webdev.asp

■ Current version of the Microsoft Platform SDK

The Microsoft Platform SDK is not required but it will come in handy as a reference to look up the definition and syntax of most of the code within this text. The platform SDK is available from the MSDN Web site at *www.microsoft.com/msdownload/platformsdk/sdkupdate/.*

Index